Current Developments in Differential Geometry and its Related Fields

Current Developments in Differential Geometry and its Related Fields

Proceedings of the 4th International Colloquium on Differential Geometry and its Related Fields

Veliko Tarnovo, Bulgaria 8 – 12 September 2014

editors

Toshiaki Adachi

Nagoya Institute of Technology, Japan

Hideya Hashimoto

Meijo University, Japan

Milen J Hristov

St Cyril and St Methodius University of Veliko Tarnovo, Bulgaria

World Scientific

EW JERSEY • LONDON • SINGAPORE • BEIJING • SHANGHAI • HONG KONG • TAIPEI • CHENNAI • TOKYO

Published by

World Scientific Publishing Co. Pte. Ltd.
5 Toh Tuck Link, Singapore 596224
USA office: 27 Warren Street, Suite 401-402, Hackensack, NJ 07601
UK office: 57 Shelton Street, Covent Garden, London WC2H 9HE

British Library Cataloguing-in-Publication Data
A catalogue record for this book is available from the British Library.

Cover images:
(front) Sokolski Monastery of the Assumption, Gabrovo
(back) The road "Alexander Stanboliyski" to the museum, Veliko Tarnovo
Photographed by T. Adachi

CURRENT DEVELOPMENTS IN DIFFERENTIAL GEOMETRY AND ITS RELATED FIELDS
Proceedings of the 4th International Colloquium on Differential Geometry and its Related Fields

ISBN 978-981-4713-78-8

Desk Editor: Lai Fun Kwong
Cover Designer: Doris Zhenghua Yu

Printed in Singapore

PREFACE

This volume contains papers written by some of the participants in the *4th International Colloquium on Differential Geometry and its Related Fields* (ICDG2014) which was held at Europe Hall in Rectral Building, St. Cyril and St. Methodius University of Veliko Tarnovo, Bulgaria, during the period of 8–12 September, 2014. With warm hospitality and assistance of the staff of the faculty of Mathematics and Informatics, Veliko Tarnovo University, participants discussed quite freely and exchanged their ideas.

The articles in this volume are mainly devoted to the study of some geometric structures on manifolds and their discretizations. These cover recent developments on Einstein metrics on Lie groups, Einstein-Weyl structures on 3-dimensional manifolds, quaternionic CR manifolds, exceptional geometry, totally geodesic surfaces in symmetric spaces, mathematical physics, lightlike submanifolds, complex statistical manifolds, new treatment of surfaces in a Euclidean 3-space, and Laplacians of graphs. We hope this volume will provide a broad overview of differential geometry and its relationship to other fields in mathematics and physics.

This academic program is based on the agreement of academic exchange and cooperation between the University of Veliko Tarnovo and Nagoya Institute of Technology. We gladly had young Bulgarian newcomers. We hope our program will encourage young scientists from Bulgaria, Japan and some other countries. Unfortunately, we have to make a solemn note. Professor Bagreliya Borisova, the former Vicerector on international activities of University of Veliko Tarnovo, passed away on December 17, 2013. We would like to pay tribute to her and her achievements.

We gratefully acknowledge a financial support for the conference activities by NITech Grant for Global Initiative Projects, Nagoya Institute of Technology. We appreciate very much the scientific reviewers who read the articles carefully and gave many important suggestions to authors. Finally, we express our hearty thanks to the local staff, especially Professor Paskal N. Piperkov, for their hospitality.

The Editors
17 May, 2015

The 4th International Colloquium on Differential Geometry and its Related Fields

8–12 September, 2014 – Veliko Tarnovo, Bulgaria

ORGANIZING COMMITTEE

T. Adachi – Nagoya Institute of Technology, Nagoya, Japan
S. H. Bouyuklieva – St. Cyril and St. Methodius University of Veliko Tarnovo, Veliko Tarnovo, Bulgaria
H. Hashimoto – Meijo University, Nagoya, Japan
M. J. Hristov – St. Cyril and St. Methodius University of Veliko Tarnovo, Veliko Tarnovo, Bulgaria

SCIENTIFIC ADVISORY COMMITTEE

G. Ganchev – Bulgarian Academy of Sciences, Sofia, Bulgaria
K. Sekigawa – Niigata University, Niigata, Japan
T. Sunada – Meiji University, Tokyo, Japan

PRESENTATIONS

at Europe Hall, Rectral Building,
St. Cyril and St. Methodius University

1. **Yusuke Sakane** (Osaka Univ.),
   New homogeneous Einstein metrics on compact simple Lie groups
2. **Galya Nakova** (Univ. Veliko Tarnovo),
   Curvature properties of some lightlike hypersurfaces of almost complex manifolds with Norden metric
3. **Fuminori Nakata** (Fukushima Univ.),
   Circle invariant indefinite Einstein-Weyl structures and the twistor correspondence
4. **Krassen Iliev** & **Milen Hristov** (Univ. Veliko Tarnovo),
   On the Laplace-matrix calculus
5. **Toshiaki Adachi** (Nagoya Inst. Tech.),
   Kähler graphs — Discrete models of manifolds equipped with magnetic fields —
6. **Yarmamat Tursun** (Nagoya Inst. Tech.),
   Laplacians for Kähler graphs
7. **Hiroshi Matsuzoe** (Nagoya Inst. Tech.),
   Geometry of deformed exponential families
8. **Nikolay Ivanov** (Univ. Veliko Tarnovo),
   Noncommutative Instantons
9. **Hiroyuki Kamada** (Miyagi Univ. of Education),
   Geometric structure modeled on a real hypersurface in a quaternionic manifold
10. **Misa Ohashi** (Nagoya Inst. Tech.),
    A method of determinig the $SO(7)$*-invariants for curves in* $\mathrm{Im}\,\mathbf{O}$ *by their* G_2*-invariants*
11. **Milen Hristov** & **Victoria Bencheva** (Univ. Veliko Tarnovo),
    On the differential geometry of Laplace-transformed curves

12. **Milen Hristov** (Univ. Veliko Tarnovo),
Matrix approach to curves in the complex plane and its conformal and Laplace transformations
13. **Toshihiro Shoda** (Saga Univ.),
On zero eigenfunctions of Jacobi operator on a minimal surface
14. **Georgi Ganchev** (Bulgarian Academy Sci.),
Special surfaces in the three or four-dimensional Euclidean and Minkowski spaces and their natural PDE's
15. **Krassimir Kanchev** (Univ. of Transport, Sofia),
Explicit solving of the system of natural PDE's of minimal surfaces in the four-dimensional Euclidean space
16. **Naoya Ando** (Kumamoto Univ.),
Local characterizations of surfaces in various spaces by induced metrics and principal distributions
17. **Mancho Manev** (Plovdiv Univ.),
On the geometry of canonical connections on manifolds with complex and contact structures and Norden-type metrics
18. **Hristro Manev** (Plovdiv Univ.) &
Miroslava Ivanova (Trakia Univ.)
On some Lie groups as 5-dimensional almost contact B-metric manifolds
19. **Hideya Hashimoto** (Meijo Univ.),
On some constructions of G_2 and Spin(7) bundles on a 3-manifold by using Clifford algebra of order 3
20. **Venelin Monev** (Univ. Veliko Tarnovo),
Maria Dzhumalieva-Stoeva (Univ. Veliko Tarnovo) &
Iliya Bouyukliev (Bulgarian Academy Sci.),
About the correctness of the classification results on binary selfdual codes of length 40

St. Cyril and St. Methodius University, September 2014

Contents

Current Developments in Differential Geometry
and its Related Fields *pp. 1 – 22*

EINSTEIN METRICS ON THE SYMPLECTIC GROUP WHICH ARE NOT NATURALLY REDUCTIVE

Andreas ARVANITOYEORGOS

University of Patras, Department of Mathematics,
GR-26500 Rion, Greece
E-mail: arvanito@math.upatras.gr

Yusuke SAKANE

Department of Pure and Applied Mathematics,
Graduate School of Information Science and Technology,
Osaka University, Toyonaka, Osaka 560-043, Japan
E-mail: sakane@math.sci.osaka-u.ac.jp

Marina STATHA

University of Patras, Department of Mathematics,
GR-26500 Rion, Greece
E-mail: statha@master.math.upatras.gr

We obtain new invariant Einstein metrics on the compact Lie groups $\mathrm{Sp}(n)$ ($n \geq 3$) which are not naturally reductive. This is achieved by imposing certain symmetry assumptions in the set of all left-invariant metrics on $\mathrm{Sp}(n)$.

Keywords: Homogeneous space, Einstein metric, isotropy representation, compact Lie group, naturally reductive metric, symplectic group, generalized Wallach space

1. Introduction

A Riemannian manifold (M, g) is called Einstein if it has constant Ricci curvature, i.e. $\mathrm{Ric}_g = \lambda \cdot g$ for some $\lambda \in \mathbb{R}$. A detailed exposition on Einstein manifolds can be found in [5], [11] and [12]. For the case of homogeneous spaces G/K the problem is to find and classify all G-invariant Einstein metrics, and for the case of a compact Lie group the problem is to find left-invariant Einstein metrics. Even for the Lie groups $\mathrm{SU}(3)$ and $\mathrm{SU}(2)\times\mathrm{SU}(2)$ the number of left-invariant Einstein metrics is still unknown. In the present paper we prove existence of left-invariant Einstein metrics on the symplectic group $\mathrm{Sp}(n)$, which are not naturally reductive.

It is known that a compact and semisimple Lie group equipped with a bi-invariant metric is Einstein. In [7] J.E. D'Atri and W. Ziller found a large number of naturally reductive metrics on the compact Lie groups. They raised the question of existence of left-invariant Einstein metrics which are not naturally reductive. A first answer was given in [8] by K. Mori who obtained non naturally reductive Einstein metrics on the Lie group $\mathrm{SU}(n)$ for $n \geq 6$. In [3] the first author, K. Mori and the second author proved existence of non naturally reductive Einstein metrics on the compact Lie groups $\mathrm{SO}(n)$ $(n \geq 11)$, $\mathrm{Sp}(n)$ $(n \geq 3)$, E_6, E_7 and E_8, by considering fibrations of these Lie groups over a generalized flag manifold. In [6] Z. Chen and K. Liang found three naturally reductive and one non naturally reductive Einstein metric on the compact Lie group F_4. Their method was based on representation theory and the study of involutions on compact simple Lie groups. Finally, in [4] the authors found new Einstein metrics on the Lie group $\mathrm{SO}(n)$ which are not naturally reductive, by imposing certain symmetry assumptions in the set of all left-invariant metrics on $\mathrm{SO}(n)$. We employ this method in the present work.

The aim of the present paper is to obtain left-invariant Einstein metrics on the compact Lie groups $\mathrm{Sp}(n)$ $(n \geq 3)$ which are not naturally reductive. We decompose the tangent space $\mathfrak{g}$ of G, via the submersion $G \to G/K$ with fiber K, into a direct sum of irreducible $\mathrm{Ad}(K)$-modules and we consider left-invariant metrics on G which are determined by diagonal $\mathrm{Ad}(K)$-invariant scalar products on $\mathfrak{g}$. By using the bracket relations of the submodules, we can see that Ricci curvature also satisfies orthogonal relations for each pair of different irreducible summands. Thus, by taking into account the diffeomorphism $G/\{e\} \cong (G \times K)/\mathrm{diag}(K)$, we can use well known formula in [10] for the Ricci curvature of such left-invariant metrics on G.

More precisely, we consider the Lie group $\mathrm{Sp}(n)$ with $n = k_1 + k_2 + k_3$ and the closed subgroup $K = \mathrm{Sp}(k_1) \times \mathrm{Sp}(k_2) \times \mathrm{Sp}(k_3)$. This determines the homogeneous space $G/K = \mathrm{Sp}(k_1+k_2+k_3)/(\mathrm{Sp}(k_1) \times \mathrm{Sp}(k_2) \times \mathrm{Sp}(k_3))$, which is an example of a generalized Wallach space according to [9]. For $k_1 \geq k_2 \geq k_3 \geq 1$ the isotropy representation of G/K does not contain equivalent summands as $\mathrm{Ad}(K)$-modules, thus we consider left-invariant metrics determined by $\mathrm{Ad}(K)$-invariant inner products of the form

$$\begin{aligned}\langle\ ,\ \rangle = {} & x_1\,(-B)|_{\mathfrak{sp}(k_1)} + x_2\,(-B)|_{\mathfrak{sp}(k_2)} + x_3\,(-B)|_{\mathfrak{sp}(k_3)} \\ & + x_{12}\,(-B)|_{\mathfrak{m}_{12}} + x_{13}\,(-B)|_{\mathfrak{m}_{13}} + x_{23}\,(-B)|_{\mathfrak{m}_{23}}.\end{aligned}$$

By using the main theorem of D'Atri and Ziller ([7]) we obtain conditions on the positive variables $x_1, x_2, x_3, x_{12}, x_{13}$ and x_{23} so that the above metric is naturally reductive (Proposition 4.1).

The main result is the following (see Theorem 6.1):

Theorem 1.1. *The compact simple Lie groups* $\mathrm{Sp}(n)$ $(n \geq 3)$ *admit left-invariant Einstein metrics which are not naturally reductive.*

We also find all Einstein metrics of the above type for the groups $\mathrm{Sp}(3)$, $\mathrm{Sp}(4)$, $\mathrm{Sp}(5)$ and $\mathrm{Sp}(6)$ in Propositions 6.1, 6.1, 6.2, 6.3, 6.4, 6.5 and 6.6. Note also that the Einstein metrics obtained are different from the ones in [3].

2. The Ricci tensor for reductive homogeneous spaces

We recall an expression for the Ricci tensor for a G-invariant Riemannian metric on a reductive homogeneous space whose isotropy representation is decomposed into a sum of non equivalent irreducible summands.

Let G be a compact semisimple Lie group, K a connected closed subgroup of G and let $\mathfrak{g}$ and $\mathfrak{k}$ be the corresponding Lie algebras. The Killing form B of $\mathfrak{g}$ is negative definite, so we can define an $\mathrm{Ad}(G)$-invariant inner product $-B$ on $\mathfrak{g}$. Let $\mathfrak{g} = \mathfrak{k} \oplus \mathfrak{m}$ be a reductive decomposition of $\mathfrak{g}$ with respect to $-B$ so that $[\,\mathfrak{k}, \mathfrak{m}\,] \subset \mathfrak{m}$ and $\mathfrak{m} \cong T_o(G/K)$. We decompose $\mathfrak{m}$ into irreducible $\mathrm{Ad}(K)$-modules as follows:

$$\mathfrak{m} = \mathfrak{m}_1 \oplus \cdots \oplus \mathfrak{m}_q. \tag{1}$$

Then for the decomposition (1) any G-invariant metric on G/K can be expressed as

$$\langle\ ,\ \rangle = x_1(-B)|_{\mathfrak{m}_1} + \cdots + x_q(-B)|_{\mathfrak{m}_q}, \tag{2}$$

for positive real numbers $(x_1, \ldots, x_q) \in \mathbb{R}^q_+$. If, for the decomposition (1) of $\mathfrak{m}$, the Ricci tensor r of a G-invariant Riemannian metric $\langle\ ,\ \rangle$ on G/K satisfies $r(\mathfrak{m}_i, \mathfrak{m}_j) = (0)$ for $i \neq j$, then the Ricci tensor r is of the same form as (2), that is

$$r = z_1(-B)|_{\mathfrak{m}_1} + \cdots + z_q(-B)|_{\mathfrak{m}_q},$$

for some real numbers $z_1, \ldots, z_q$.

Let $\{e_\alpha\}$ be a $(-B)$-orthonormal basis adapted to the decomposition of $\mathfrak{m}$, i.e. $e_\alpha \in \mathfrak{m}_i$ for some i, and $\alpha < \beta$ if $i < j$. We put $A^\gamma_{\alpha\beta} = -B\left([e_\alpha, e_\beta], e_\gamma\right)$ so that $[e_\alpha, e_\beta] = \sum_\gamma A^\gamma_{\alpha\beta} e_\gamma$ and set $\begin{bmatrix} k \\ ij \end{bmatrix} = \sum (A^\gamma_{\alpha\beta})^2$, where the sum is taken over all indices α, β, γ with $e_\alpha \in \mathfrak{m}_i,\ e_\beta \in \mathfrak{m}_j,\ e_\gamma \in \mathfrak{m}_k$ (cf. [13]). Then the positive numbers $\begin{bmatrix} k \\ ij \end{bmatrix}$ are independent of the $(-B)$-orthonormal bases chosen for $\mathfrak{m}_i, \mathfrak{m}_j, \mathfrak{m}_k$, and $\begin{bmatrix} k \\ ij \end{bmatrix} = \begin{bmatrix} k \\ ji \end{bmatrix} = \begin{bmatrix} j \\ ki \end{bmatrix}$.

Let $d_k = \dim \mathfrak{m}_k$. Then we have the following:

Lemma 2.1. ([10]) *The components $r_1, \dots, r_q$ of the Ricci tensor r of the metric $\langle\ ,\ \rangle$ of the form* (2) *on G/K are given by*

$$r_k = \frac{1}{2x_k} + \frac{1}{4d_k}\sum_{j,i}\frac{x_k}{x_j x_i}\begin{bmatrix}k\\ ji\end{bmatrix} - \frac{1}{2d_k}\sum_{j,i}\frac{x_j}{x_k x_i}\begin{bmatrix}j\\ ki\end{bmatrix} \quad (k = 1, \dots, q), \tag{3}$$

where the sum is taken over $i, j = 1, \dots, q$.

If, for the decomposition (1) of $\mathfrak{m}$, the Ricci tensor r of a G-invariant Riemannian metric $\langle\ ,\ \rangle$ on G/K satisfies $r(\mathfrak{m}_i, \mathfrak{m}_j) = 0$ for $i \neq j$, then, by Lemma 2.1, it follows that G-invariant Einstein metrics on $M = G/K$ are exactly the positive real solutions $(x_1, \dots, x_q) \in \mathbb{R}^q_+$ of the system of equations $\{r_1 = \lambda,\ r_2 = \lambda,\ \dots,\ r_q = \lambda\}$, where $\lambda \in \mathbb{R}_+$ is the Einstein constant.

3. A class of left-invariant metrics on $\mathrm{Sp}(n) = \mathrm{Sp}(k_1+k_2+k_3)$

We will describe a decomposition of the tangent space of the Lie group $\mathrm{Sp}(n)$ which will be convenient for our study. We consider the closed subgroup $K = \mathrm{Sp}(k_1)\times\mathrm{Sp}(k_2)\times\mathrm{Sp}(k_3)$ of $G = \mathrm{Sp}(n) = \mathrm{Sp}(k_1+k_2+k_3)$ $(k_1, k_2, k_3 \geq 1)$, where the embedding of K in G is diagonal. Then the tangent space $\mathfrak{sp}(k_1+k_2+k_3)$ of the symplectic group $G = \mathrm{Sp}(k_1+k_2+k_3)$ can be written as a direct sum of $\mathrm{Ad}(K)$-invariant modules as

$$\mathfrak{sp}(k_1+k_2+k_3) = \mathfrak{sp}(k_1)\oplus\mathfrak{sp}(k_2)\oplus\mathfrak{sp}(k_3)\oplus\mathfrak{m}, \tag{4}$$

where $\mathfrak{m}$ corresponds to the tangent space of G/K.

For $i = 1, 2, 3$, we embed the Lie subalgebras

$$\mathfrak{sp}(k_i) = \left\{\begin{pmatrix}X_i & -\bar{Y}_i\\ Y_i & \bar{X}_i\end{pmatrix}\ \middle|\ \begin{array}{l}X_i \in \mathfrak{u}(k_i),\\ Y_i\ \text{ is a } k_i\times k_i \text{ complex symmetric matrix}\end{array}\right\}$$

in the Lie algebra $\mathfrak{sp}(k_1+k_2+k_3)$ as follows:

$$\left\{\left(\begin{array}{ccc|ccc} X_1 & 0 & 0 & -\bar{Y}_1 & 0 & 0\\ 0 & 0 & 0 & 0 & 0 & 0\\ 0 & 0 & 0 & 0 & 0 & 0\\ \hline Y_1 & 0 & 0 & \bar{X}_1 & 0 & 0\\ 0 & 0 & 0 & 0 & 0 & 0\\ 0 & 0 & 0 & 0 & 0 & 0\end{array}\right)\right\}, \left\{\left(\begin{array}{ccc|ccc} 0 & 0 & 0 & 0 & 0 & 0\\ 0 & X_2 & 0 & 0 & -\bar{Y}_2 & 0\\ 0 & 0 & 0 & 0 & 0 & 0\\ \hline 0 & 0 & 0 & 0 & 0 & 0\\ 0 & Y_2 & 0 & 0 & \bar{X}_2 & 0\\ 0 & 0 & 0 & 0 & 0 & 0\end{array}\right)\right\}, \left\{\left(\begin{array}{ccc|ccc} 0 & 0 & 0 & 0 & 0 & 0\\ 0 & 0 & 0 & 0 & 0 & 0\\ 0 & 0 & X_3 & 0 & 0 & -\bar{Y}_3\\ \hline 0 & 0 & 0 & 0 & 0 & 0\\ 0 & 0 & 0 & 0 & 0 & 0\\ 0 & 0 & Y_3 & 0 & 0 & \bar{X}_3\end{array}\right)\right\}.$$

Then the tangent space $\mathfrak{m}$ of G/K is given by $\mathfrak{k}^\perp$ in $\mathfrak{g} = \mathfrak{sp}(k_1+k_2+k_3)$ with respect to the $\mathrm{Ad}(G)$-invariant inner product $-B$. If we denote by

$M(p,q)$ the set of all $p \times q$ matrices, then we see that $\mathfrak{m}$ is given by

$$\mathfrak{m} = \left\{ \begin{pmatrix} 0 & A_{12} & A_{13} & 0 & -\bar{B}_{12} & -\bar{B}_{13} \\ -{}^t\bar{A}_{12} & 0 & A_{23} & -{}^t\bar{B}_{12} & 0 & -\bar{B}_{23} \\ -{}^t\bar{A}_{13} & -{}^t\bar{A}_{23} & 0 & -{}^t\bar{B}_{13} & -{}^t\bar{B}_{23} & 0 \\ 0 & B_{12} & B_{13} & 0 & \bar{A}_{12} & \bar{A}_{13} \\ {}^tB_{12} & 0 & B_{23} & -{}^tA_{12} & 0 & \bar{A}_{23} \\ {}^tB_{13} & {}^tB_{23} & 0 & -{}^tA_{13} & -{}^tA_{23} & 0 \end{pmatrix} \;\middle|\; \begin{array}{l} A_{ij}, B_{ij} \in M(k_i, k_j) \\ (1 \leq i < j \leq 3) \end{array} \right\}.$$

Now we set

$$\mathfrak{m}_{12} = \left\{ \begin{pmatrix} 0 & A_{12} & 0 & 0 & -\bar{B}_{12} & 0 \\ -{}^t\bar{A}_{12} & 0 & 0 & -{}^t\bar{B}_{12} & 0 & 0 \\ 0 & 0 & 0 & 0 & 0 & 0 \\ 0 & B_{12} & 0 & 0 & \bar{A}_{12} & 0 \\ {}^tB_{12} & 0 & 0 & -{}^tA_{12} & 0 & 0 \\ 0 & 0 & 0 & 0 & 0 & 0 \end{pmatrix} \;\middle|\; A_{12}, B_{12} \in M(k_1, k_2) \right\},$$

$$\mathfrak{m}_{13} = \left\{ \begin{pmatrix} 0 & 0 & A_{13} & 0 & 0 & -\bar{B}_{13} \\ 0 & 0 & 0 & 0 & 0 & 0 \\ -{}^t\bar{A}_{13} & 0 & 0 & -{}^t\bar{B}_{13} & 0 & 0 \\ 0 & 0 & B_{13} & 0 & 0 & \bar{A}_{13} \\ 0 & 0 & 0 & 0 & 0 & 0 \\ {}^tB_{13} & 0 & 0 & -{}^tA_{13} & 0 & 0 \end{pmatrix} \;\middle|\; A_{13}, B_{13} \in M(k_1, k_3) \right\},$$

$$\mathfrak{m}_{23} = \left\{ \begin{pmatrix} 0 & 0 & 0 & 0 & 0 & 0 \\ 0 & 0 & A_{23} & 0 & 0 & -\bar{B}_{23} \\ 0 & -{}^t\bar{A}_{23} & 0 & 0 & -{}^t\bar{B}_{23} & 0 \\ 0 & 0 & 0 & 0 & 0 & 0 \\ 0 & 0 & B_{23} & 0 & 0 & \bar{A}_{23} \\ 0 & {}^tB_{23} & 0 & 0 & -{}^tA_{23} & 0 \end{pmatrix} \;\middle|\; A_{23}, B_{23} \in M(k_2, k_3) \right\}.$$

Note that the subspaces $\mathfrak{m}_{12}, \mathfrak{m}_{13}, \mathfrak{m}_{23}$ are the irreducible $\mathrm{Ad}(K)$-submodules whose dimensions are $\dim \mathfrak{m}_{12} = 4k_1k_2$, $\dim \mathfrak{m}_{13} = 4k_1k_3$ and $\dim \mathfrak{m}_{23} = 4k_2k_3$.

The modules $\mathfrak{m}_{12}, \mathfrak{m}_{13}$ and $\mathfrak{m}_{23}$ are given as orthogonal complements, with respect to the negative of Killing form, of $\mathfrak{sp}(k_1) \oplus \mathfrak{sp}(k_2)$ in $\mathfrak{sp}(k_1+k_2)$, of $\mathfrak{sp}(k_1) \oplus \mathfrak{sp}(k_3)$ in $\mathfrak{sp}(k_1+k_3)$, and of $\mathfrak{sp}(k_2) \oplus \mathfrak{sp}(k_3)$ in $\mathfrak{sp}(k_2+k_3)$ respectively. The space $\mathrm{Sp}(k_1+k_2+k_3)/(\mathrm{Sp}(k_1) \times \mathrm{Sp}(k_2) \times \mathrm{Sp}(k_3))$ is an example of a generalized Wallach space (cf. [9]). Note that the irreducible submodules $\mathfrak{m}_{ij}$ are mutually non equivalent, so any G-invariant metric on G/K is determined by an $\mathrm{Ad}(K)$-invariant scalar product $x_{12}\,(-B)|_{\mathfrak{m}_{12}} +$

$x_{13}\,(-B)|_{\mathfrak{m}_{13}} + x_{23}\,(-B)|_{\mathfrak{m}_{23}}$. We also set $\mathfrak{m}_1 = \mathfrak{sp}(k_1)$, $\mathfrak{m}_2 = \mathfrak{sp}(k_2)$ and $\mathfrak{m}_3 = \mathfrak{sp}(k_3)$.

Therefore, the decomposition (4) of the tangent space of the symplectic group $G = \mathrm{Sp}(k_1 + k_2 + k_3)$ takes the form

$$\mathfrak{sp}(k_1 + k_2 + k_3) = \mathfrak{m}_1 \oplus \mathfrak{m}_2 \oplus \mathfrak{m}_3 \oplus \mathfrak{m}_{12} \oplus \mathfrak{m}_{13} \oplus \mathfrak{m}_{23}. \tag{5}$$

Then we see that the following relations hold:

Lemma 3.1. *The submodules in the decomposition* (5) *satisfy the following bracket relations*:

$$\begin{array}{lll}
[\mathfrak{m}_1, \mathfrak{m}_1] = \mathfrak{m}_1, & [\mathfrak{m}_2, \mathfrak{m}_2] = \mathfrak{m}_2, & [\mathfrak{m}_3, \mathfrak{m}_3] = \mathfrak{m}_3, \\
[\mathfrak{m}_1, \mathfrak{m}_{12}] = \mathfrak{m}_{12}, & [\mathfrak{m}_1, \mathfrak{m}_{13}] = \mathfrak{m}_{13}, & [\mathfrak{m}_2, \mathfrak{m}_{12}] = \mathfrak{m}_{12}, \\
[\mathfrak{m}_2, \mathfrak{m}_{23}] = \mathfrak{m}_{23}, & [\mathfrak{m}_3, \mathfrak{m}_{13}] = \mathfrak{m}_{13}, & [\mathfrak{m}_3, \mathfrak{m}_{23}] = \mathfrak{m}_{23}, \\
[\mathfrak{m}_{12}, \mathfrak{m}_{23}] \subset \mathfrak{m}_{13}, & [\mathfrak{m}_{13}, \mathfrak{m}_{23}] \subset \mathfrak{m}_{12}, & [\mathfrak{m}_{12}, \mathfrak{m}_{13}] \subset \mathfrak{m}_{23}, \\
[\mathfrak{m}_{12}, \mathfrak{m}_{12}] \subset \mathfrak{m}_1 + \mathfrak{m}_2, & [\mathfrak{m}_{13}, \mathfrak{m}_{13}] \subset \mathfrak{m}_1 + \mathfrak{m}_3, & [\mathfrak{m}_{23}, \mathfrak{m}_{23}] \subset \mathfrak{m}_2 + \mathfrak{m}_3,
\end{array}$$

and the other bracket relations are zero.

Then by taking into account the diffeomorphism

$$G/\{e\} \cong (G \times \mathrm{Sp}(k_1) \times \mathrm{Sp}(k_2) \times \mathrm{Sp}(k_3))/\mathrm{diag}(\mathrm{Sp}(k_1) \times \mathrm{Sp}(k_2) \times \mathrm{Sp}(k_3))$$

we consider left-invariant metrics on G which are determined by the $\mathrm{Ad}(\mathrm{Sp}(k_1) \times \mathrm{Sp}(k_2) \times \mathrm{Sp}(k_3))$-invariant scalar products on $\mathfrak{sp}(k_1 + k_2 + k_3)$ given by

$$\begin{aligned}\langle\ ,\ \rangle = {} & x_1\,(-B)|_{\mathfrak{sp}(k_1)} + x_2\,(-B)|_{\mathfrak{sp}(k_2)} + x_3\,(-B)|_{\mathfrak{sp}(k_3)} \\ & + x_{12}\,(-B)|_{\mathfrak{m}_{12}} + x_{13}\,(-B)|_{\mathfrak{m}_{13}} + x_{23}\,(-B)|_{\mathfrak{m}_{23}}.\end{aligned} \tag{6}$$

Then we see that the only non zero symbols (up to permutation of indices) are

$$\begin{aligned}&\begin{bmatrix}1\\11\end{bmatrix}, \begin{bmatrix}2\\22\end{bmatrix}, \begin{bmatrix}3\\33\end{bmatrix}, \begin{bmatrix}(12)\\1(12)\end{bmatrix}, \begin{bmatrix}(13)\\1(13)\end{bmatrix}, \begin{bmatrix}(12)\\2(12)\end{bmatrix}, \begin{bmatrix}(23)\\2(23)\end{bmatrix}, \\ &\begin{bmatrix}(13)\\3(13)\end{bmatrix}, \begin{bmatrix}(23)\\3(23)\end{bmatrix}, \begin{bmatrix}(13)\\(12)(23)\end{bmatrix}.\end{aligned} \tag{7}$$

Denote by d_i and d_{ij} the dimensions of the modules $\mathfrak{m}_i$ and $\mathfrak{m}_{ij}$ respectively. Then it is $d_i = 2k_i^2 + k_i$, $d_{ij} = 4k_ik_j$.

4. Naturally reductive metrics on the compact Lie group Sp(n)

We recall the basic result of D'Atri and Ziller in [7], where they had investigated naturally reductive metrics among left-invariant metrics on compact Lie groups and gave a complete classification in the case of simple Lie groups. Let G be a compact, connected semisimple Lie group, L a closed subgroup of G and let $\mathfrak{g}$ be the Lie algebra of G and $\mathfrak{l}$ the subalgebra corresponding to L. We denote by Q the negative of the Killing form of $\mathfrak{g}$. Then Q is an $\mathrm{Ad}(G)$-invariant inner product on $\mathfrak{g}$. Let $\mathfrak{m}$ be an orthogonal complement of $\mathfrak{l}$ with respect to Q. Then we have

$$\mathfrak{g} = \mathfrak{l} \oplus \mathfrak{m}, \quad \mathrm{Ad}(L)\mathfrak{m} \subset \mathfrak{m}.$$

Let $\mathfrak{l} = \mathfrak{l}_0 \oplus \mathfrak{l}_1 \oplus \cdots \oplus \mathfrak{l}_p$ be a decomposition of $\mathfrak{l}$ into ideals, where $\mathfrak{l}_0$ is the center of $\mathfrak{l}$ and $\mathfrak{l}_i$ $(i = 1, \ldots, p)$ are simple ideals of $\mathfrak{l}$. Let $A_0|_{\mathfrak{l}_0}$ be an arbitrary metric on $\mathfrak{l}_0$.

Theorem 4.1. ([7, Theorem 1, p. 92]) *Under the notations above a left-invariant metric on G of the form*

$$\langle\ ,\ \rangle = x \cdot Q|_{\mathfrak{m}} + A_0|_{\mathfrak{l}_0} + u_1 \cdot Q|_{\mathfrak{l}_1} + \cdots + u_p \cdot Q|_{\mathfrak{l}_p}, \quad (x, u_1, \ldots, u_p > 0) \quad (8)$$

is naturally reductive with respect to $G \times L$, where $G \times L$ acts on G by $(g, l)y = gyl^{-1}$.

Moreover, if a left-invariant metric $\langle\ ,\ \rangle$ on a compact simple Lie group G is naturally reductive, then there is a closed subgroup L of G and the metric $\langle\ ,\ \rangle$ is given by the form (8).

For the Lie group $\mathrm{Sp}(n)$ we consider $\mathrm{Ad}(\mathrm{Sp}(k_1) \times \mathrm{Sp}(k_2) \times \mathrm{Sp}(k_3))$-invariant metrics of the form (6). Recall that $K = \mathrm{Sp}(k_1) \times \mathrm{Sp}(k_2) \times \mathrm{Sp}(k_3)$ with Lie algebra $\mathfrak{k}$.

Proposition 4.1. *If a left invariant metric $\langle\ ,\ \rangle$ of the form* (6) *on* $\mathrm{Sp}(n)$ *is naturally reductive with respect to* $\mathrm{Sp}(n) \times L$ *for some closed subgroup L of* $\mathrm{Sp}(n)$, *then one of the following holds:*

1) $x_1 = x_2 = x_{12}$, $x_{13} = x_{23}$
2) $x_2 = x_3 = x_{23}$, $x_{12} = x_{13}$
3) $x_1 = x_3 = x_{13}$, $x_{12} = x_{23}$,
4) $x_{12} = x_{13} = x_{23}$.

Conversely, if one of the conditions 1), 2), 3), 4) *is satisfied, then the metric $\langle\ ,\ \rangle$ of the form* (6) *is naturally reductive with respect to* $\mathrm{Sp}(n) \times L$ *for some closed subgroup L of* $\mathrm{Sp}(n)$.

Proof. Let $\mathfrak{l}$ be the Lie algebra of L. Then we have either $\mathfrak{l} \subset \mathfrak{k}$ or $\mathfrak{l} \not\subset \mathfrak{k}$. First we consider the case of $\mathfrak{l} \not\subset \mathfrak{k}$. Let $\mathfrak{h}$ be the subalgebra of $\mathfrak{g}$ generated by $\mathfrak{l}$ and $\mathfrak{k}$. Since $\mathfrak{sp}(k_1 + k_2 + k_3) = \mathfrak{m}_1 \oplus \mathfrak{m}_2 \oplus \mathfrak{m}_3 \oplus \mathfrak{m}_{12} \oplus \mathfrak{m}_{13} \oplus \mathfrak{m}_{23}$ is an irreducible decomposition as $\mathrm{Ad}(K)$-modules, we see that the Lie algebra $\mathfrak{h}$ contains at least one of $\mathfrak{m}_{12}$, $\mathfrak{m}_{13}$, $\mathfrak{m}_{23}$. We first consider the case that $\mathfrak{h}$ contains $\mathfrak{m}_{12}$. Note that $[\mathfrak{m}_{12}, \mathfrak{m}_{12}] \subset \mathfrak{m}_1 \oplus \mathfrak{m}_2$ and $\mathfrak{m}_1 \oplus \mathfrak{m}_2 \oplus \mathfrak{m}_{12}$ is a subalgebra $\mathfrak{sp}(k_1 + k_2)$. Thus we see that $\mathfrak{h}$ contains $\mathfrak{sp}(k_1 + k_2) \oplus \mathfrak{sp}(k_3)$. If $\mathfrak{h} \neq \mathfrak{sp}(k_1 + k_2) \oplus \mathfrak{sp}(k_3)$, then we see that $\mathfrak{h} = \mathfrak{sp}(k_1 + k_2 + k_3)$ and the metric $\langle\ ,\ \rangle$ is bi-invariant. If $\mathfrak{h} = \mathfrak{sp}(k_1 + k_2) \oplus \mathfrak{sp}(k_3)$, then we obtain an irreducible decomposition $\mathfrak{sp}(k_1 + k_2 + k_3) = \mathfrak{h} \oplus \mathfrak{n}$, where $\mathfrak{n} = \mathfrak{m}_{13} \oplus \mathfrak{m}_{23}$. Thus the metric $\langle\ ,\ \rangle$ of the form (6) satisfies $x_1 = x_2 = x_{12}$ and $x_{13} = x_{23}$, so we obtain case 1). Cases 2) and 3) are obtained by the similar way.

Now we consider the case $\mathfrak{l} \subset \mathfrak{k}$. Since the orthogonal complement $\mathfrak{l}^{\perp}$ of $\mathfrak{l}$ with respect to $-B$ contains the orthogonal complement $\mathfrak{k}^{\perp}$ of $\mathfrak{k}$, we see that $\mathfrak{l}^{\perp} \supset \mathfrak{m}_{12} \oplus \mathfrak{m}_{13} \oplus \mathfrak{m}_{23}$. Since the invariant metric $\langle\ ,\ \rangle$ is naturally reductive with respect to $G \times L$, it follows that $x_{12} = x_{13} = x_{23}$ by Theorem 4.1. The converse is a direct consequence of Theorem 4.1. □

5. The Ricci tensor for a class of left-invariant metrics on $\mathrm{Sp}(n) = \mathrm{Sp}(k_1 + k_2 + k_3)$

We will compute the Ricci tensor for the left-invariant metrics on $\mathrm{Sp}(n) = \mathrm{Sp}(k_1 + k_2 + k_3)$, determined by the $\mathrm{Ad}(K) = \mathrm{Ad}(\mathrm{Sp}(k_1) \times \mathrm{Sp}(k_2) \times \mathrm{Sp}(k_3))$-invariant scalar products of the form (6). Note that the Ricci tensor r of the metric (6) is also $\mathrm{Ad}(K)$-invariant.

For a moment we write the $\mathrm{Ad}(K)$-invariant irreducible decomposition (5) as

$$\mathfrak{sp}(n) = \mathfrak{sp}(k_1 + k_2 + k_3) = \mathfrak{w}_1 \oplus \mathfrak{w}_2 \oplus \mathfrak{w}_3 \oplus \mathfrak{w}_4 \oplus \mathfrak{w}_5 \oplus \mathfrak{w}_6,$$

where $\mathfrak{w}_i = \mathfrak{m}_i$ for $i = 1, 2, 3$ and $\mathfrak{w}_4 = \mathfrak{m}_{12}$, $\mathfrak{w}_5 = \mathfrak{m}_{13}$ and $\mathfrak{w}_6 = \mathfrak{m}_{23}$. Then by Lemma 3.1 it is easy to see that

Lemma 5.1. *For an $\mathrm{Ad}(K)$-invariant symmetric 2-tensor ρ on $\mathfrak{sp}(k_1 + k_2 + k_3)$, we have $\rho(\mathfrak{w}_i, \mathfrak{w}_j) = (0)$ for $i \neq j$. In particular, for the Ricci tensor r of the metric (6), we have $r(\mathfrak{w}_i, \mathfrak{w}_j) = (0)$ for $i \neq j$.*

By Lemma 2.1 and by taking into account (7) we obtain the following:

Proposition 5.1. *The components of the Ricci tensor r for the left-invariant metric $\langle\ ,\ \rangle$ on $\mathrm{Sp}(n)$ defined by (6) are given as follows:*

$$r_1 = \frac{1}{2x_1} + \frac{1}{4d_1}\left(\begin{bmatrix}1\\11\end{bmatrix}\frac{1}{x_1} + \begin{bmatrix}1\\(12)(12)\end{bmatrix}\frac{x_1}{x_{12}{}^2} + \begin{bmatrix}1\\(13)(13)\end{bmatrix}\frac{x_1}{x_{13}{}^2}\right)$$
$$-\frac{1}{2d_1}\left(\begin{bmatrix}1\\11\end{bmatrix}\frac{1}{x_1} + \begin{bmatrix}(12)\\1(12)\end{bmatrix}\frac{1}{x_1} + \begin{bmatrix}(13)\\1(13)\end{bmatrix}\frac{1}{x_1}\right),$$

$$r_2 = \frac{1}{2x_2} + \frac{1}{4d_2}\left(\begin{bmatrix}2\\22\end{bmatrix}\frac{1}{x_2} + \begin{bmatrix}2\\(12)(12)\end{bmatrix}\frac{x_2}{x_{12}{}^2} + \begin{bmatrix}2\\(23)(23)\end{bmatrix}\frac{x_2}{x_{23}{}^2}\right)$$
$$-\frac{1}{2d_2}\left(\begin{bmatrix}2\\22\end{bmatrix}\frac{1}{x_2} + \begin{bmatrix}(12)\\2(12)\end{bmatrix}\frac{1}{x_2} + \begin{bmatrix}(23)\\2(23)\end{bmatrix}\frac{1}{x_2}\right),$$

$$r_3 = \frac{1}{2x_3} + \frac{1}{4d_3}\left(\begin{bmatrix}3\\33\end{bmatrix}\frac{1}{x_3} + \begin{bmatrix}3\\(13)(13)\end{bmatrix}\frac{x_3}{x_{13}{}^2} + \begin{bmatrix}3\\(23)(23)\end{bmatrix}\frac{x_3}{x_{23}{}^2}\right)$$
$$-\frac{1}{2d_3}\left(\begin{bmatrix}3\\33\end{bmatrix}\frac{1}{x_3} + \begin{bmatrix}(13)\\3(13)\end{bmatrix}\frac{1}{x_3} + \begin{bmatrix}(23)\\3(23)\end{bmatrix}\frac{1}{x_3}\right),$$

$$r_{12} = \frac{1}{2x_{12}} + \frac{1}{4d_{12}}\left(\begin{bmatrix}(12)\\1(12)\end{bmatrix}\frac{1}{x_1}\times 2 + \begin{bmatrix}(12)\\2(12)\end{bmatrix}\frac{1}{x_2}\times 2 + \begin{bmatrix}(12)\\(13)(23)\end{bmatrix}\frac{x_{12}}{x_{13}x_{23}}\times 2\right)$$
$$-\frac{1}{2d_{12}}\left(\begin{bmatrix}1\\(12)(12)\end{bmatrix}\frac{x_1}{x_{12}{}^2} + \begin{bmatrix}(12)\\(12)1\end{bmatrix}\frac{1}{x_1} + \begin{bmatrix}2\\(12)(12)\end{bmatrix}\frac{x_2}{x_{12}{}^2} + \begin{bmatrix}(12)\\(12)2\end{bmatrix}\frac{1}{x_2}\right.$$
$$\left.+\begin{bmatrix}(13)\\(12)(23)\end{bmatrix}\frac{x_{13}}{x_{12}x_{23}} + \begin{bmatrix}(23)\\(12)(13)\end{bmatrix}\frac{x_{23}}{x_{12}x_{13}}\right),$$

$$r_{13} = \frac{1}{2x_{13}} + \frac{1}{4d_{13}}\left(\begin{bmatrix}(13)\\1(13)\end{bmatrix}\frac{1}{x_1}\times 2 + \begin{bmatrix}(13)\\2(13)\end{bmatrix}\frac{1}{x_2}\times 2 + \begin{bmatrix}(13)\\(12)(23)\end{bmatrix}\frac{x_{13}}{x_{12}x_{23}}\times 2\right)$$
$$-\frac{1}{2d_{13}}\left(\begin{bmatrix}1\\(13)(13)\end{bmatrix}\frac{x_1}{x_{13}{}^2} + \begin{bmatrix}(13)\\(13)1\end{bmatrix}\frac{1}{x_1} + \begin{bmatrix}3\\(13)(13)\end{bmatrix}\frac{x_3}{x_{13}{}^2} + \begin{bmatrix}(13)\\(13)3\end{bmatrix}\frac{1}{x_3}\right.$$
$$\left.+\begin{bmatrix}(12)\\(13)(23)\end{bmatrix}\frac{x_{12}}{x_{13}x_{23}} + \begin{bmatrix}(23)\\(13)(12)\end{bmatrix}\frac{x_{23}}{x_{13}x_{12}}\right),$$

$$r_{23} = \frac{1}{2x_{23}} + \frac{1}{4d_{23}}\left(\begin{bmatrix}(23)\\2(23)\end{bmatrix}\frac{1}{x_2}\times 2 + \begin{bmatrix}(23)\\3(23)\end{bmatrix}\frac{1}{x_3}\times 2 + \begin{bmatrix}(23)\\(12)(13)\end{bmatrix}\frac{x_{23}}{x_{12}x_{13}}\times 2\right)$$

$$-\frac{1}{2d_{23}}\left(\begin{bmatrix}2\\(23)(23)\end{bmatrix}\frac{x_2}{{x_{23}}^2}+\begin{bmatrix}(23)\\(23)2\end{bmatrix}\frac{1}{x_2}+\begin{bmatrix}3\\(23)(23)\end{bmatrix}\frac{x_3}{{x_{23}}^2}+\begin{bmatrix}(23)\\(23)3\end{bmatrix}\frac{1}{x_3}\right.$$

$$\left.+\begin{bmatrix}(12)\\(23)(13)\end{bmatrix}\frac{x_{12}}{x_{23}x_{13}}+\begin{bmatrix}(13)\\(23)(12)\end{bmatrix}\frac{x_{13}}{x_{23}x_{12}}\right),$$

where $n = k_1 + k_2 + k_3$.

We recall the following result by A. Arvanitoyeorgos, V.V. Dzhepko and Yu.G. Nikonorov:

Lemma 5.2. ([1], [2]) *For $a, b, c = 1, 2, 3$ and $(a-b)(b-c)(c-a) \neq 0$ the following relations hold:*

$$\begin{bmatrix}a\\aa\end{bmatrix}=\frac{k_a(k_a+1)(2k_a+1)}{n+1},\quad \begin{bmatrix}a\\(ab)(ab)\end{bmatrix}=\frac{k_ak_b(2k_a+1)}{(n+1)},$$
$$\begin{bmatrix}(ac)\\(ab)(bc)\end{bmatrix}=\frac{2k_ak_bk_c}{n+1}.$$

By using the above lemma, we can now obtain the components of the Ricci tensor for the metrics (6).

Proposition 5.2. *The components of the Ricci tensor r for the left-invariant metric $\langle\ ,\ \rangle$ on $\mathrm{Sp}(n)$ defined by (6) are given as follows:*

$$r_1 = \frac{k_1+1}{4(n+1)x_1}+\frac{k_2}{4(n+1)}\frac{x_1}{{x_{12}}^2}+\frac{k_3}{4(n+1)}\frac{x_1}{{x_{13}}^2},$$

$$r_2 = \frac{k_2+1}{4(n+1)x_2}+\frac{k_1}{4(n+1)}\frac{x_2}{{x_{12}}^2}+\frac{k_3}{4(n+1)}\frac{x_2}{{x_{23}}^2},$$

$$r_3 = \frac{k_3+1}{4(n+1)x_3}+\frac{k_2}{4(n+1)}\frac{x_1}{{x_{12}}^2}+\frac{k_3}{4(n+1)}\frac{x_1}{{x_{13}}^2},$$

$$r_{12} = \frac{1}{2x_{12}}+\frac{k_3}{4(n+1)}\left(\frac{x_{12}}{x_{13}x_{23}}-\frac{x_{13}}{x_{12}x_{23}}-\frac{x_{23}}{x_{12}x_{13}}\right)$$
$$-\frac{2k_1+1}{8(n+1)}\frac{x_1}{{x_{12}}^2}-\frac{2k_2+1}{8(n+1)}\frac{x_2}{{x_{12}}^2},$$

$$
\left.\begin{aligned}
r_{13} &= \frac{1}{2x_{13}} + \frac{k_2}{4(n+1)}\left(\frac{x_{13}}{x_{12}x_{23}} - \frac{x_{12}}{x_{13}x_{23}} - \frac{x_{23}}{x_{12}x_{13}}\right) \\
&\quad - \frac{2k_1+1}{8(n+1)}\frac{x_1}{x_{13}^2} - \frac{2k_3+1}{8(n+1)}\frac{x_3}{x_{13}^2}, \\
r_{23} &= \frac{1}{2x_{23}} + \frac{k_1}{4(n+1)}\left(\frac{x_{23}}{x_{13}x_{12}} - \frac{x_{13}}{x_{12}x_{23}} - \frac{x_{12}}{x_{23}x_{13}}\right) \\
&\quad - \frac{2k_2+1}{8(n+1)}\frac{x_2}{x_{23}^2} - \frac{2k_3+1}{8(n+1)}\frac{x_3}{x_{23}^2}.
\end{aligned}\right. \tag{9}
$$

6. Left-invariant Einstein metrics on Sp(n)

In this section we consider first the case where $k_1 = n-2, k_2 = 1, k_3 = 1$.

Theorem 6.1. *The Lie group* Sp(n) *($n \geq 3$) admits at least one* Ad(Sp$(n-2) \times$ Sp$(1) \times$ Sp(1))*-invariant Einstein metric of the form* (6)*, which is not naturally reductive.*

Proof. We consider the system of equations

$$r_1 = r_2, \quad r_2 = r_3, \quad r_3 = r_{12}, \quad r_{12} = r_{13}, \quad r_{13} = r_{23}. \tag{10}$$

Then finding Einstein metrics of the form (6) reduces to finding positive solutions of the system (10).

We consider our equations by putting

$$x_{12} = x_{13} = 1, \quad x_2 = x_3.$$

Then the system of equations (10) reduces to the following system of algebraic equations:

$$
\left.\begin{aligned}
f_1 &= -nx_1{x_2}^2{x_{23}}^2 + nx_2{x_{23}}^2 + 2{x_1}^2x_2{x_{23}}^2 \\
&\quad +2x_1{x_2}^2{x_{23}}^2 - x_1{x_2}^2 - 2x_1{x_{23}}^2 - x_2{x_{23}}^2 = 0, \\
f_2 &= 2nx_1x_2{x_{23}}^2 + 2n{x_2}^2{x_{23}}^2 - 4nx_2{x_{23}}^2 - 3x_1x_2{x_{23}}^2 \\
&\quad -{x_2}^2{x_{23}}^2 + 2{x_2}^2 + 2x_2{x_{23}}^3 - 4x_2{x_{23}}^2 + 4{x_{23}}^2 = 0, \\
f_3 &= -2nx_1{x_{23}}^2 - 2n{x_{23}}^3 + 4n{x_{23}}^2 + 3x_1{x_{23}}^2 \\
&\quad -3x_2{x_{23}}^2 + 6x_2 + 2{x_{23}}^3 + 4{x_{23}}^2 - 12x_{23} = 0.
\end{aligned}\right\} \tag{11}
$$

We consider the polynomial ring $R = \mathbb{Q}[z, x_2, x_1, x_{23}]$ and the ideal I generated by $\{f_1, f_2, f_3, z(x_2 - x_{23})\, x_1\, x_2\, x_{23} - 1\}$ and we need to find non zero solutions of equations (11) with $x_2 \neq x_{23}$. We take a lexicographic order $>$ with $z > x_2 > x_1 > x_{23}$ for a monomial ordering on R. Then, by the aid of computer, we see that a Gröbner basis for the ideal I contains the polynomials $\{g(x_{23}), p_1(x_{23}, x_1), p_2(x_{23}, x_2)\}$, where $g(x_{23})$ is a polynomial of x_{23} given by

$$\begin{aligned}
&g(x_{23}) = 2(n-2)^2(n-1)\left(2n^2-3n+2\right){x_{23}}^8 \\
&-4(n-2)^2(n+1)\left(2n^2+n-2\right){x_{23}}^7 \\
&+(n-2)\left(4n^4+88n^3-159n^2+47n-114\right){x_{23}}^6 \\
&-8(n-2)(n+1)\left(14n^2+7n-26\right){x_{23}}^5 \\
&+8\left(4n^4+76n^3-141n^2+11n-81\right){x_{23}}^4 \\
&-8(n+1)\left(58n^2+29n-178\right){x_{23}}^3 \\
&+4\left(16n^3+424n^2-74n+47\right){x_{23}}^2 \\
&-288(n+1)(2n+5)x_{23}+648(2n+1),
\end{aligned}$$

$p_1(x_{23}, x_1)$ is a polynomial of x_1 and x_{23} given by

$$\begin{aligned}
&p_1(x_{23}, x_1) = -18(2n-3)(2n+1)\left(2n^2-3n+6\right)x_1 \\
&+2(n-2)^2(n-1)n(2n-3)\left(2n^2-3n+2\right){x_{23}}^7 \\
&-4(n-2)^2n(n+1)(2n-3)\left(2n^2+n-2\right){x_{23}}^6 \\
&+(n-2)\left(8n^6+140n^5-486n^4+445n^3-315n^2+366n-24\right){x_{23}}^5 \\
&-4(n-2)(n+1)\left(44n^4-44n^3-113n^2+144n-12\right){x_{23}}^4 \\
&+\left(40n^6+724n^5-2694n^4+2879n^3-1623n^2+1104n-588\right){x_{23}}^3 \\
&-16(n+1)\left(28n^4-28n^3-103n^2+183n-66\right){x_{23}}^2 \\
&+4\left(8n^5+272n^4-610n^3+598n^2+225n+246\right)x_{23} \\
&-288(n+1)\left(2n^2-3n+6\right)
\end{aligned}$$

and $p_2(x_{23}, x_2)$ is a polynomial of x_2 and x_{23} given by

$$\begin{aligned}
&p_2(x_{23}, x_2) = 36n\left(2n^2-3n+6\right)x_2 \\
&+2(n-2)^3(n-1)\left(2n^2-3n+2\right){x_{23}}^7-4(n-2)^3(n+1) \\
&\times\left(2n^2+n-2\right){x_{23}}^6+(n-2)^2\left(4n^4+72n^3-119n^2+7n-98\right){x_{23}}^5 \\
&-8(n-2)^2(n+1)\left(10n^2+5n-22\right){x_{23}}^4 \\
&+4(n-2)\left(4n^4+80n^3-163n^2+15n-64\right){x_{23}}^3 \\
&-72(n-2)^2(n+1)(2n+5){x_{23}}^2+36(2n-3)\left(4n^2-9n-6\right)x_{23}.
\end{aligned}$$

Thus we see that, if there exists a real root $x_{23} = \alpha_{23}$ of $g(x_{23}) = 0$, then there are also a real solution $x_1 = \alpha_1$ of $p_1(\alpha_{23}, x_1) = 0$ and a real solution $x_2 = \alpha_2$ of $p_2(\alpha_{23}, x_2) = 0$.

Now we have $g(0) = 648(2n+1) > 0$ for $n > 0$,

$$\begin{aligned} g(2) &= 8(32n^5 - 256n^4 + 760n^3 - 960n^2 + 366n + 103) \\ &= 8(32(n-3)^5 + 224(n-3)^4 + 568(n-3)^3 \\ &\quad + 696(n-3)^2 + 438(n-3) + 121) > 0 \end{aligned}$$

for $n \geq 3$ and $g(1) = -3(n-3)n^2(2n+1) < 0$ for $n > 3$. Note that $g(3/2) = -11565/64 < 0$ for $n = 3$. Thus we see that the equation $g(x_{23}) = 0$ has at least two positive roots $x_{23} = \alpha_{23}, \beta_{23}$ with $0 < \alpha_{23} < 1 < \beta_{23} < 2$ for $n > 3$. For $n = 3$ we have the roots $x_{23} = 1, \beta_{23}$ with $3/2 < \beta_{23} < 2$.

Now we have to show that the real solutions $x_1 = \alpha_1$ of $p_1(\alpha_{23}, x_1) = 0$ and $x_2 = \alpha_2$ of $p_2(\alpha_{23}, x_2) = 0$ are positive. We take a lexicographic order $>$ with $z > x_2 > x_{23} > x_1$ for a monomial ordering on R. Then, by the aid of computer, we see that a Gröbner basis for the ideal I contains the polynomial $g_1(x_1)$ of x_1 given by

$$\begin{aligned} & g_1(x_1) = (2n+1)(2n^2-3n+2)(2n^2-3n+6)^2{x_1}^8 \\ & -16\left(4(n-2)^4 + 20(n-2)^3 + 39(n-2)^2 + 35(n-2) + 8\right)(n+1) \\ & \times(2n^2-3n+6){x_1}^7 + 2\left(224(n-2)^7 + 2424(n-2)^6 + 11068(n-2)^5\right. \\ & \left.+27998(n-2)^4 + 42247(n-2)^3 + 36635(n-2)^2 + 15596(n-2) + 2224\right){x_1}^6 \\ & -8\left(112(n-2)^6 + 892(n-2)^5 + 2836(n-2)^4 + 4623(n-2)^3 + 3947(n-2)^2\right. \\ & \left.+1468(n-2) + 192\right)(n+1){x_1}^5 + 4\left(280(n-2)^7 + 3060(n-2)^6\right. \\ & +13442(n-2)^5 + 30443(n-2)^4 + 37842(n-2)^3 + 25212(n-2)^2 \\ & \left.+7913(n-2) + 936\right){x_1}^4 - 16\left(56(n-2)^5 + 376(n-2)^4 + 890(n-2)^3\right. \\ & \left.+886(n-2)^2 + 351(n-2) + 48\right)(n-1)(n+1){x_1}^3 + 8\left(56(n-2)^5\right. \\ & \left.+466(n-2)^4 + 1393(n-2)^3 + 1781(n-2)^2 + 891(n-2) + 131\right)(n-1)^2{x_1}^2 \\ & -32(n-1)^3n(n+1)(4n^2-9n+4)x_1 + 16(n-1)^5n^2. \end{aligned}$$

Thus we see that for $n \geq 3$, the coefficients of the polynomial $g_1(x_1)$ are positive for even degree terms and negative for odd degree terms, so if the equation $g_1(x_1) = 0$ has real solutions, then these are all positive.

We take a lexicographic order $>$ with $z > x_1 > x_{23} > x_2$ for a monomial ordering on R. Then, by the aid of computer, we see that a Gröbner basis for the ideal I contains the polynomial $g_2(x_2)$ of x_2 given by

$$\begin{aligned} & g_2(x_2) = 81(n-2)^2n^2\left(2n^2-3n+6\right)^2{x_2}^8 - 432(n-2)^2n(n+1) \\ & \times(2n^2-3n+6)\left(2n^2+n+6\right){x_2}^7 + 18\left(192(n-2)^7 + 3056(n-2)^6\right. \end{aligned}$$

$$
\begin{aligned}
&+20988(n-2)^5+78896(n-2)^4+173419(n-2)^3+217840(n-2)^2\\
&+126704(n-2)+7168\Big)(n-2){x_2}^6-24\Big(128(n-2)^6+2476(n-2)^5\\
&+18356(n-2)^4+66255(n-2)^3+126232(n-2)^2+108688(n-2)+17408\Big)\\
&\times(n-2)(n+1){x_2}^5+16\Big(64(n-2)^8+2512(n-2)^7+31752(n-2)^6\\
&+193768(n-2)^5+645578(n-2)^4+1170181(n-2)^3+1013926(n-2)^2\\
&+273104(n-2)+12544\Big){x_2}^4-32\Big(272(n-2)^6+5260(n-2)^5+36640(n-2)^4\\
&+115467(n-2)^3+157086(n-2)^2+69824(n-2)+7168\Big)(n+1){x_2}^3\\
&+8\Big(3464(n-2)^6+53316(n-2)^5+319130(n-2)^4+907499(n-2)^3\\
&+1223519(n-2)^2+679136(n-2)+110272\Big){x_2}^2-576\Big(68(n-2)^4\\
&+508(n-2)^3+1197(n-2)^2+1072(n-2)+288\Big)(n+1)x_2\\
&+5184(n-1)(2n+1)\Big(2n^2-3n+2\Big)
\end{aligned}
$$

Thus we see that, for $n \geq 3$, the coefficients of the polynomial $g_2(x_2)$ are positive for even degree terms and negative for odd degree terms so, if the equation $g_2(x_2) = 0$ has real solutions, then these are all positive. Since the obtained solutions $\{x_1, x_2, x_3, x_{12}, x_{13}, x_{23}\}$ of the system (10) have the property that

$$x_{12} = x_{13} = 1, \quad x_2 = x_3, \quad x_2 \neq x_{23}, \quad x_{23} \neq 1,$$

we see that these metrics are not naturally reductive by Proposition 4.1. □

Now our main Theorem 1.1 follows from Theorem 6.1.

Remark 6.1. We see that, for $n \geq 3$,

$$
\begin{aligned}
g(9/(2n)) = &-\frac{81}{128n^8}\big(5120n^8+56n^7-187384n^6+466920n^5+1638063n^4\\
&-10288377n^3+20273490n^2-16769916n+4251528\big)\\
=&-\frac{81}{128n^8}\big(5120(n-3)^8+122936(n-3)^7+1104032(n-3)^6\\
&+4846032(n-3)^5+12428343(n-3)^4+20035539(n-3)^3\\
&+19325547(n-3)^2+9757665(n-3)+1873530\big)<0
\end{aligned}
$$

and, for $n \geq 28$,

$$
\begin{aligned}
g(1/2) = &\frac{1}{128}\big(2n^5-25n^4-905n^3-168n^2+46380n+13216\big)\\
&+245252(n-28)^2+1859772(n-28)+367920\big)>0.
\end{aligned}
$$

Thus we see that, for $n \geq 28$, the equation $g(x_{23}) = 0$ has four positive roots $x_{23} = \alpha_{23}, \beta_{23}, \gamma_{23}, \delta_{23}$ with $0 < \gamma_{23} < 9/(2n) < \delta_{23} < 1/2 < \alpha_{23} < 1 < \beta_{23} < 2$. Thus there exist at least four Einstein metrics of the form (6) on $\mathrm{Sp}(n)$ for $n \geq 28$.

We now consider the system of equations

$$r_1 = r_2, \quad r_2 = r_3, \quad r_3 = r_{12}, \quad r_{12} = r_{13}, \quad r_{13} = r_{23}, \tag{12}$$

to find more Einstein metrics on $\mathrm{Sp}(n)$.

Proposition 6.1. *The Lie group* $\mathrm{Sp}(3)$ *admits exactly four* $\mathrm{Ad}(\mathrm{Sp}(1) \times \mathrm{Sp}(1) \times \mathrm{Sp}(1))$*-invariant Einstein metrics of the form* (6) *up to isometry and scalar. One of them is not naturally reductive and the other three are naturally reductive.*

Proof. By putting $x_{13} = 1$, the system of equations (12) reduces to the system of algebraic equations:

$$\left.\begin{aligned}
f_1 &= {x_1}^2{x_{12}}^2x_2{x_{23}}^2 + {x_1}^2x_2{x_{23}}^2 - x_1{x_{12}}^2{x_2}^2 \\
&\quad - 2x_1{x_{12}}^2{x_{23}}^2 - x_1{x_2}^2{x_{23}}^2 + 2{x_{12}}^2x_2{x_{23}}^2 = 0, \\
f_2 &= {x_{12}}^2{x_2}^2x_3 - {x_{12}}^2x_2{x_{23}}^2{x_3}^2 - 2{x_{12}}^2x_2{x_{23}}^2 - {x_{12}}^2x_2{x_3}^2 \\
&\quad + 2{x_{12}}^2{x_{23}}^2x_3 + {x_2}^2{x_{23}}^2x_3 = 0, \\
f_3 &= 3x_1{x_{23}}^2x_3 - 2{x_{12}}^3x_{23}x_3 + 2{x_{12}}^2{x_{23}}^2{x_3}^2 + 4{x_{12}}^2{x_{23}}^2 \\
&\quad + 2{x_{12}}^2{x_3}^2 + 2x_{12}{x_{23}}^3x_3 - 16x_{12}{x_{23}}^2x_3 + 2x_{12}x_{23}x_3 \\
&\quad + 3x_2{x_{23}}^2x_3 = 0, \\
f_4 &= -3x_1{x_{23}}^2 + 4{x_{12}}^3x_{23} + 3{x_{12}}^2x_2 - 16{x_{12}}^2x_{23} + 3{x_{12}}^2x_3 \\
&\quad - 4x_{12}{x_{23}}^3 + 16x_{12}{x_{23}}^2 - 3x_2{x_{23}}^2 = 0, \\
f_5 &= 3x_1x_{12}{x_{23}}^2 - 3x_{12}x_2 + 3x_{12}{x_{23}}^2x_3 - 16x_{12}{x_{23}}^2 \\
&\quad + 16x_{12}x_{23} - 3x_{12}x_3 + 4{x_{23}}^3 - 4x_{23} = 0.
\end{aligned}\right\} \tag{13}$$

We consider the polynomial ring $R = \mathbb{Q}[z, x_1, x_2, x_3, x_{12}, x_{23}]$ and the ideal I generated by $\{f_1, f_2, f_3, f_4, f_5, \ z\,x_1\,x_2\,x_3\,x_{12}\,x_{23} - 1\}$. We take a lexicographic order $>$ with $z > x_1 > x_2 > x_3 > x_{12} > x_{23}$ for a monomial ordering on R. Then, by the aid of computer, we see that a Gröbner basis for the ideal I contains a polynomial $g(x_{23})$ of x_{23} of the form:

$$\begin{aligned}
g(x_{23}) &= (x_{23} - 1)(119{x_{23}}^3 - 377{x_{23}}^2 + 402x_{23} - 126)\times \\
&\quad (126{x_{23}}^3 - 402{x_{23}}^2 + 377x_{23} - 119)\times \\
&\quad (11{x_{23}}^7 - 65{x_{23}}^6 + 259{x_{23}}^5 - 709{x_{23}}^4 + 1409{x_{23}}^3 - 2039{x_{23}}^2 \\
&+ 2034x_{23} - 1134)(1134{x_{23}}^7 - 2034{x_{23}}^6 + 2039{x_{23}}^5 - 1409{x_{23}}^4 \\
&+ 709{x_{23}}^3 - 259{x_{23}}^2 + 65x_{23} - 11)h(x_{23}),
\end{aligned}$$

where $h(x_{23})$ is a polynomial of degree 72. We see that the equation $h(x_{23}) = 0$ has no real solutions and each of the other factors of $g(x_{23})$ has only one positive solution.

We consider the ideals J_1 generated by

$$\{f_1, f_2, f_3, f_4, f_5, 11{x_{23}}^7 - 65{x_{23}}^6 + 259{x_{23}}^5 - 709{x_{23}}^4 + 1409{x_{23}}^3 \\ -2039{x_{23}}^2 + 2034x_{23} - 1134,\ z\,x_1\,x_2\,x_3\,x_{12}\,x_{23} - 1\},$$

J_2 generated by

$$\{f_1, f_2, f_3, f_4, f_5, 1134{x_{23}}^7 - 2034{x_{23}}^6 + 2039{x_{23}}^5 - 1409{x_{23}}^4 \\ +709{x_{23}}^3 - 259{x_{23}}^2 + 65x_{23} - 11,\ z\,x_1\,x_2\,x_3\,x_{12}\,x_{23} - 1\},$$

J_3 generated by

$$\{f_1, f_2, f_3, f_4, f_5, 119{x_{23}}^3 - 377{x_{23}}^2 + 402x_{23} - 126, \\ z\,x_1\,x_2\,x_3\,x_{12}\,x_{23} - 1\},$$

J_4 generated by

$$\{f_1, f_2, f_3, f_4, f_5, 126{x_{23}}^3 - 402{x_{23}}^2 + 377x_{23} - 119, \\ z\,x_1\,x_2\,x_3\,x_{12}\,x_{23} - 1\}$$

and J_5 generated by

$$\{f_1, f_2, f_3, f_4, f_5, x_{23} - 1,\ z\,x_1\,x_2\,x_3\,x_{12}\,x_{23} - 1\}$$

of the polynomial ring $R = \mathbb{Q}[z, x_1, x_2, x_3, x_{12}, x_{23}]$.

We first consider the case of the system of equations defined by the ideal J_1. We take a lexicographic order $>$ with $z > x_1 > x_2 > x_3 > x_{12} > x_{23}$ for a monomial ordering on $R = \mathbb{Q}[z, x_1, x_2, x_3, x_{12}, x_{23}]$. Then, by the aid of computer, we see that a Gröbner basis for the ideal J_1 is given by

$$\{11{x_{23}}^7 - 65{x_{23}}^6 + 259{x_{23}}^5 - 709{x_{23}}^4 + 1409{x_{23}}^3 - 2039{x_{23}}^2 + 2034x_{23} \\ -1134,\ x_{12} - 1,\ 11{x_{23}}^6 - 21{x_{23}}^5 - 45{x_{23}}^4 + 411{x_{23}}^3 - 1247{x_{23}}^2 \\ +1953x_{23} - 405x_3 - 1134,\ 405x_2 - 11{x_{23}}^6 + 21{x_{23}}^5 + 45{x_{23}}^4 - 411{x_{23}}^3 \\ +1247{x_{23}}^2 - 1953x_{23} + 1134,\ 945x_1 + 66{x_{23}}^6 - 236{x_{23}}^5 + 490{x_{23}}^4 \\ -334{x_{23}}^3 - 842{x_{23}}^2 + 2998x_{23} - 3924,\ -717827{x_{23}}^6 + 3259724{x_{23}}^5 \\ -12321400{x_{23}}^4 + 28797880{x_{23}}^3 - 51686081{x_{23}}^2 + 59349464x_{23} \\ +7715736z - 51640839\}.$$

Thus we see that there is only one real solution of the system of equations defined from the ideal J_1 which is approximately given by

$$(z, x_1, x_2, x_3, x_{12}, x_{23}) \approx (4.2933, 0.55597, 0.49417, 0.49417, 1, 1.7155).$$

We substitute the values $(x_1, x_2, x_3, x_{12}, x_{23}, x_{13}) \approx (0.55597,\ 0.494172,\ 0.494172, 1, 1.71553, 1)$ into the system (9). Then we obtain that $r_1 = r_2 = r_3 = r_{12} = r_{23} = r_{13} \approx 0.294328$. We multiply this solution by a scale factor and we obtain the solution

$$(x_1,\ x_2,\ x_3,\ x_{12},\ x_{23}, x_{13}) \approx \\ (0.163638, 0.145449, 0.145449, 0.294328, 0.50493, 0.294328)$$

for the system of equations $r_1 = r_2 = r_3 = r_{12} = r_{23} = r_{13} = 1$.

Next, we consider the cases of the system of equations defined by the ideals J_2, J_3 and J_4. By the same method as in the case of the ideal J_1 for the system of equations $r_1 = r_2 = r_3 = r_{12} = r_{23} = r_{13} = 1$, we obtain the solutions

$$(x_1,\ x_2,\ x_3,\ x_{12},\ x_{23}, x_{13}) \approx \\ (0.145449, 0.163638, 0.145449, 0.294328, 0.294328, 0.50493),$$

$$(x_1,\ x_2,\ x_3,\ x_{12},\ x_{23}, x_{13}) \approx \\ (0.142567, 0.205804, 0.205804, 0.380298, 0.205804, 0.380298)$$

and

$$(x_1,\ x_2,\ x_3,\ x_{12},\ x_{23}, x_{13}) \approx \\ (0.142567, 0.205804, 0.205804, 0.380298, 0.205804, 0.380298)$$

respectively.

We now consider the case of the system of equations defined by the ideal J_5. By substituting the value $x_{23} = 1$ into $f_5 = 0$ of (13), we obtain that $3x_{12}(x_1 - x_2) = 0$ and thus we obtain the system of equations from (13):

$$\left.\begin{aligned} g_1 &= {x_1}^2{x_{12}}^2x_3 + {x_1}^2x_3 - 2x_1{x_{12}}^2{x_3}^2 - 2x_1{x_{12}}^2 + 2{x_{12}}^2x_3 = 0, \\ g_2 &= -3x_1x_3 + {x_{12}}^3x_3 - 2{x_{12}}^2{x_3}^2 - 2{x_{12}}^2 + 6x_{12}x_3 = 0, \\ g_3 &= 3x_1{x_{12}}^2 - 6x_1 + 4{x_{12}}^3 + 3{x_{12}}^2x_3 - 16{x_{12}}^2 + 12x_{12} = 0. \end{aligned}\right\} \quad (14)$$

We consider the polynomial ring $R = \mathbb{Q}[z, x_1, x_3, x_{12}]$ and the ideal I_5 generated by $\{g_1,\ g_2,\ g_3, x_1x_3x_{12}z - 1\}$. We take a lexicographic order $>$ with $z > x_1 > x_3 > x_{12}$ for a monomial ordering on $R = \mathbb{Q}[z, x_1, x_3, x_{12}]$. Then, by the aid of computer, we see that a Gröbner basis for the ideal I_5 contains a polynomial of x_{12}:

$$(x_{12} - 1)(119{x_{12}}^3 - 377{x_{12}}^2 + 402x_{12} - 126)(11{x_{12}}^7 - 65{x_{12}}^6 + 259{x_{12}}^5 \\ -709{x_{12}}^4 + 1409{x_{12}}^3 - 2039{x_{12}}^2 + 2034x_{12} - 1134).$$

From $x_{12} - 1 = 0$, we obtain the real solutions

$$(x_1, x_3, x_{12}) = (1, 1, 1) \text{ and } (2/5, 2/5, 1),$$

from $119{x_{12}}^3 - 377{x_{12}}^2 + 402x_{12} - 126 = 0$, we obtain the real solution which is approximately given by

$$(x_1, x_3, x_{12}) \approx (0.541165, 0.374883, 0.541165)$$

and, from $11{x_{12}}^7 - 65{x_{12}}^6 + 259{x_{12}}^5 - 709{x_{12}}^4 + 1409{x_{12}}^3 - 2039{x_{12}}^2 + 2034x_{12} - 1134 = 0$, we obtain the real solution which is approximately given by

$$(x_1, x_3, x_{12}) \approx (0.494172, 0.55597, 1.71553).$$

We now substitute the values $(x_1, x_2, x_3, x_{12}, x_{23}, x_{13}) = (2/5, 2/5, 2/5, 1, 1, 1)$ into the system (9). Then we obtain that $r_1 = r_2 = r_3 = r_{12} = r_{23} = r_{13} = 29/80$. We multiply this solution by a scale factor and we obtain the solution

$$(x_1, x_2, x_3, x_{12}, x_{23}, x_{13}) = (29/200, 29/200, 29/200, 29/80, 29/80, 29/80)$$

for the system of equations $r_1 = r_2 = r_3 = r_{12} = r_{23} = r_{13} = 1$.

By the same method as above, we obtain a solution given by

$$(x_1, x_2, x_3, x_{12}, x_{23}, x_{13}) = (1/4, 1/4, 1/4, 1/4, 1/4, 1/4),$$

and solutions which are approximately given by

$$\begin{aligned}(x_1, x_2, x_3, x_{12}, x_{23}, x_{13}) &\approx \\ &(0.205804, 0.205804, 0.142567, 0.205804, 0.380298, 0.380298), \\ &(0.145449, 0.145449, 0.163638, 0.50493, 0.294328, 0.294328),\end{aligned}$$

for the system of equations $r_1 = r_2 = r_3 = r_{12} = r_{23} = r_{13} = 1$.

Note that last two metrics are isometric to the metrics obtained by the ideals J_3 and J_2, respectively. □

By the same method we obtain the following propositions.

Proposition 6.2. *The Lie group* Sp(4) *admits exactly six* Ad(Sp(2) × Sp(1) × Sp(1))*-invariant Einstein metrics of the form* (6) *up to isometry and scalar. Three of them are not naturally reductive and the other* *three are naturally reductive. The solutions for the non naturally reductive metrics are approximately given by*

$$
\begin{aligned}
&(x_1, x_2, x_3, x_{12}, x_{23}, x_{13}) \approx \\
&\qquad (0.180564, 0.114935, 0.114935, 0.326608, 0.508812, 0.326608),\\
&\qquad (0.169957, 0.116403, 0.116403, 0.380445, 0.310184, 0.380445),\\
&\qquad (0.170185, 0.117632, 0.131837, 0.489011, 0.241674, 0.320149)
\end{aligned}
$$

and the solutions for the naturally reductive metrics are approximately given by

$$
\begin{aligned}
&(x_1, x_2, x_3, x_{12}, x_{23}, x_{13}) \approx \\
&\qquad (0.216504, 0.216504, 0.113641, 0.216504, 0.376842, 0.376842),\\
&\qquad (33/196, 33/196, 33/196, 11/28, 33/196, 11/28),\\
&\qquad (1/4, 1/4, 1/4, 1/4, 1/4, 1/4).
\end{aligned}
$$

Proposition 6.3. *The Lie group* Sp(5) *admits exactly six* Ad(Sp(3) × Sp(1) × Sp(1))*-invariant Einstein metrics of the form* (6) *up to isometry and scalar. Three of them are not naturally reductive and the other three are naturally reductive. The solutions for the non naturally reductive metrics are approximately given by*

$$
\begin{aligned}
&(x_1, x_2, x_3, x_{12}, x_{23}, x_{13}) \approx \\
&\qquad (0.186299, 0.0981436, 0.107724, 0.469801, 0.207125, 0.332506),\\
&\qquad (0.194296, 0.0947096, 0.0947096, 0.337434, 0.494427, 0.337434),\\
&\qquad (0.186037, 0.0970275, 0.0970275, 0.385871, 0.260268, 0.385871)
\end{aligned}
$$

and the solutions for the naturally reductive metrics are approximately given by

$$
\begin{aligned}
&(x_1, x_2, x_3, x_{12}, x_{23}, x_{13}) \approx \\
&\qquad (0.185009, 0.140975, 0.140975, 0.394342, 0.140975, 0.394342),\\
&\qquad (0.223412, 0.223412, 0.0940135, 0.223412, 0.371385, 0.371386),\\
&\qquad (1/4, 1/4, 1/4, 1/4, 1/4, 1/4).
\end{aligned}
$$

Proposition 6.4. *The Lie group* Sp(5) *admits exactly six* Ad(Sp(2) × Sp(2) × Sp(1))*-invariant Einstein metrics of the form* (6) *up to isometry and scalar. Three of them are not naturally reductive and the other three are naturally reductive. The solutions for the non naturally reductive* *metrics are approximately given by*

$$\begin{aligned}(x_1, x_2, x_3, x_{12}, x_{23}, x_{13}) \approx\\ &(0.143750, 0.143750, 0.109487, 0.479841, 0.276389, 0.276389),\\ &(0.143036, 0.143036, 0.0967338, 0.398959, 0.341148, 0.341148),\\ &(0.142057, 0.153625, 0.0961501, 0.349437, 0.505958, 0.280275)\end{aligned}$$

and the solutions for the naturally reductive metrics are approximately given by

$$\begin{aligned}(x_1, x_2, x_3, x_{12}, x_{23}, x_{13}) \approx\\ &(0.140975, 0.185009, 0.185009, 0.394342, 0.185009, 0.394342),\\ &(0.223412, 0.223412, 0.0967338, 0.223412, 0.71386, 0.71386),\\ &(1/4, 1/4, 1/4, 1/4, 1/4, 1/4).\end{aligned}$$

Proposition 6.5. *The Lie group* Sp(6) *admits exactly six* Ad(Sp(4) × Sp(1) × Sp(1))*-invariant Einstein metrics of the form* (6) *up to isometry and scalar. Three of them are not naturally reductive and the other three are naturally reductive. The solutions for the non naturally reductive metrics are approximately given by*

$$\begin{aligned}(x_1, x_2, x_3, x_{12}, x_{23}, x_{13}) \approx\\ &(0.197228, 0.0831302, 0.0831302, 0.385909, 0.220579, 0.385909),\\ &(0.20424, 0.0803836, 0.0803836, 0.340661, 0.480162, 0.340661),\\ &(0.197486, 0.0838840, 0.0901956, 0.451246, 0.182822, 0.339648)\end{aligned}$$

and the solutions for the naturally reductive metrics are approximately given by

$$\begin{aligned}(x_1, x_2, x_3, x_{12}, x_{23}, x_{13}) \approx\\ &(0.228189, 0.0799654, 0.228189, 0.365730, 0.365730, 0.228189),\\ &(0.196534, 0.120691, 0.120691, 0.391909, 0.120691, 0.391909),\\ &(1/4, 1/4, 1/4, 1/4, 1/4, 1/4).\end{aligned}$$

Proposition 6.6. *The Lie group* Sp(6) *admits exactly eight* Ad(Sp(3) × Sp(2) × Sp(1))*-invariant Einstein metrics of the form* (6) *up to isometry and scalar. Four of them are not naturally reductive and four others are naturally reductive. Solutions for the non naturally reductive metrics are* *approximately given by*

$$(x_1, x_2, x_3, x_{12}, x_{23}, x_{13}) \approx$$
$$(0.160905, 0.131760, 0.0824922, 0.360299, 0.245488, 0.493148)$$
$$(0.162065, 0.123756, 0.0913345, 0.465574, 0.245943, 0.298142),$$
$$(0.161502, 0.123065, 0.0827579, 0.406554, 0.297661, 0.355452),$$
$$(0.170401, 0.121360, 0.0812213, 0.360818, 0.500972, 0.298828)$$

and solutions for the naturally reductive metrics are given by

$$(x_1, x_2, x_3, x_{12}, x_{23}, x_{13}) =$$
$$(0.228189, 0.228189, 0.0799654, 0.228189, 0.36573, 0.36573),$$
$$(0.196534, 0.120691, 0.196534, 0.391909, 0.391909, 0.196534),$$
$$(4/25, 4/25, 4/25, 2/5, 4/25, 2/5),$$
$$(1/4, 1/4, 1/4, 1/4, 1/4, 1/4).$$

Proposition 6.7. *The Lie group* Sp(6) *admits exactly four* Ad(Sp(2) × Sp(2) × Sp(2))*-invariant Einstein metrics of the form* (6) *up to isometry and scalar. One of them is not naturally reductive and the other three are naturally reductive. The solution for the non naturally reductive metric is approximately given by*

$$(x_1, x_2, x_3, x_{12}, x_{23}, x_{13}) \approx$$
$$(0.122619, 0.122619, 0.134157, 0.506100, 0.308511, 0.308511)$$

and the solutions for the naturally reductive metrics are approximately given by

$$(x_1, x_2, x_3, x_{12}, x_{23}, x_{13}) \approx$$
$$(0.120691, 0.196534, 0.196534, 0.391909, 0.196534, 0.391909),$$
$$(31/252, 31/252, 31/252, 31/84, 31/84, 31/84),$$
$$(1/4, 1/4, 1/4, 1/4, 1/4, 1/4).$$

Acknowledgments

The work was supported by Grant #*E*.037 from the Research Committee of the University of Patras (Programme K. Karatheodori) and JSPS KAKENHI Grant Number 25400071. It was developed and completed while the first author was a visiting scholar at Tufts University during Fall 2014.

References

1. A. Arvanitoyeorgos, V. V. Dzhepko and Yu. G. Nikonorov, Invariant Einstein metrics on some homogeneous spaces of classical Lie groups, Canad. J. Math. **61**(6), 1201–1213 (2009).
2. A. Arvanitoyeorgos, V. V. Dzhepko and Yu. G. Nikonorov, Invariant Einstein metrics on quaternionic Stiefel manifolds, Bull. Greek Math. Soc. **53**, 1–14 (2007).
3. A. Arvanitoyeorgos, K. Mori and Y. Sakane, *Einstein metrics on compact Lie groups which are not naturally reductive*, Geom. Dedicata **160**(1), 261–285 (2012).
4. A. Arvanitoyeorgos, Y. Sakane and M. Statha, *New Einstein metrics on the Lie group $SO(n)$ which are not naturally reductive*, preprint.
5. A. L. Besse, *Einstein Manifolds*, Springer-Verlag, Berlin, 1986.
6. Z. Chen and K. Liang, Non-naturally reductive Einstein metrics on the compact simple Lie group F_4, Ann. Glob. Anal. Geom. **46**, 103–115 (2014).
7. J. E. D'Atri and W. Ziller, Naturally reductive metrics and Einstein metrics on compact Lie groups, Memoirs A.M.S. **19** (215) (1979).
8. K. Mori, Left Invariant Einstein Metrics on $SU(N)$ that are not Naturally Reductive, Master Thesis (in Japanese) Osaka University 1994, English Translation: Osaka University RPM 96010 (preprint series) 1996.
9. Yu. G. Nikonorov, E. D. Rodionov and V. V. Slavskii, Geometry of Homogeneous Riemannian Manifolds, J. Math. Sci. (N.Y.) **146**(6), 6313–6390 (2007).
10. J-S. Park and Y. Sakane, Invariant Einstein metrics on certain homogeneous spaces, Tokyo J. Math. **20**(1), 51–61 (1997).
11. M. Wang, Einstein metrics from symmetry and bundle constructions, in *Surveys in Differential Geometry: Essays on Einstein Manifolds*, Surv. Diff. Geom. VI, Int. Press, Boston, MA 1999.
12. M. Wang, Einstein metrics from symmetry and bundle constructions: A sequel, in *Differential Geometry: Under the Influence of S.-S. Chern*, Advanced Lectures in Mathematics, vol. 22, 253–309, Higher Education Press/International Press, 2012.
13. M. Wang and W. Ziller, Existence and non-existence of homogeneous Einstein metrics, Invent. Math. **84**, 177–194 (1986).

Received January 6, 2015

Current Developments in Differential Geometry
and its Related Fields pp. 23 – 44

LAPLACIANS FOR FINITE REGULAR KÄHLER GRAPHS AND FOR THEIR DUAL GRAPHS

Toshiaki ADACHI *

Department of Mathematics, Nagoya Institute of Technology,
Nagoya 466-8555, Japan
E-mail: adachi@nitech.ac.jp

In memory of Professor Rugreliya Borisova,
the former Vicerector on international activities,
University of Veliko Tarnovo

A Kähler graph is a compound of two graphs having the same set of vertices. Paths on this graph are said to be bicolored paths if two kinds of edges appear reciprocally. As generating operators for random walks of bicolored paths, we define its Laplacians. Being different from ordinary discrete Laplacians for graphs, these operators are not selfadjoint in general. In this paper we show that Laplacians for regular Kähler graphs are adjoint to those for their dual Kähler graphs. Also, we give a condition on Kähler graphs that their (p, q)-Laplacians are selfadjoint.

Keywords: Regular Kähler graphs, bicolored paths, (p, q)-Laplacians, adjacency and transition operators, probabilistic weights, Kähler magnetic fields, magnetic means, Kähler-dual, selfadjoint.

1. Introduction

The aim of this paper is to study adjointness of (p, q)-Laplacians for regular Kähler graphs and those for their dual graphs by comparing transition operators of Kähler graphs with magnetic mean operators associated with Kähler magnetic fields on Kähler manifolds. A *magnetic field* on a Riemannian manifold M is a closed 2-form. Through the identification of vector fields on a Euclidean 3-space $\mathbb{R}^3$ with 2-forms on this space, we regard a closed 2-form on a Riemannian manifold as a generalization of a static magnetic field on $\mathbb{R}^3$. Under the influence of this magnetic field, a motion of a charged particle of unit speed on this manifold is said to be a trajectory. When there are no influence of magnetic fields, the motion of a charged particle is a geodesic. We may therefore say that trajectories for magnetic fields are generalizations of geodesics.

*The author is partially supported by Grant-in-Aid for Scientific Research (C) (No. 24540075) Japan Society for the Promotion of Science.

We here consider discretizations of Riemannian manifolds admitting magnetic fields. It is said that graphs are discrete models of Riemannian manifolds. A graph consists of a set of vertices and a set of edges, and forms a 1-dimensional CW-complex. Paths, which are chains of edges, on a graph correspond to geodesics. As a discrete model of a Riemannian manifold admitting magnetic fields, the author introduced the notion of Kähler graphs in [3]. He calls a graph Kähler if it is a compound of a principal and an auxiliary graphs (see §3 for more detail). He considered that paths which alternatively consist of principal edges and auxiliary edges correspond to trajectories.

In their papers [12, 13, 14], Yaermaimaiti and the author defined Laplacians for Kähler graphs as generating operators for random walks with specified paths, and studied their eigenvalues of Laplacians for some kinds of Kähler graphs. It is well known that Laplacians for Riemannian manifolds and for ordinary graphs are selfadjoint operators. But our Laplacians for Kähler graphs are not selfadjoint in general. In this paper we explain this non-symmetric property by comparing Laplacians for regular Kähler graphs with magnetic mean operators for manifolds admitting magnetic fields. Also, we give a condition for our (p,q)-Laplacians to be selfadjoint.

2. Magnetic fields

We shall start by recalling the definition of trajectories for magnetic fields briefly. For a magnetic field $\mathbb{B}$, which is a closed 2-form, on a Riemannian manifold M, we take a skew-symmetric endomorphism $\Omega_{\mathbb{B}}$ of the tangent bundle TM of M defined by $\mathbb{B}(v,w) = \langle v, \Omega_{\mathbb{B}}(w)\rangle$ for all tangent vectors $v, w \in T_pM$ at an arbitrary point $p \in M$. This shows the action of the Lorentz force on moving electric charged particles. When $\Omega_{\mathbb{B}}$ is parallel (i.e. $\nabla\Omega_{\mathbb{B}} = 0$), we call $\mathbb{B}$ *uniform*.

A smooth curve γ parameterized by its arclength is said to be a trajectory if it satisfies the differential equation $\nabla_{\dot\gamma}\dot\gamma = \Omega_{\mathbb{B}}(\dot\gamma)$. When there are no influence of magnetic fields, which is the case that $\mathbb{B}$ is the null 2-form 0, trajectories are geodesics because Ω_0 is the null endomorphism. Given a magnetic field $\mathbb{B}$ we get a family of magnetic fields $\{\mathbb{B}_\kappa = \kappa\mathbb{B}\}_{\kappa\in\mathbb{R}}$ by multiplying constants to it. Since we have $\Omega_{\mathbb{B}_\kappa} = \kappa\Omega_{\mathbb{B}}$, we can consider that trajectories for such magnetic fields to be perturbations of geodesics. A typical example of uniform magnetic fields is the Kähler form $\mathbb{B}_J$ on a Kähler manifold (M,J) with complex structure J. For a Kähler magnetic field $\mathbb{B}_\kappa = \kappa\mathbb{B}_J$, which is a constant multiple of $\mathbb{B}_J$, its trajectories are curves satisfying the differential equation $\nabla_{\dot\gamma}\dot\gamma = \kappa J\dot\gamma$, and are circles of constant geodesic curvature $|\kappa|$. As Kähler magnetic fields are closely related with complex structure on a underlying manifold, the author considers that properties of trajectories for Kähler magnetic fields have an interaction to properties of the underlying manifold (see [1 ,4]).

3. Kähler graphs

In this section we give a discrete model of a manifold admitting uniform magnetic fields. A pair (V, E) of a set V of vertices and a set E of edges is said to be a graph. We suppose that each edge does not have its orientation. We can regard a graph to be a 1-dimensional CW-complex. In this paper we suppose that every graph does not have loops, hairs and multiple edges. Here, a loop is an edge joining a vertex to itself, a hair is an edge which is a blind alley, and multiple edges are edges joining the same pair of vertices. We say two vertices $v, v' \in V$ to be adjacent to each other if there is an edge joining them. In this case we denote as $v \sim v'$. We denote by $d_G(v)$ the cardinality of the set of vertices which are adjacent to v, and call it the degree at v. For example, for a Petersen graph, which consists of 10 vertices and 15 edges, the degree at each vertex is 3 (see Fig. 1). A chain of edges $\gamma = (v_0, v_1, \ldots, v_n)$ ($\{v_{i-1}, v_i\} \in E$ for $i = 1, \ldots, n$) is called a path, or more precisely an n-step path. When there is j_0 with $v_{j_0-1} = v_{j_0+1}$, we say a path γ contains a backtracking. It is considered that paths without backtracking on a graph correspond to geodesics on a Riemannian manifold. For a path γ we denote its origin v_0 by $o(\gamma)$ and its terminus v_n by $t(\gamma)$. Though a 1-step path, which is an edge, does not have direction, we use these notations. Given two paths $\sigma = (v_0, \ldots, v_p)$, $\rho = (w_0, \ldots, w_q)$ with $v_p = w_0$, we set paths $\sigma \cdot \rho$, σ^{-1} by $\sigma \cdot \rho = (v_0, \ldots, v_p, w_1 \ldots, w_q)$ and $\sigma^{-1} = (v_p, \ldots, v_0)$. We call a graph connected if for arbitrary distinct vertices there is a path joining them. That is, a graph is said to be connected if it is connected as a CW-complex.

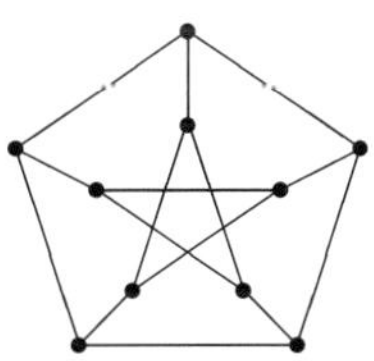

Fig. 1. Petersen graph

We call a graph $G = (V, E)$ *Kähler* if the set E of edges is divided into two subsets as $E = E^{(p)} \cup E^{(a)}$ ($E^{(p)} \cap E^{(a)} = \emptyset$) so that the cardinalities of the sets $\{e \in E^{(p)} \mid o(e) = v\}$ and $\{e \in E^{(a)} \mid o(e) = v\}$ are not smaller than 2 for each vertex $v \in V$. We denote these cardinalities by $d_G^{(p)}(v)$ and $d_G^{(a)}(v)$ and call them the principal and the auxiliary degrees, respectively. We say two graphs $G^{(p)} = (V, E^{(p)})$, $G^{(a)} = (V, E^{(a)})$ to be the principal and the auxiliary graphs of G. For two vertices $v, v' \in V$ we denote by $v \sim_p v'$ if they are adjacent to each other in $G^{(p)}$ and by $v \sim_a v'$ if they are adjacent to each other in $G^{(a)}$. We call a Kähler graph *regular* if both of its principal and auxiliary graphs are regular. That is, both the principal and the auxiliary degrees are constant functions on the set of vertices. We call a Kähler graph *complete* if it is regular as a Kähler graph and is complete as an ordinary graph, that is two distinct vertices are adjacent to each other either in the principal graph or in the auxiliary graph.

Example 3.1. We take the set $V = \{z = x + \sqrt{-1}y \mid x, y \in \mathbb{Z}\}$ of lattice points in a complex line $\mathbb{C}$.

(1) For two $z = x + \sqrt{-1}y$, $z' = x' + \sqrt{-1}y' \in V$, we define $z \sim_p z'$ if and only if $z' - z = \pm 1$ and define $z \sim_a z'$ if and only if $z' - z = \pm\sqrt{-1}$. We then get a Kähler graph as in Fig. 2. We call this a Kähler graph of complex lattice.
(2) We define $z \sim_p z'$ if and only if either $z' - z = \pm 1$ or $z' - z = \pm\sqrt{-1}$ holds, and define $z \sim_a z'$ if and only if either $z' - z = \pm(1+\sqrt{-1})$ or $z' - z = \pm(1-\sqrt{-1})$ holds. We then get another Kähler graph as in Fig. 3. We shall call this a complex line of Cartesian-tensor product type.
(3) We define $z \sim_p z'$ if and only if either $z' - z = \pm 1$ or $z' - z = \pm(1 + \sqrt{-1})$ holds, and define $z \sim_a z'$ if and only if either $z' - z = \pm\sqrt{-1}$ or $z' - z = \pm(1 - \sqrt{-1})$ holds. Then we get the third Kähler graph as in Fig. 4. We shall call this a Cayley complex line.

These three Kähler graphs are regular. For the first Kähler graph, we have $d^{(p)} = d^{(a)} = 2$ and both the principal and the auxiliary graphs are not connected. For the second, we have $d^{(p)} = d^{(a)} = 4$ and its principal graph is connected but not for its auxiliary graph. It has two connected components. For the third, we have $d^{(p)} = d^{(a)} = 4$ and both of the principal and the auxiliary graphs are connected. In the following figures, we show principal edges by lines and auxiliary edges by dotted lines.

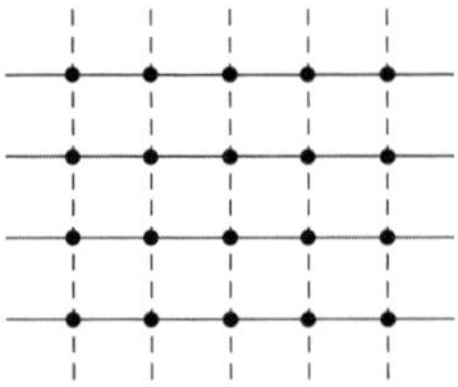

Fig. 2. Complex lattice

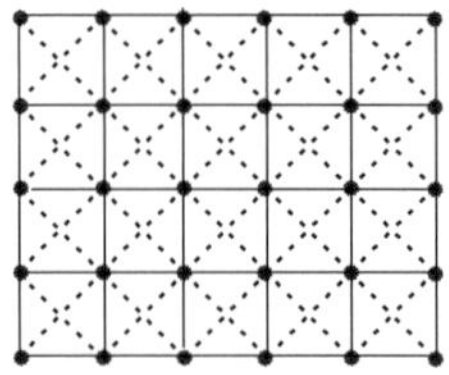

Fig. 3. Complex line of Cartesian-tensor product type

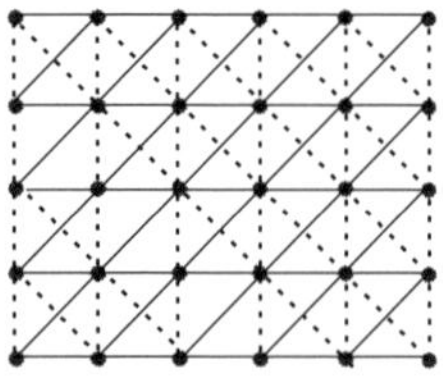

Fig. 4. Cayley complex line

Example 3.2. Fig. 5 shows a regular Kähler graph of $d_G^{(p)} = d_G^{(a)} = 3$ having ten vertices. Both of its principal and its auxiliary graphs are Petersen graphs. We shall call this graph a Petersen Kähler graph.

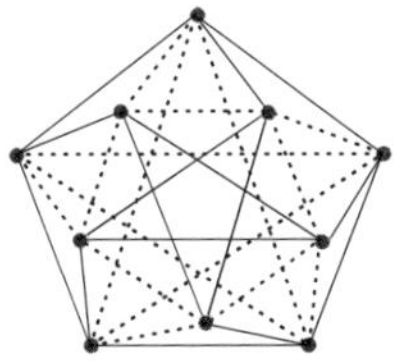

Fig. 5. Petersen Kähler graph

Example 3.3. We take a set

$$V = \{(i,j) \mid i = 0,1,2,3,\ j = 0,1,2,3,4\}.$$

We define $E^{(p)}$ as

i) $(i,j) \sim_p (i,j+1)$ for $i = 0,2$,
ii) $(i,j) \sim_p (i,j+2)$ for $i = 1,3$,
iii) $(0,j) \sim_p (1,j)$ and $(2,j) \sim_p (3,j)$,

and define $E^{(a)}$ as

i) $(i,j) \sim_a (i,j+1)$ for $i = 1,3$,
ii) $(i,j) \sim_a (i,j+2)$ for $i = 0,2$,
iii) $(1,j) \sim_a (2,j)$ and $(3,j) \sim_a (0,j)$,

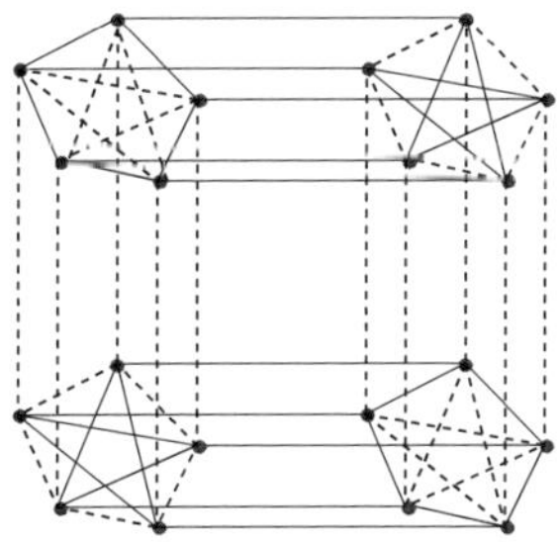

Fig. 6. Complexified Petersen graph

where we consider the second index by modulo 5. We then get a regular Kähler graph of $d_G^{(p)} = d_G^{(a)} = 3$ having 20 vertices. We shall call this graph a complexified Petersen graph.

Example 3.4. Let $G = (V, E)$ be an ordinary graph. We define its complement graph $G^c = (V, E^c)$ so that two distinct vertices $v, w \in V$ are adjacent to each other in G^c if and only if they are not adjacent to each other in G. Then $G^K = (V, E \cup E^c)$ is a Kähler graph. We call this the complement-filled Kähler graph of G.

We can construct many examples of Kähler graphs by taking products and by considering Cayley graphs (see [10, 11, 12, 14]).

Since we have graphs admitting a "structure", we here introduce objects corresponding to trajectories for magnetic fields. Since every trajectory for a Kähler magnetic field is a curve of constant geodesic curvature, we need to define "bended" paths on a Kähler graph. As paths on an ordinary graph are considered to correspond to geodesics, we consider paths on the principal graph of a Kähler graph show motions of charged particles without the influence of magnetic fields. We use the auxiliary graph to show how trajectories are bended. Given a pair (p, q) of relatively prime positive integers, we say a $(p+q)$-step path $\gamma = (v_0, \ldots, v_{p+q})$ in a Kähler graph G

to be a *(p,q)-primitive bicolored path* if it does not have backtracking, the subpath $(v_0,\dots,v_p)$ is a path in $G^{(p)}$, and $(v_p,\dots,v_{p+q})$ is a path in $G^{(a)}$. An $m(p+q)$-step path $\gamma=(v_0,\dots,v_{m(p+q)})$ with positive integer m, we say it is a (p,q)-bicolored path if its subpaths $\gamma_i=(v_{(i-1)(p+q)},\dots,v_{i(p+q)})$, $i=1,\dots,m$, are (p,q)-primitive bicolored paths. We consider (p,q)-bicolored paths are motions of charged particles under a magnetic field of strength q/p. Under the influence of a magnetic field of strength q/p, a p-step path on the principal graph is bended and its terminus reaches to the terminus of a (p,q)-primitive bicolored path whose first p-step coincides with the given path.

On a Riemannian manifold the direction of the action of the Lorentz force of a magnetic field $\mathbb{B}$ is given by the skew-symmetric endomorphism $\Omega_{\mathbb{B}}$. But on a Kähler graph we do not have such direction of the action of magnetic fields. We hence can not determine which direction paths on the principal graph are bended to. Therefore we treat bicolored paths probabilistically. Given a $(p+q)$-primitive bicolored path $\gamma=(v_0,\dots,v_{p+q})$ on a Kähler graph G, we define its probabilistic weight $\omega_G(\gamma)$ by

$$\omega_G(\gamma)=\bigl\{d_G^{(a)}(v_p)\bigl(d_G^{(a)}(v_{p+1})-1\bigr)\cdots\bigl(d_G^{(a)}(v_{p+q-1})-1\bigr)\bigr\}^{-1}.$$

For a $(p+q)$-bicolored path $\gamma=\gamma_1\cdots\gamma_m$, we set $\omega_G(\gamma)=\omega_G(\gamma_1)\times\cdots\times\omega_G(\gamma_m)$. We denote by $\mathfrak{P}_q^{(a)}(v;G)$ the set of all q-step paths without backtracking in the auxiliary graph $G^{(a)}$ of G whose origin is a vertex v in G. For $\rho\in\mathfrak{P}_q^{(a)}(v;G)$ we define its probabilistic weight $\omega_{G^{(a)}}(\rho)$ by just the same way as above. That is, we set $\omega_{G^{(a)}}(\rho)=\bigl\{d_G^{(a)}(v_0)\bigl(d_G^{(a)}(v_1)-1\bigr)\cdots\bigl(d_G^{(a)}(v_{q-1})-1\bigr)\bigr\}^{-1}$. We should note that when $q\geq 2$ the probabilistic weight $\omega_{G^{(a)}}(\rho)$ does not coincide with the reciprocal of the cardinality $\sharp\bigl(\mathfrak{P}_q^{(a)}(v;G)\bigr)$ in general. Only in the case that $G^{(a)}$ is regular they coincide with each other. Also we note that $\omega_{G^{(a)}}(\rho^{-1})d_G^{(a)}(t(\rho))=\omega_{G^{(a)}}(\rho)d_G^{(a)}(o(\rho))$ holds.

4. Duality of adjacency and transition operators

For a finite Kähler graph $G=(V,E^{(p)}\cup E^{(a)})$, we denote by $\mathfrak{P}_{p,q}(v;G)$ the set of all (p,q)-primitive bicolored paths whose origin is $v\in V$. We define the adjacency operator $\mathcal{A}_G^{(p,q)}:C(V)\to C(V)$ and the probabilistic transition operator $\mathcal{Q}_G^{(p,q)}:C(V)\to C(V)$ acting on the set $C(V)$ of all complex-valued functions by

$$\mathcal{A}_G^{(p,q)}f(v)=\sum_{\gamma\in\mathfrak{P}_{p,q}(v;G)}\omega_G(\gamma)f\bigl(t(\gamma)\bigr),$$

$$\mathcal{Q}_G^{(p,q)} f(v) = \frac{1}{\sum_{\gamma \in \mathfrak{P}_{p,q}(v;G)} \omega_G(\gamma)} \sum_{\gamma \in \mathfrak{P}_{p,q}(v;G)} \omega_G(\gamma) f\bigl(t(\gamma)\bigr).$$

First we decompose these operators into operators for the principal and the auxiliary graphs. Given an ordinary finite graph $H = (W, F)$ and a positive integer k, we denote by $\mathfrak{P}_k(w; H)$ the set of all k-step paths without backtracking whose origin is w. We define the k-step adjacency operator $\mathcal{A}_H^{(k)}$ acting on $C(W)$ by $\mathcal{A}_H^{(k)} f(w) = \sum_{\sigma \in \mathfrak{P}_k(w;H)} f\bigl(t(\sigma)\bigr)$.

Coming back to our Kähler graph G we put $\mathfrak{P}_p^{(p)}(v; G) = \mathfrak{P}_p(v; G^{(p)})$. We define operators $\mathcal{A}_G^{(p,0)}, \mathcal{P}_G^{(p,0)}, \mathcal{Q}_G^{(0,q)}$ acting on $C(V)$ by $\mathcal{A}_G^{(p,0)} = \mathcal{A}_{G^{(p)}}^{(p)}$ and

$$\mathcal{P}_G^{(p,0)} f(v) = \frac{1}{\sharp\bigl(\mathfrak{P}_p^{(p)}(v; G)\bigr)} \sum_{\sigma \in \mathfrak{P}_p(v; G^{(p)})} f\bigl(t(\sigma)\bigr),$$

$$\mathcal{Q}_G^{(0,q)} f(v) = \sum_{\rho \in \mathfrak{P}_q^{(a)}(v;G)} \omega_G(\rho) f\bigl(t(\rho)\bigr).$$

When $p = 1$, these operators $\mathcal{A}_G^{(1,0)}$, $\mathcal{P}_G^{(1,0)}$ are the adjacency and the transition operators $\mathcal{A}_{G^{(p)}}$ and $\mathcal{P}_{G^{(p)}}$ of the principal graph $G^{(p)}$, respectively, and when $q = 1$ the operator $\mathcal{Q}_G^{(0,1)}$ is the transition operator $\mathcal{P}_{G^{(a)}}$ of the auxiliary graph $G^{(a)}$. For a (p,q)-primitive bicolored path $\gamma = \sigma \cdot \rho$ with a p-step path σ in $G^{(p)}$ and a q-step path ρ in $G^{(a)}$ which satisfy $t(\sigma) = o(\rho)$, we have $\omega_G(\gamma) = \omega_{G^{(a)}}(\rho)$ by definitions. As we have $\sum_{\gamma \in \mathfrak{P}_{p,q}(v;G)} \omega(\gamma) = \sharp\bigl(\mathfrak{P}_p(v; G^{(p)})\bigr)$, we get the following.

Lemma 4.1. *For an arbitrary pair (p,q) of positive integers, we have $\mathcal{A}_G^{(p,q)} = \mathcal{A}_G^{(p,0)} \circ \mathcal{Q}_G^{(0,q)}$ and $\mathcal{P}_G^{(p,q)} = \mathcal{P}_G^{(p,0)} \circ \mathcal{Q}_G^{(0,q)}$.*

By this lemma one can easily see that $\mathcal{A}_G^{(p,q)}$ and $\mathcal{Q}_G^{(p,q)}$ are not self-adjoint operators in general. In order to justify this we consider the duals of Kähler graphs. For a Kähler graph $G = (V, E^{(p)} \cup E^{(a)})$, we put $F^{(p)} = E^{(a)}$, $F^{(a)} = E^{(p)}$ and call $G^* = (V, F^{(p)} \cup F^{(a)})$ the *dual* Kähler graph of G. Operators for G and G^* are related with each other in the following manner.

Theorem 4.1. *Let $G = (V, E^{(p)} \cup E^{(a)})$ be a finite regular Kähler graph. Then, for an arbitrary pair (p,q) of positive integers, its probabilistic transition operator $\mathcal{Q}_G^{(p,q)}$ and the probabilistic transition operator $\mathcal{Q}_{G^*}^{(q,p)}$ of its dual Kähler graph G^* are adjoint to each other with respect to the ordinary*

(Hermitian) inner product. That is, we have

$$\langle \mathcal{Q}_G^{(p,q)} f, g\rangle = \langle f, \mathcal{Q}_{G^*}^{(q,p)} g\rangle \qquad \textit{for all} \quad f, g \in C(V). \tag{4.1}$$

Proof. Since G is regular, we see that the cardinality $\sharp\big(\mathfrak{P}_q^{(a)}(v;G)\big)$ does not depend on the choice of $v \in V$, and we have

$$\omega(\gamma) = (d_G^{(a)})^{-1}(d_G^{(a)} - 1)^{1-q} = \big\{\sharp\big(\mathfrak{P}_q^{(a)}(\star;G)\big)\big\}^{-1}$$

for an arbitrary (p,q)-primitive bicolored path γ on G. It is clear that the path γ^{-1} is a (q,p)-primitive bicolored path on G^* if and only if γ is a (p,q)-primitive bicolored path on G. Therefore if we put $\mathfrak{P}_{p,q}(G) = \bigcup_{v\in V} \mathfrak{P}_{p,q}(v;G)$, we find that there is a bijective correspondence between $\mathfrak{P}_{p,q}(G)$ and $\mathfrak{P}_{q,p}(G^*)$. We hence have

$$\begin{aligned}
\langle \mathcal{Q}_G^{(p,q)} f, g\rangle &= \sum_{v\in V} \sum_{\gamma\in\mathfrak{P}_{p,q}(v;G)} \frac{1}{\sharp\big(\mathfrak{P}_p^{(p)}(v;G)\big)}\, \omega(\gamma) f\big(t(\gamma)\big)\, \overline{g(v)} \\
&= \frac{1}{\sharp\big(\mathfrak{P}_p^{(p)}(\star;G)\big)} \frac{1}{\sharp\big(\mathfrak{P}_q^{(a)}(\star;G)\big)} \sum_{v\in V} \sum_{\gamma\in\mathfrak{P}_{p,q}(v;G)} f\big(t(\gamma)\big)\, \overline{g(v)} \\
&= \frac{1}{\sharp\big(\mathfrak{P}_q^{(p)}(\star;G^*)\big)} \frac{1}{\sharp\big(\mathfrak{P}_p^{(a)}(\star;G^*)\big)} \sum_{v\in V} \sum_{\gamma\in\mathfrak{P}_{p,q}(v;G)} f\big(o(\gamma^{-1})\big)\, \overline{g\big(t(\gamma^{-1})\big)} \\
&= \sum_{w\in V} \sum_{\eta\in\mathfrak{P}_{q,p}(w;G^*)} \frac{1}{\sharp\big(\mathfrak{P}_q^{(p)}(w;G^*)\big)}\, \omega_{G^*}(\eta) f(w)\, \overline{g\big(t(\eta)\big)} \\
&= \langle f, \mathcal{Q}_{G^*}^{(q,p)} g\rangle,
\end{aligned}$$

and get the conclusion. □

Remark 4.1. For a finite Kähler graph $G = (V, E^{(p)} \cup E^{(a)})$ and a positive integer p, we define an inner product $(\ ,\ )_{p;G}$ on $C(V)$ by

$$(f, g)_{p,G} = \sum_{v\in V} \big\{\sharp\big(\mathfrak{P}_p^{(p)}(v;G)\big)\big\}^{-1} f(v)\, \overline{g(v)}.$$

When G is regular, we have

$$\big(\mathcal{A}_G^{(p,q)} f, g\big)_{p,G} = \big(f, \mathcal{A}_{G^*}^{(q,p)} g\big)_{q,G^*} \quad \text{for all} \quad f, g \in C(V).$$

We now restrict ourselves to $(1,1)$-adjacency operators and the $(1,1)$-transition operators for Kähler graphs.

Theorem 4.2. *Let $G = (V, E^{(p)} \cup E^{(a)})$ be a finite Kähler graph. Suppose the principal and the auxiliary degrees satisfy $d_G^{(p)}(v) = d_G^{(a)}(v)$ at each* *vertex $v \in V$. Then $\mathcal{A}_G^{(1,1)}$ and $\mathcal{A}_{G^*}^{(1,1)}$ are adjoint to each other with respect to the ordinary inner product.*

Proof. By the condition, for each $\gamma = (v_0, v_1, v_2) \in \mathfrak{P}_{1,1}(G)$ we have

$$\omega_G(\gamma) = \left\{d_G^{(a)}(v_1)\right\}^{-1} = \left\{d_G^{(p)}(v_1)\right\}^{-1} = \left\{d_{G^*}^{(a)}(v_1)\right\}^{-1} = \omega_{G^*}(\gamma^{-1}).$$

Therefore we obtain

$$\begin{aligned}\langle \mathcal{A}_G^{(1,1)} f, g\rangle &= \sum_{v\in V} \sum_{\gamma\in\mathfrak{P}_{1,1}(G)} \omega_G(\gamma) f\big(t(\gamma)\big)\,\overline{g(v)} \\ &= \sum_{v\in V} \sum_{\gamma\in\mathfrak{P}_{1,1}(G)} \omega_{G^*}(\gamma^{-1}) f\big(o(\gamma^{-1})\big)\,\overline{g\big(t(\gamma^{-1})\big)} = \langle f, \mathcal{A}_{G^*}^{(1,1)} g\rangle,\end{aligned}$$

which shows the conclusion. □

Remark 4.2. For a finite Kähler graph $G = (V, E^{(p)} \cup E^{(a)})$, we define an inner product $\langle\ ,\ \rangle_G$ on $C(V)$ by $\langle f, g\rangle_G = \sum_{v\in V} d_G^{(p)}(v) f(v)\,\overline{g(v)}$. When $d^{(p)}(v) = d^{(a)}(v)$ at each vertex $v \in V$, we have

$$\big\langle \mathcal{Q}_G^{(1,1)} f, g\big\rangle_G = \big\langle f, \mathcal{Q}_{G^*}^{(1,1)} g\big\rangle_{G^*} \quad \text{for all} \quad f, g \in C(V).$$

Being different from the regularity condition, coincidence of the principal and the auxiliary degrees does not guarantee the adjointness of arbitrary (p,q)-adjacency and (p,q)-transition operators.

Example 4.1. We consider the Kähler graph in Fig. 7. We see that this graph satisfies $d^{(p)}(v) = d^{(a)}(v)$ at an arbitrary vertex v and that it is not regular. If we represent these operators $\mathcal{A}_G^{(1,1)}$, $\mathcal{A}_{G^*}^{(1,1)}$, $\mathcal{Q}_G^{(2,1)}$, $\mathcal{Q}_{G^*}^{(1,2)}$ by matrices $A_G^{(1,1)}$, $A_{G^*}^{(1,1)}$, $Q_G^{(2,1)}$, $Q_{G^*}^{(1,2)}$, then we find that ${}^tA_G^{(1,1)} = A_{G^*}^{(1,1)}$ and ${}^tQ_G^{(2,1)} \neq Q_{G^*}^{(1,2)}$, where for a matrix M we denote by ${}^tM$ its transposed matrix. In fact we have

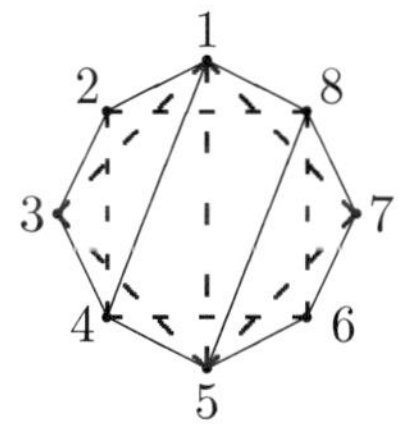

Fig. 7.

$$Q_G^{(2,1)} = \begin{pmatrix} \frac{13}{30} & 0 & \frac{2}{15} & 0 & \frac{3}{10} & 0 & \frac{2}{15} & 0 \\ 0 & \frac{1}{3} & 0 & \frac{1}{9} & 0 & \frac{1}{3} & 0 & \frac{2}{9} \\ \frac{1}{9} & 0 & \frac{1}{3} & 0 & \frac{2}{9} & 0 & \frac{1}{3} & 0 \\ 0 & \frac{2}{15} & 0 & \frac{13}{30} & 0 & \frac{2}{15} & 0 & \frac{3}{10} \\ \frac{3}{10} & 0 & \frac{2}{15} & 0 & \frac{13}{30} & 0 & \frac{2}{15} & 0 \\ 0 & \frac{1}{3} & 0 & \frac{2}{9} & 0 & \frac{1}{3} & 0 & \frac{1}{9} \\ \frac{2}{9} & 0 & \frac{1}{3} & 0 & \frac{1}{9} & 0 & \frac{1}{3} & 0 \\ 0 & \frac{2}{15} & 0 & \frac{3}{10} & 0 & \frac{2}{15} & 0 & \frac{13}{30} \end{pmatrix}, \quad Q_{G^*}^{(1,2)} = \begin{pmatrix} \frac{4}{9} & 0 & \frac{1}{18} & 0 & \frac{1}{3} & 0 & \frac{1}{6} & 0 \\ 0 & \frac{1}{3} & 0 & \frac{1}{6} & 0 & \frac{1}{3} & 0 & \frac{1}{6} \\ \frac{1}{6} & 0 & \frac{1}{3} & 0 & \frac{1}{6} & 0 & \frac{1}{3} & 0 \\ 0 & \frac{1}{18} & 0 & \frac{4}{9} & 0 & \frac{1}{6} & 0 & \frac{1}{3} \\ \frac{1}{3} & 0 & \frac{1}{6} & 0 & \frac{4}{9} & 0 & \frac{1}{18} & 0 \\ 0 & \frac{1}{3} & 0 & \frac{1}{6} & 0 & \frac{1}{3} & 0 & \frac{1}{6} \\ \frac{1}{6} & 0 & \frac{1}{3} & 0 & \frac{1}{6} & 0 & \frac{1}{3} & 0 \\ 0 & \frac{1}{6} & 0 & \frac{1}{3} & 0 & \frac{1}{18} & 0 & \frac{4}{9} \end{pmatrix}.$$

When a Kähler graph satisfies neither the regularity condition nor the condition on principal and auxiliary degrees, its adjacency and its transition operators are not adjoint to those for its dual Kähler graph, in general.

Example 4.2. We consider the Kähler graph in Fig. 8. Its auxiliary graph is regular but not for its principal graph. The principal and the auxiliary degrees do not coincide at the vertices 1 and 4. By using the same notations as in Example 4.1, we find that ${}^tA_G^{(1,1)} \neq A_{G^*}^{(1,1)}$ and ${}^tQ_G^{(1,1)} \neq Q_{G^*}^{(1,1)}$. In fact we have

$$A_G^{(1,1)} = \begin{pmatrix} 0 & 1 & 0 & 1 & 0 & 1 \\ \frac{1}{2} & 0 & \frac{1}{2} & 0 & 1 & 0 \\ 0 & \frac{1}{2} & 0 & \frac{1}{2} & 0 & 1 \\ 1 & 0 & 1 & 0 & 1 & 0 \\ 0 & 1 & 0 & \frac{1}{2} & 0 & \frac{1}{2} \\ \frac{1}{2} & 0 & 1 & 0 & \frac{1}{2} & 0 \end{pmatrix}, \quad A_{G^*}^{(1,1)} = \begin{pmatrix} 0 & \frac{1}{2} & 0 & 1 & 0 & \frac{1}{2} \\ \frac{5}{6} & 0 & \frac{1}{3} & 0 & \frac{5}{6} & 0 \\ 0 & \frac{1}{3} & 0 & \frac{5}{6} & 0 & \frac{5}{6} \\ 1 & 0 & \frac{1}{2} & 0 & \frac{1}{2} & 0 \\ 0 & \frac{5}{6} & 0 & \frac{5}{6} & 0 & \frac{1}{3} \\ \frac{5}{6} & 0 & \frac{5}{6} & 0 & \frac{1}{3} & 0 \end{pmatrix}.$$

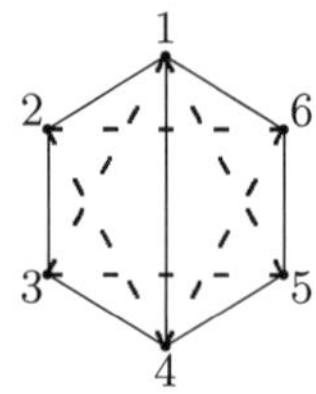

Fig. 8.

For a Kähler graph G we define its combinatorial (p,q)-Laplacian $\Delta_{\mathcal{A}_G}^{(p,q)}$ and its probabilistic transitional (p,q)-Laplacian $\Delta_{\mathcal{Q}_G}^{(p,q)}$ acting on $C(V)$ by

$$\Delta_{\mathcal{A}_G}^{(p,q)} = \mathcal{D}_G^{(p,0)} - \mathcal{A}_G^{(p,q)}, \quad \Delta_{\mathcal{Q}_G}^{(p,q)} = I - \mathcal{Q}_G^{(p,q)},$$

where the operator $\mathcal{D}_G^{(p,0)}$ is defined by $\mathcal{D}_G^{(p,0)} f(v) = \sharp\big(\mathfrak{P}_p^{(p)}(v;G)\big) f(v)$ for $f \in C(V)$. As a direct consequences of Theorems 4.1, 4.2, we get the following.

Corollary 4.1. *For a regular finite Kähler graph G, we have*

$$\langle \Delta_{\mathcal{Q}_G}^{(p,q)} f, g\rangle = \langle f, \Delta_{\mathcal{Q}_{G^*}}^{(q,p)} g\rangle \quad \textit{and} \quad \big(\Delta_{\mathcal{A}_G}^{(p,q)} f, g\big)_{p,G} = \big(f, \Delta_{\mathcal{A}_{G^*}}^{(q,p)} g\big)_{q,G^*}$$

for all $f, g \in C(V)$.

Corollary 4.2. *For a finite Kähler graph G satisfying $d_G^{(p)}(v) = d_G^{(a)}(v)$ at each vertex v, we have*

$$\langle \Delta_{\mathcal{A}_G}^{(1,1)} f, g\rangle = \langle f, \Delta_{\mathcal{A}_{G^*}}^{(1,1)} g\rangle \quad \textit{and} \quad \langle \Delta_{\mathcal{Q}_G}^{(1,1)} f, g\rangle_G = \langle f, \Delta_{\mathcal{Q}_{G^*}}^{(1,1)} g\rangle_{G^*}$$

for all $f, g \in C(V)$.

5. Magnetic means

In this section we make mention of the duality of probabilistic transition operators for Kähler graphs and their duals from the viewpoint of magnetic fields on a manifold. For a constant multiple $\mathbb{B}_\kappa = \kappa\mathbb{B}$ of a magnetic field $\mathbb{B}$

on a manifold M, we define the magnetic exponential map $\mathbb{B}_\kappa\exp_p : T_pM \to M$ at a point $p \in M$ by

$$\mathbb{B}_\kappa\exp_p(v) = \begin{cases} \gamma_{v/\|v\|}(\|v\|), & \text{if } v \neq 0_p, \\ p, & \text{if } v = 0_p, \end{cases}$$

where γ_w denotes the trajectory for $\mathbb{B}_\kappa$ with initial unit tangent vector $w \in UM$. We define the magnetic mean $\mathcal{M}_{r,\kappa}$ of radius r for $\mathbb{B}_\kappa$ which acts on the set $C^0(M)$ of continuous functions on M by

$$\mathcal{M}_{r,\kappa}f(p) = \int_{UM} f\big(\mathbb{B}_\kappa\exp_p(rw)\big)\, dS_p(w),$$

where dS_p denotes the normalized canonical density of the unit sphere U_pM. This operator is extended onto the set $L^2(M)$ of square integrable functions. When $\kappa = 0$, this operator is called a spherical mean. Its expansion with respect to the radius r is given as

$$\mathcal{M}_{r,0}f = f - \frac{r^2}{2m}\Delta f + \frac{r^4}{4!}P_4 f + \cdots + \frac{r^{2k}}{(2k)!}P_{2k}f + o\big(r^{2(k+1)}\big),$$

where m denotes the real dimension of M and P_{2j} is a formally selfadjoint operator (see [8]). When $\kappa \neq 0$ we have a similar expression

$$\mathcal{M}_{r,\kappa}f = f - \frac{r^2}{2m}\Delta f + \frac{r^3}{3!}Q_3 f + \cdots + \frac{r^k}{k!}Q_k f + o\big(r^{k+1}\big),$$

with some operators $Q_3, \ldots, Q_k$ which are not selfadjoint in general (see [2]). Thus we may consider that probabilistic transition operators correspond to magnetic mean operators.

Since Liouville's theorem guarantees that the magnetic flow $\mathbb{B}_\kappa\varphi_t$ on UM defined by $\mathbb{B}_\kappa\varphi_t(w) = \dot\gamma_w(t)$ preserves the density dS on UM, we find

$$\langle \mathcal{M}_{r,\kappa}f, g\rangle_{L^2} = \langle f, \mathcal{M}_{r,-\kappa}g\rangle_{L^2} \quad \text{for all} \quad f, g \in L^2(M) \tag{5.1}$$

(see p.32 of [2]). We should note that if γ is a trajectory for $\mathbb{B}_\kappa$ then the curve γ^{-1} defined by $\gamma^{-1}(t) = \gamma(-t)$ is a trajectory for $\mathbb{B}_{-\kappa}$ because

$$\nabla_{(\gamma^{-1})'}(\gamma^{-1})' = \nabla_{-\dot\gamma}(-\dot\gamma) = \kappa\dot\gamma = -\kappa(\gamma^{-1})'.$$

Thus we can regard that the adjointness (4.1) of probabilistic transition operators corresponds to the adjointness (5.1) of magnetic mean operators. Therefore, for a regular Kähler graph G and for a negative $-q/p$ with relatively prime positive integers p, q, we consider (q,p)-bicolored paths on the dual Kähler graph G^* and regard them as trajectories for a magnetic field on G of strength $-q/p$. Since the adjointness (5.1) holds on an arbitrary Riemannian manifold, we are interested in studying adjointness of probabilistic transition operators on a given Kähler graph.

6. Kähler graphs having selfadjoint (p, q)-Laplacians

Let M be a quotient of a complex space form, which is one of a complex projective space, a complex Euclidean space and a complex hyperbolic space. For a Kähler magnetic field $\mathbb{B}_\kappa$, we denote by $\iota_\kappa(p)$ the $\mathbb{B}_\kappa$-injectivity radius at $p \in M$. It is defined as $\iota_\kappa(p) = \sup\{r \mid \mathbb{B}_\kappa\exp_p|_{B_r(0_p)} \text{ is injective}\}$, where $B_r(0_p) \subset T_pM$ denotes a Euclidean closed ball of radius r. On this manifold M, it is known that $\iota_\kappa(p) = \iota_{-\kappa}(p)$ and that a trajectory-ball $B_r(p;\kappa) = \mathbb{B}_\kappa\exp_p\big(B_r(0_p)\big)$ coincides with $B_r(p;-\kappa)$ at each point p for an arbitrary r with $0 < r < \iota_\kappa(p)$. Therefore on M we have $\mathcal{M}_{r;\kappa} = \mathcal{M}_{r;-\kappa}$ for an arbitrary r with $0 < r < \iota_\kappa(M) = \inf\{\iota_\kappa(p) \mid p \in M\}$. In other words, by (5.1) we can say that $\mathcal{M}_{r;\kappa}$ is a selfadjoint operator of $L^2(M)$. We are therefore interested in Kähler graphs having selfadjoint (p, q)-Laplacians.

Example 6.1. For the regular Kähler graphs G_1, G_2 given in Figs. 9, 10, we find that their $(1, 1)$-probabilistic transition operators are selfadjoint but not for the Petersen Kähler graph. The adjacency operators of G_1, G_2 are simultaneously diagonalizable ($\mathcal{A}_{G_i^{(p)}}\mathcal{A}_{G_i^{(a)}} = \mathcal{A}_{G_i^{(a)}}\mathcal{A}_{G_i^{(p)}}$). On the other hand, the Petersen Kähler graph G is regular with $d_G^{(p)} = d_G^{(a)} = 3$, but its adjacency operators are not simultaneously diagonalizable. For the Petersen Kähler graph G in Fig. 5, we label the outside five vertices as 1,2,3,4,5 anticlockwisely, and label inside five vertices as 6,5,7,8,9,10 anticlockwisely so that 1 and 6 are joined by a principal edge. By using the similar notations as in Example 4.1, we have

$$4Q_{G_1}^{(1,1)} = \begin{pmatrix} 0&1&0&2&0&1 \\ 1&0&1&0&2&0 \\ 0&1&0&1&0&2 \\ 2&0&1&0&1&0 \\ 0&2&0&1&0&1 \\ 1&0&2&0&1&0 \end{pmatrix}, \quad 8Q_{G_2}^{(1,1)} = \begin{pmatrix} 0&1&2&1&1&2&1 \\ 1&0&1&2&1&1&2 \\ 2&1&0&1&2&1&1 \\ 1&2&1&0&1&2&1 \\ 1&1&2&1&0&1&2 \\ 2&1&1&2&1&0&1 \\ 1&2&1&1&2&1&0 \end{pmatrix},$$

$$9Q_G^{(1,1)} = \begin{pmatrix} 0&1&1&1&2&1&1&1&0&1 \\ 2&0&1&1&1&1&1&1&1&0 \\ 1&2&0&1&1&0&1&1&1&1 \\ 1&1&2&0&1&1&0&1&1&1 \\ 1&1&1&2&0&1&1&0&1&1 \\ 0&1&2&1&0&0&2&1&1&1 \\ 0&0&1&2&1&1&0&2&1&1 \\ 1&0&0&1&2&1&1&0&2&1 \\ 2&1&0&0&1&1&1&1&0&2 \\ 1&2&1&0&0&2&1&1&1&0 \end{pmatrix}.$$

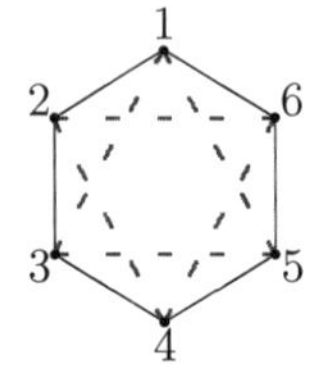

Fig. 9. G_1

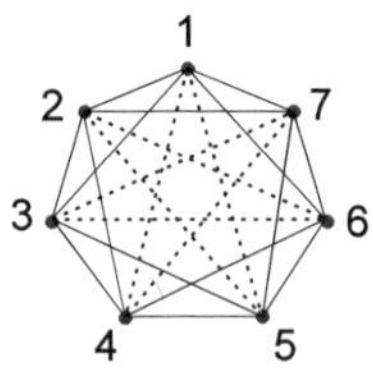

Fig. 10. G_2

In view of the above examples, we consider the case that the adjacency operators $\mathcal{A}_{G^{(p)}}$, $\mathcal{A}_{G^{(a)}}$ for the principal and the auxiliary graphs of a Kähler graph G are commutative. For example, finite complement-filled Kähler graphs satisfy this condition. When G is regular, it is clear that

$$\mathcal{Q}_G^{(1,1)} = \big(d_G^{(p)} d_G^{(a)}\big)^{-1} \mathcal{A}_{G^{(p)}} \mathcal{A}_{G^{(a)}} = \big(d_G^{(a)} d_G^{(p)}\big)^{-1} \mathcal{A}_{G^{(a)}} \mathcal{A}_{G^{(p)}} = \mathcal{Q}_{G^*}^{(1,1)}.$$

This shows that when G is regular the probabilistic transitional $(1,1)$-Laplacian $\Delta_{\mathcal{Q}_G}^{(1,1)}$ is selfadjoint with respect to the ordinary inner product $\langle\ ,\ \rangle$ on $C(V)$.

Next we study $\mathcal{Q}_G^{(p,q)}$ for a general pair (p,q). The regularity condition guarantees that higher step adjacency operators are given as $\mathcal{A}_{G^{(p)}}^{(p)} = F_p(\mathcal{A}_{G^{(p)}}; d^{(p)})$ and $\mathcal{A}_{G^{(a)}}^{(q)} = F_q(\mathcal{A}_{G^{(a)}}; d^{(a)})$ with the polynomials $F_n(t;d)$ which are inductively defined by the following relations

$$\begin{cases} F_{n+1}(t;d) = tF_n(t;d) - (d-1)F_{n-1}(t;d) & (n \geq 2), \\ F_0(t;d) = 1, \quad F_1(t;d) = t, \quad F_2(t;d) = t^2 - d, \end{cases}$$

(see [13]). This shows that $\mathcal{A}_{G^{(p)}}^{(p)} \mathcal{A}_{G^{(a)}}^{(q)} = \mathcal{A}_{G^{(a)}}^{(q)} \mathcal{A}_{G^{(p)}}^{(p)}$ and that

$$\mathcal{Q}_G^{(p,q)} = \frac{1}{d_G^{(p)}\big(d_G^{(p)}-1\big)^{p-1} d_G^{(a)}\big(d_G^{(a)}-1\big)^{q-1}} \mathcal{A}_{G^{(p)}}^{(p)} \mathcal{A}_{G^{(a)}}^{(q)} = \mathcal{Q}_{G^*}^{(q,p)}.$$

We therefore get the following.

Theorem 6.1. *Let G be a finite regular Kähler graph. We suppose the adjacency operators $\mathcal{A}_{G^{(p)}}$, $\mathcal{A}_{G^{(a)}}$ for its principal and its auxiliary graphs are commutative.*

(1) *The probabilistic transitional Laplacian $\Delta_{\mathcal{Q}_G}^{(p,q)}$ is selfadjoint with respect to the ordinary inner product on $C(V)$.*
(2) *Two probabilistic transitional Laplacians $\Delta_{\mathcal{Q}_G}^{(p,q)}$ and $\Delta_{\mathcal{Q}_{G^*}}^{(q,p)}$ acting on $C(V)$ coincide with each other.*
(3) *When $d_G^{(p)} = d_G^{(a)}$, the combinatorial Laplacian $\Delta_{\mathcal{A}_G}^{(p,q)}$ is selfadjoint with respect to the ordinary inner product on $C(V)$.*

We have many examples satisfying the assumption in Theorem 6.1. If we take two regular finite graphs, then their Kähler graphs of product types have commutative adjacency operators (see [12, 13, 14]). Moreover we can construct many such Kähler graphs. For two Kähler graphs $G = (V, E^{(p)} \cup E^{(a)})$, $H = (W, F^{(p)} \cup F^{(a)})$, we call a map $f : V \to W$ a homomorphism if it satisfies $f(v) \sim_p f(v')$ when $v \sim_p v'$ and $f(v) \sim_a f(v')$ when $v \sim_a v'$. A bijection $f : V \to W$ is said to be an isomorphism if f and f^{-1} are homomorphism of Kähler graphs. When we have an isomorphism of G to H, we say G is isomorphic to H. We say a Kähler graph G to be vertex-transitive if for an arbitrary pair (v, v') of distinct vertices there is an isomorphism f of G with $f(v) = v'$. Clearly, a vertex-transitive Kähler graph is regular.

Proposition 6.1 ([15]). *Let $n\,(\geq 5)$ be an integer which satisfies $n \equiv 1$ modulo 4. Then there exists a complete vertex-transitive Kähler graph $G = (V, E^{(p)} \cup E^{(a)})$ satisfying the following conditions:*

i) *the adjacency operators $\mathcal{A}_{G^{(p)}}$ and $\mathcal{A}_{G^{(a)}}$ are commutative;*
ii) *$n_G = n$ and $d_G^{(p)} = d_G^{(a)} = (n-1)/2$, where n_G denotes the cardinality of the set V.*

Though we have many examples of regular Kähler graphs whose adjacency operators are commutative, the commutativity condition seems a bit strong. Since Laplacians for a finite regular Kähler graph and those for its dual Kähler graph are adjoint to each other, we weaken the condition in the following manner. We shall say that the Laplacian $\Delta_{\mathcal{A}_G}^{(p,q)}$ (resp. $\Delta_{\mathcal{Q}_G}^{(p,q)}$) for a Kähler graph G has a dual correspondence if the following conditions hold:

i) the dual Kähler graph G^* is isomorphic to G;
ii) there is an isomorphism $\varphi = \varphi_{p,q} : G \to G^*$ which satisfies $\Delta_{\mathcal{A}_G}^{(p,q)} \circ \varphi^* = \varphi^* \circ \Delta_{\mathcal{A}_{G^*}}^{(q,p)}$ (resp. $\Delta_{\mathcal{Q}_G}^{(p,q)} \circ \varphi^* = \varphi^* \circ \Delta_{\mathcal{Q}_{G^*}}^{(q,p)}$).

Here, we define $\varphi^* : C(V) \to C(V)$ by $\varphi^*(f) = f \circ \varphi$ for each $f \in C(V)$. In view of Theorem 4.1, for a regular Kähler graph G we say $\Delta_{\mathcal{Q}_G}^{(p,q)}$ is *quasi-selfadjoint* with respect to the ordinary inner product if it has a dual correspondence. Similarly, in view of Theorem 4.2, for a Kähler graph G satisfying $d^{(p)}(v) = d^{(a)}(v)$ at each its vertex v, we say $\Delta_{\mathcal{A}_G}^{(1,1)}$ is quasi-selfadjoint with respect to the ordinary inner product if it has a dual correspondence. We note that the eigenvalues of (p,q)-Laplacians are not necessarily real even if they have dual correspondences (see [10]). By the definition we get the following.

Theorem 6.2. *Let $G = (V, E^{(p)} \cup E^{(a)})$ be a finite Kähler graph which is isomorphic to its dual Kähler graph G^*.*

(1) *If $d_G^{(p)}(v) = d_G^{(a)}(v)$ at each $v \in V$, then $\Delta_{\mathcal{A}_G}^{(1,1)}$ is quasi-selfadjoint with respect to the ordinary inner product.*
(2) *If G is regular, then $\Delta_{\mathcal{Q}_G}^{(1,1)}$ is quasi-selfadjoint with respect to the ordinary inner product.*

Proof. Let $\varphi : G \to G^*$ be an isomorphism. We then have

$$\mathcal{A}_{G^{(p)}} = \varphi^* \circ \mathcal{A}_{(G^*)^{(p)}} \circ (\varphi^*)^{-1} = \varphi^* \circ \mathcal{A}_{G^{(a)}} \circ (\varphi^*)^{-1},$$
$$\mathcal{P}_{G^{(a)}} = \varphi^* \circ \mathcal{P}_{(G^*)^{(a)}} \circ (\varphi^*)^{-1} = \varphi^* \circ \mathcal{P}_{G^{(p)}} \circ (\varphi^*)^{-1},$$

and obtain

$$\mathcal{A}_G^{(1,1)} = \mathcal{A}_{G^{(p)}}\mathcal{P}_{G^{(a)}} = \varphi^* \circ \mathcal{A}_{G^{(a)}} \circ \mathcal{P}_{G^{(p)}} \circ (\varphi^*)^{-1}.$$

By the proof of Theorem 4.2 we can show our assertions. But in order to make more clearer, we show them directly. We denote by $\widehat{\mathcal{D}}^{(p)}, \widehat{\mathcal{D}}^{(a)}$ the operators acting on $C(V)$ which are given by

$$\widehat{\mathcal{D}}^{(p)} f(v) = \frac{1}{d_G^{(p)}(v)} f(v), \qquad \widehat{\mathcal{D}}^{(a)} f(v) = \frac{1}{d_G^{(a)}(v)} f(v).$$

We then have

$$\mathcal{P}_{G^{(p)}} = \widehat{\mathcal{D}}^{(p)} \circ \mathcal{A}_{G^{(p)}}, \quad \mathcal{P}_{G^{(a)}} = \widehat{\mathcal{D}}^{(a)} \circ \mathcal{A}_{G^{(a)}}, \quad \mathcal{Q}_G^{(1,1)} = \widehat{\mathcal{D}}^{(a)} \circ \mathcal{A}_G^{(1,1)}.$$

We represent $\mathcal{A}_{G^{(p)}}, \mathcal{A}_{G^{(a)}}, \widehat{\mathcal{D}}^{(p)}, \widehat{\mathcal{D}}^{(a)}, \varphi^*$ by matrices $A_{G^{(p)}}, A_{G^{(a)}}, \widehat{D}^{(p)}$, $\widehat{D}^{(a)}, \Phi$, respectively. As $\mathcal{A}_G^{(1,1)} = \varphi^* \circ \mathcal{A}_{G^{(a)}} \circ \mathcal{P}_{G^{(p)}} \circ (\varphi^*)^{-1}$, the matrix representation $A_G^{(1,1)}$ of $\mathcal{A}_G^{(1,1)}$ is given by $A_G^{(1,1)} = \Phi A_{G^{(a)}} \widehat{D}^{(p)} A_{G^{(p)}} \Phi^{-1}$. Thus we have

$${}^tA_G^{(1,1)} = {}^t\Phi^{-1}\, {}^tA_{G^{(p)}}\, {}^t\widehat{D}^{(p)}\, {}^tA_{G^{(a)}}\, {}^t\Phi = \Phi A_{G^{(p)}} \widehat{D}^{(p)} A_{G^{(a)}} \Phi^{-1},$$

because $A_{G^{(a)}}, A_{G^{(p)}}, \widehat{D}^{(p)}$ are symmetric and ${}^t\Phi = \Phi^{-1}$. On the other hand, as $\mathcal{A}_G^{(1,1)} = \mathcal{A}_{G^{(p)}}\mathcal{P}_{G^{(a)}}$, we have $A_G^{(1,1)} = A_{G^{(p)}} \widehat{D}^{(a)} A_{G^{(a)}}$ hence have

$${}^tA_G^{(1,1)} = A_{G^{(a)}} \widehat{D}^{(a)} A_{G^{(p)}} = A_{(G^*)^{(p)}} \widehat{D}^{(a)} A_{(G^*)^{(a)}},$$

where $A_{(G^*)^{(p)}}$ and $A_{(G^*)^{(a)}}$ denote the matrix representations of $\mathcal{A}_{(G^*)^{(p)}}$ and $\mathcal{A}_{(G^*)^{(a)}}$, respectively.

Under the assumption in the assertion (1), we have $\widehat{D}^{(p)} = \widehat{D}^{(a)}$. If we denote by $A_{G^*}^{(1,1)}$ the matrix representation of $\mathcal{A}_{G^*}^{(1,1)}$, we get

$$A_{G^*}^{(1,1)} = A_{(G^*)^{(p)}} \widehat{D}^{(p)} A_{(G^*)^{(a)}} = {}^tA_G^{(1,1)} = \Phi A_{G^{(p)}} \widehat{D}^{(a)} A_{G^{(a)}} \Phi^{-1} = \Phi A_G^{(1,1)} \Phi^{-1}.$$

We hence obtain $\mathcal{A}_{G^*}^{(1,1)} = \varphi^* \circ \mathcal{A}_G^{(1,1)} \circ (\varphi^*)^{-1}$.

Under the assumption in the assertion (2), we have $\widehat{D}^{(p)} = d_G^{(p)} I$ and $\widehat{D}^{(a)} = d_G^{(a)} I$. Hence we obtain $\mathcal{Q}_{G^*}^{(1,1)} = \varphi^* \circ \mathcal{Q}_G^{(1,1)} \circ (\varphi^*)^{-1}$. Therefore, we get the conclusions. □

Remark 6.1. When G is regular and is isomorphic to its dual G^*, our proof shows that $\Delta_{\mathcal{Q}_G}^{(p,q)} \circ \varphi^* = \varphi^* \circ \Delta_{\mathcal{Q}_{G^*}}^{(p,q)}$. But this does not guarantee that $\Delta_{\mathcal{Q}_G}^{(p,q)}$ is quasi-selfadjoint.

Example 6.2. For a complexified Petersen graph G in Example 3.3, we see it is isomorphic to G^* through the map $(i,j) \mapsto (i+1,j)$. The adjacency operators $\mathcal{A}_{G^{(p)}}, \mathcal{A}_{G^{(a)}}$ and the 2-step adjacency operators $\mathcal{A}^{(2)}_{G^{(p)}}, \mathcal{A}^{(2)}_{G^{(a)}}$ are represented by the matrices

$$A_{G^{(p)}} = \begin{pmatrix} B & I & O & O \\ I & C & O & O \\ O & O & B & I \\ O & O & I & C \end{pmatrix}, \qquad A_{G^{(a)}} = \begin{pmatrix} C & O & O & I \\ O & B & I & O \\ O & I & C & O \\ I & O & O & B \end{pmatrix},$$

$$A^{(2)}_{G^{(p)}} = \begin{pmatrix} C & M-I & O & O \\ M-I & B & O & O \\ O & O & C & M-I \\ O & O & M-I & B \end{pmatrix}, \quad A^{(2)}_{G^{(p)}} = \begin{pmatrix} B & O & O & M-I \\ O & C & M-I & O \\ O & M-I & B & O \\ M-I & O & O & C \end{pmatrix},$$

with symmetric matrices

$$B = \begin{pmatrix} 0&1&0&0&1 \\ 1&0&1&0&0 \\ 0&1&0&1&0 \\ 0&0&1&0&1 \\ 1&0&0&1&0 \end{pmatrix}, \qquad C = \begin{pmatrix} 0&0&1&1&0 \\ 0&0&0&1&1 \\ 1&0&0&0&1 \\ 1&1&0&0&0 \\ 0&1&1&0&0 \end{pmatrix}, \qquad M = \begin{pmatrix} 1&1&1&1&1 \\ 1&1&1&1&1 \\ 1&1&1&1&1 \\ 1&1&1&1&1 \\ 1&1&1&1&1 \end{pmatrix}.$$

One can easily check that $\mathcal{A}_{G^{(p)}}$ and $\mathcal{A}_{G^{(a)}}$ are not commutative.

If we study $\mathcal{A}^{(1,1)}_G$, $\mathcal{Q}^{(2,1)}_G$, $\mathcal{Q}^{(1,2)}_G$, they are expressed by the matrices

$$A^{(1,1)}_G = A_{G^{(p)}} A_{G^{(a)}} = \frac{1}{3}\begin{pmatrix} M-I & B & I & B \\ C & M-I & C & I \\ I & B & M-I & B \\ C & I & C & M-I \end{pmatrix},$$

$$Q^{(2,1)}_G = A^{(2)}_{G^{(p)}} A_{G^{(a)}} = \frac{1}{18}\begin{pmatrix} B+2I & 2M-B & M-I & C \\ 2M-C & C+2I & B & M-I \\ M-I & C & B+2I & 2M-B \\ B & M-I & 2M-C & C+2I \end{pmatrix},$$

$$Q^{(1,2)}_G = A_{G^{(p)}} A^{(2)}_{G^{(a)}} = \frac{1}{18}\begin{pmatrix} C+2I & C & M-I & 2M-B \\ B & B+2I & 2M-C & M-I \\ M-I & 2M-B & C+2I & C \\ 2M-C & M-I & B & B+2I \end{pmatrix}.$$

Since $A^{(1,1)}_{G^*}$, $\mathcal{Q}^{(1,2)}_{G^*}$, $\mathcal{Q}^{(2,1)}_{G*}$ are expressed by the transposed matrices ${}^tA^{(1,1)}_G$ ${}^tQ^{(2,1)}_G$, ${}^tQ^{(1,2)}_G$, we see that $\Delta^{(1,1)}_{\mathcal{A}_G}$ is quasi-selfadjoint through the isomorphism $(i,j) \mapsto (i+1,j)$. And through this isomorphism φ we find that $\Delta^{(2,1)}_{\mathcal{Q}_G} \circ \varphi^* \neq \varphi^* \circ \Delta^{(1,2)}_{\mathcal{Q}_{G^*}}$ and $\Delta^{(1,2)}_{\mathcal{Q}_G} \circ \varphi^* \neq \varphi^* \circ \Delta^{(2,1)}_{\mathcal{Q}_{G^*}}$. This example shows also that even if G is isomorphic to G^* its Laplacians are not selfadjoint.

We can construct many examples of Kähler graphs satisfying the condition in Theorem 6.2. There are many pairs of finite ordinary graphs $H_p = (W, F_p)$ and $H_a = (W, F_a)$ satisfying that $d_{H_p}(w) = d_{H_a}(w)$ for each $w \in W$, and that H_a is a subgraph of the complement graph $H_p^c = (W, F_p^c)$ of H_p (i.e. $F_a \subset F_p^c$). If we define a Kähler graph as $G = (W, F_p \cup F_a)$, then it satisfies the condition in Theorem 6.2. For example, we take H_p, H_a as in Figs. 11, 12. Their adjacency operators are not commutative.

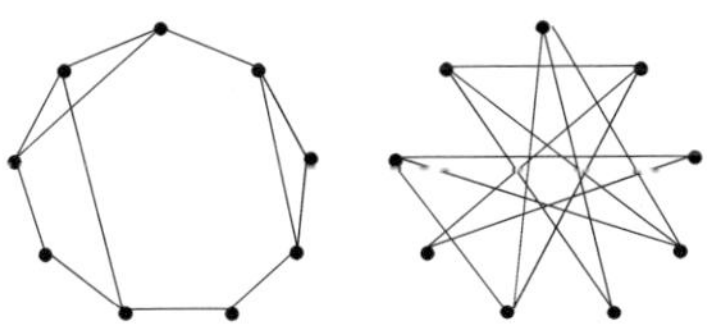

Fig. 11.

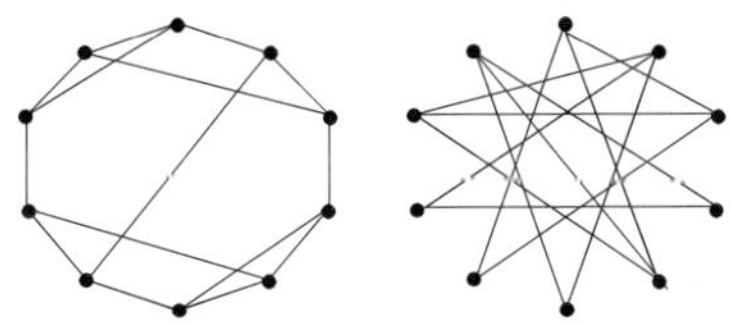

Fig. 12.

We here show some ways of constructing Kähler graphs which satisfy the condition in Theorem 6.2. We take a set $W = \{0, 1, \ldots, 2km-1\}$ with positive integers k, m satisfying $m \geq 2$. We define principal and auxiliary edges in the following manner and make a vertex-transitive regular Kähler graph $H_{k,m} = (W, F^{(p)} \cup F^{(a)})$ of $d_H^{(p)} = d_H^{(a)} = k$.

i) When i is even, $i \sim_p i'$ if and only if either $i' = i+4j-3$ for $j = 1, \ldots, S_k$ or $i' = i-4j+1$ for $j = 1, \ldots, T_k$, and $i \sim_a i'$ if and only if either $i' = i-4j+3$ for $j = 1, \ldots, T_k$ or $i' = i+4j-1$ for $j = 1, \ldots, S_k$;

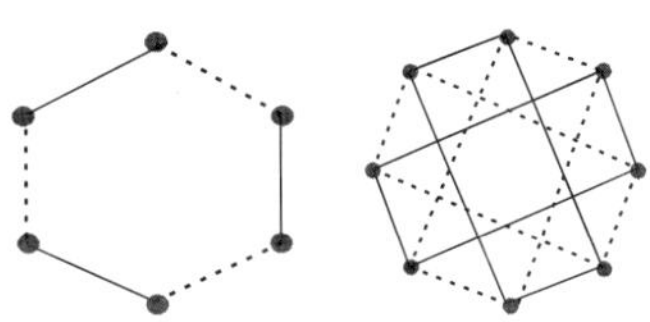

Fig. 13. $H_{1,3}$ and $H_{2,2}$

ii) When i is odd, $i \sim_p i'$ if and only if either $i' = i-4j+3$ for $j = 1, \ldots, T_k$ or $i' = i+4j-1$ for $j = 1, \ldots, S_k$, and $i \sim_a i'$ if and only if either $i' = i+4j-3$ for $j = 1, \ldots, S_k$ or $i' = i-4j+1$ for $j = 1, \ldots, T_k$.

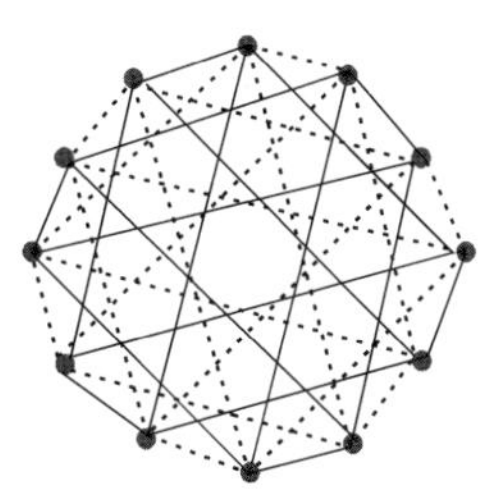

Fig. 14. $H_{3,2}$

Here, $S_k = T_k = k/2$ when k is even, and $S_k = (k+1)/2$, $T_k = (k-1)/2$ when k is odd.

We consider indices of vertices by modulo $2km$. One can easily check that $H_{k,m}$ is isomorphic to its dual through the rotation $i \mapsto i+1$, and also is isomorphic to its dual through the reflection $i \mapsto 2km-i$. Also, we can see that the adjacency operators $\mathcal{A}^{(p)}_{H_{k,m}}$ and $\mathcal{A}^{(a)}_{H_{k,m}}$ of its principal and its auxiliary graphs are not commutative if $m \geq 3$.

By taking a Kähler graph $G = (V, E^{(p)} \cup E^{(a)})$, we define a new Kähler graph $K = K(G; H_{k,m})$ in the following manner.

i) The set of vertices is $\{(i, v) \mid i = 0, 1, \ldots, 2km-1,\ v \in V\}$;

ii) $(i, v) \sim_p (i', v')$ if and only if they satisfy one of the following conditions;
   (a) $i = i'$ and are even, and $v \sim_p v'$ in G,
   (b) $i = i'$ and are odd, and $v \sim_a v'$ in G,
   (c) $v = v'$, and $i \sim_p i'$ in $H_{k,m}$:

iii) $(i, v) \sim_a (i', v')$ if and only if they satisfy one of the following conditions;
   (a) $i = i'$ and are odd, and $v \sim_p v'$ in G,
   (b) $i = i'$ and are even, and $v \sim_a v'$ in G,
   (c) $v = v'$, and $i \sim_a i'$ in $H_{k,m}$.

The complexified Petersen graph is obtained by this way. It is clear that K is isomorphic to its dual through the block rotation $(i, v) \mapsto (i+1, v)$. As we have

$$d_K^{(p)}(i, v) = d_G^{(p)}(v)+k, \quad d_K^{(a)}(i, v) = d_G^{(a)}(v)+k, \qquad \text{when } i \text{ is even,}$$
$$d_K^{(p)}(i, v) = d_G^{(a)}(v)+k, \quad d_K^{(a)}(i, v) = d_G^{(p)}(v)+k, \qquad \text{when } i \text{ is odd,}$$

if we take G so that $d_G^{(p)}(v) = d_G^{(a)}(v)$ at each vertex $v \in V$ we get a Kähler graph K satisfying $d_K^{(p)}(i, v) = d_K^{(a)}(i, v)$ at each vertex (i, v) of K.

Let $G_j = (V, E_j^{(p)} \cup E_j^{(a)})$ $(j = 0, \ldots, km)$ be Kähler graphs having the common sets of vertices. We suppose there are isomorphisms $\varphi_0 : G_0 \to G_0^*$ and $\varphi_{km} : G_{km} \to G_{km}^*$. We can define another Kähler graph $L = L(G_0, \ldots, G_{km}; H_{k,m})$ in the following manner.

i) The set of vertices is $\{(i, v) \mid i = 0, 1, \ldots, 2km-1,\ v \in V\}$;

ii) $(i, v) \sim_p (i', v')$ if and only if they satisfy one of the following conditions;
   (a) $0 \leq i = i' \leq km$, and $v \sim_p v'$ in G_i,
   (b) $km < i = i' < 2km$, and $v \sim_a v'$ in G_{2km-i},
   (c) $v = v'$, and $i \sim_p i'$ in $H_{k,m}$:

iii) $(i, v) \sim_a (i', v')$ if and only if they satisfy one of the following conditions;
   (a) $0 \leq i = i' \leq km$, and $v \sim_a v'$ in G_i,
   (b) $km < i = i' < 2km$, and $v \sim_p v'$ in G_{2km-i},
   (c) $v = v'$, and $i \sim_a i'$ in $H_{k,m}$.

It is clear that L is isomorphic to its dual through the block reflection defined by

$$(0,v)\mapsto\big(0,\varphi_0(v)\big),\ (km,v)\mapsto\big(km,\varphi_{km}(v)\big),\ (i,v)\mapsto(2km-i,v)$$

for $i=1,\dots,km-1$. As we have

$$\begin{aligned}
&d_L^{(p)}(i,v)=d_{G_i}^{(p)}(v)+k, && d_L^{(a)}(i,v)=d_{G_i}^{(a)}(v)+k, && \text{for } 0\le i\le km,\\
&d_L^{(p)}(i,v)=d_{G_{2km-i}}^{(a)}(v)+k, && d_L^{(a)}(i,v)=d_{G_{2km-i}}^{(p)}(v)+k, && \text{for } km<i<2km,
\end{aligned}$$

if we take G_i so that $d_{G_i}^{(p)}(v)=d_{G_i}^{(a)}(v)$ at each vertex $v\in V$, we get a Kähler graph L satisfying $d_L^{(p)}(i,v)=d_L^{(a)}(i,v)$ at each vertex (i,v) of L.

We give another series of examples of Kähler graphs satisfying the condition in Theorem 6.2. We take a finite Kähler graph $H=(W,F^{(p)}\cup F^{(a)})$ which satisfies $d_H^{(p)}(w)=d_H^{(a)}(w)$ at each $w\in W$. We denote the set of vertices as $W=\{0,1,\dots,n-1\}$ with $n\ge 5$. We choose a set I of indices satisfying the condition that I and $-I$ are disjoint. Here, $-I=\{-i\mid i\in I\}$, and we consider indices by modulo n. We define a Kähler graphs $G_I(H)$ in the following manner.

i) The set of vertices is $V=\{(0,i),(1,i)\mid i=0,\dots,n-1\}$.
ii) We define $(\epsilon,i)\sim_p(\delta,j)$ if and only if they satisfy one of the following conditions:
   (a) $\epsilon=\delta=0$ and $i\sim_p j$ in H;
   (b) $\epsilon=\delta=1$ and $i\sim_a j$ in H;
   (c) $\epsilon=0,\delta=1$ and $j=i-k$, or $\epsilon=1,\delta=0$ and $i=j+k$ for some $k\in I$.
iii) We define $(\epsilon,i)\sim_a(\delta,j)$ if and only if they satisfy one of the following conditions:
   (a) $\epsilon=\delta=0$ and $i\sim_a j$ in H;
   (b) $\epsilon=\delta=1$ and $i\sim_p j$ in H;
   (c) $\epsilon=0,\delta=1$ and $j=i+k$, or $\epsilon=1,\delta=0$ and $i=j-k$ for some $k\in I$.

We here consider second components by modulo n. For example, the Petersen Kähler graph (Fig. 5) is obtained by this way. By the construction of G, if we put n_I the cardinality of the set I, we find that

$$\begin{aligned}
d_G^{(p)}(0,i)&=d_H^{(p)}(i)+n_I, & d_G^{(a)}(0,i)&=d_H^{(a)}(i)+n_I,\\
d_G^{(p)}(1,i)&=d_H^{(a)}(i)+n_I, & d_G^{(a)}(1,i)&=d_H^{(p)}(i)+n_I,
\end{aligned}$$

hence find that $d_G^{(p)}(\epsilon,i) = d_G^{(a)}(\epsilon,i)$ at each vertex $(\epsilon,i)\in V$. If we take a map $f: V\to V$ given by $(\epsilon,i)\mapsto(\epsilon+1,i)$, where the first indices are considered by modulo 2, we find that it gives an isomorphism of $G_I(H)$ to its dual $G_I(H)^*$. This Kähler graph $G_I(H)$ does not have commutative adjacency operators of the principal and the auxiliary graphs, in general.

We construct $G_I(H)$ by taking H and H^* and joining them "symmetrically". We can construct such type of Kähler graphs by turning over dual graphs. When the cardinality n_H of the set W of vertices of a Kähler graph H is even ($n_H = 2m$), we denote W as $W = \{(\rho,i)\mid \rho=0,1;\ i = 0,1,\dots,m-1\}$. We take the disjoint union V of the sets of vertices of H and H^* which is denote as $V = \{(\epsilon,(\rho,i))\mid \epsilon=0,1;\ \rho=0,1;\ i=0,\dots,m-1\}$. Here a vertex of H is represented as $\big(0,(\rho,i)\big)$ and that of H^* is represented as $\big(1,(\rho,i)\big)$. We take a set J ($\subset\{0,\dots,m-1\}$) of indices which satisfies the condition that J and $-J$ are disjoint, where we consider the indices in J by modulo m, and join H and H^* in the following manner:

i) $\big(0,(\rho,i)\big)\sim_p\big(1,(\tau,i)\big)$ if and only if either $\rho=0,\tau=1$ and $j=i-k$, or $\rho=1,\tau=0$ and $j=i+k$, with some $k\in J$;
ii) $\big(0,(\rho,i)\big)\sim_a\big(1,(\tau,i)\big)$ if and only if either $\rho=0,\tau=1$ and $j=i+k$, or $\rho=1,\tau=0$ and $j=i-k$, with some $k\in J$.

Then we get a Kähler graph $G_J^t(H)$ whose set of vertices is V. The principal and the auxiliary degrees of $G_J^t(H)$ coincide with each other at each its vertex, and it is isomorphic to its dual $G_J^t(H)^*$ through the map $\big(\epsilon,(\rho,i)\big)\mapsto\big(\epsilon+1,(\rho,i)\big)$, where the first component is considered by modulo 2.

When n_H is odd ($n_H = 2m+1$), we denote W as $W=\{(\rho,i)\mid\rho= 0,1;\ i=0,1,\dots,m-1\}\cup\{\star\}$. By the same way as in the case that n_H is even where we ignore the element $\star$ in joining H and H^*, we get a Kähler graph $G_J^t(H)$ also in this case. Therefore we can easily construct many examples of Kähler graphs satisfying the condition in Theorem 6.2.

We here give some examples of Kähler graphs constructed by the above ways. In Fig. 15, we take H as the complement-filled Kähler graph of 5-circuit, and set $W=\{0=\star,\ 1=(0,0),\ 2=(0,1),\ 3=(1,0),\ 4=(1,1)\}$ and $J=\{1\}$. In Figs. 16, 17 we take H as the Kähler graph whose principal graph is a 6-circuit and whose auxiliary graph is the union of two 3-circuits (see Fig. 9). We set $W=\{0=(0,0),\ 1=(0,1),\ 2=(0,2),\ 3=(1,0),\ 4= (1,1),\ 5=(1,2)\}$ and $I=J=\{1\}$. If one represents them in $\mathbb{R}^3$, he can get their properties more clearly. The adjacency operators of their principal and auxiliary graphs are not commutative.

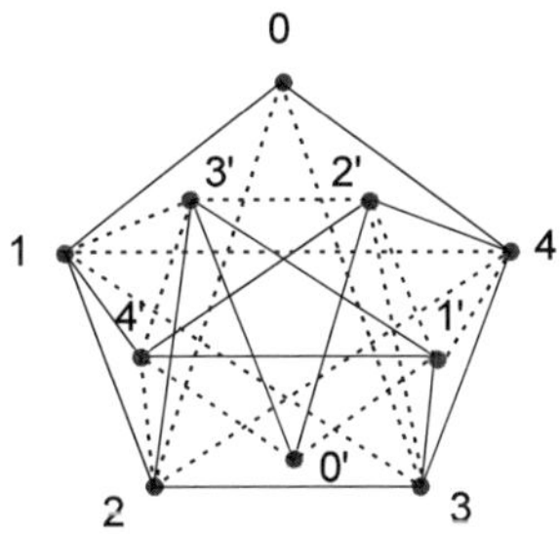

Fig. 15. $G^t_J(H)$

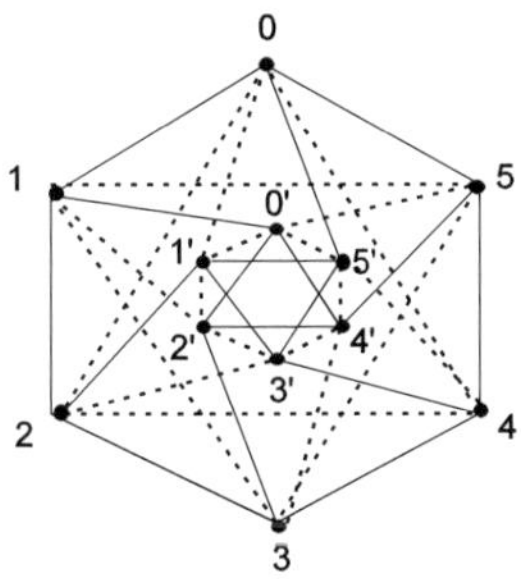

Fig. 16. $G_I(H)$

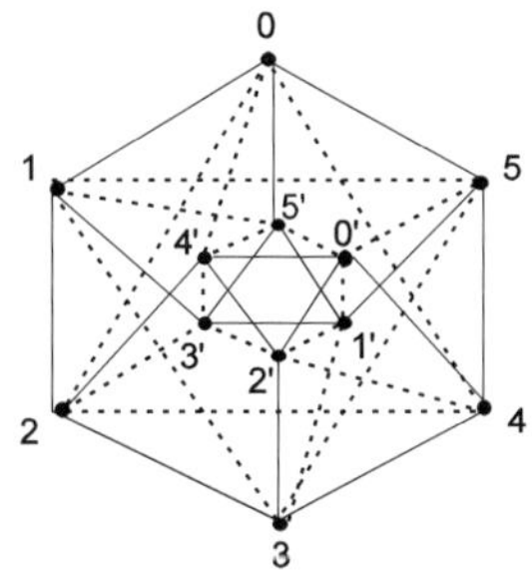

Fig. 17. $G^t_J(H)$

At last, we note that our argument goes through to combinatorial and probabilistic transitional (p,q)-Laplacians of locally finite Kähler graphs which act on the set of square summable functions on the set of vertices. Also, if we replace probabilistic weights by cardinalities of auxiliary paths in the definitions of adjacency and probabilistic transition operators, we can get similar results.

In the forthcoming paper, we study first eigenvalue of (p,q)-Laplacians corresponding to the results in [6 ,7 ,9] and in their references.

References

1. T. Adachi, Kähler magnetic flows on a manifold of constant holomorphic sectional curvature, *Tokyo J. Math.* **18**, 473–483 (1995).
2. ______, Magnetic mean operators on a Kähler manifold, in *Topics in Almost Hermitian geometry and related fields*, Y. Matsushita, E.G. Rio, H. Hashimoto, T. Koda & T. Oguro eds., 30–40, (World Scientific, Singapore, 2005).
3. ______, A discrete model for Kähler magnetic fields on a complex hyperbolic space, in *Trends in Differential Geometry, Complex Analysis and Mathematical Physics*, K. Sekigawa, V.S. Gerdjikov & S. Dimiev eds, 1–9 (World Scientific, Singapore, 2009).
4. ______, A theorem of Hadamard-Cartan type for Kähler magnetic fields, *J. Math. Soc. Japan* **64**, 969–984 (2012).
5. T. Adachi & Yaermaimaiti T., Zeta functions for a Kähler graph, preprint (2013).

6. A. Katsuda & H. Urakawa, The first eigenvalue of the discrete Dirichlet problem for a graph, *J. Combin. Math. Combin. Comput.* **27**, 217–225 (1998).
7. ———, The Faber-Kraph type isoperimetric inequalities for a graph, *Tohoku Math. J.* **51**, 267–281 (1999).
8. T. Sunada, Spherical means and geodesic chains on a Riemannian manifold, *Trans. A.M.S.* **267**, 483–501 (1981).
9. H. Urakawa, Eigenvalue comparison theorems of the discrete Laplacian for a graph, *Goem. Dedicata* **74**, 95–112 (1999).
10. Yaermaimaiti T., Laplacians of Kähler graphs, this volume.
11. Yaermaimaiti T. & T. Adachi, A note on vertex-transitive Kähler graphs, to appear in *Hokkaido Math. J.*
12. ———, Isospectral Kähler graphs, *Kodai Math. J.* **38**, 560–580 (2015).
13. ———, Laplacians for derived graphs of a regular Kähler graph, *C. R. Math. Acad. Sci. Soc. R. Canada* **37**, 141–156 (2015).
14. ———, Kähler graphs of connected product type, *Applied Mathematics and Information Sciences* **9**, 2767–2773 (2015).
15. ———, Complete vertex-transitive Kähler graphs whose adjacency operators of principal and auxiliary graphs are commutative, preprint (2015).

Received November 13, 2014
Revised January 3, 2015

Current Developments in Differential Geometry
and its Related Fields pp. 45 – 56

S^1-INVARIANT EINSTEIN-WEYL STRUCTURE AND TWISTOR CORRESPONDENCE

Fuminori NAKATA
Faculty of Human Development and Culture, Fukushima University,
Kanayagawa, Fukushima, Japan
E-mail: fnakata@educ.fukushima-u.ac.jp

A twistor correspondence for three-dimensional S^1-invariant indefinite Einstein-Weyl structures is explicitly established. The correspondence is given by using an integral transform related to the wave equation with cylindrical symmetry.

Keywords: Twistor theory, pseudo-Riemannian geometry, wave equation, integral transform.

1. Introduction

LeBrun and Mason established three types of twistor theory concerning holomorphic disks [7–9]. In particular, in [9], they investigate the twistor correspondence for indefinite Einstein-Weyl structures, and established a correspondence between

- smooth, space-time-oriented, conformally compact, globally hyperbolic, Lorentzian Einstein-Weyl 3-manifolds, and
- orientation-reversing diffeomorphisms $\psi : \mathbb{C}P^1 \to \mathbb{C}P^1$.

The model case occurs when ψ is the complex conjugation, then the Einstein-Weyl 3-manifold is the class of the de-Sitter 3-space which is $\mathbb{R} \times S^2$ with an indefinite Einstein metric. In general, the above conditions for Einstein-Weyl 3-manifolds give a strong restriction to the topology of Einstein-Weyl spaces. In fact, such an Einstein-Weyl space is diffeomorhpic to $\mathbb{R} \times S^2$. The author also studied this twistor correspondence in an explicit mannar [10].

In this article, we establish another kind of twistor correspondence for indefinite Einstein-Weyl 3-spaces. The model case is the flat Lorentz space $\mathbb{R}^{1,2}$. We notice an S^1-symmetry on $\mathbb{R}^{1,2}$ and deform the structure preserving this symmetry. We will give the correspondence in an explicit mannar, and will see that this correspondence is related to an old result [1] for the solutions to the wave equation with a cylindrical symmetry.

2. Preliminaries

Here we review some definitions and basic results, and fix some notations.

Einstein-Weyl structures. A *Weyl structure* on a smooth manifold M is the pair $([g], \nabla)$ of a conformal structure $[g]$ and an affine connection ∇ on M satisfying

$$\nabla g = -2\omega \otimes g \tag{1}$$

for some 1-form ω. Though the conformal structure $[g]$ can be either Riemannian or psuedo-Riemannian, our main interest is in the case of indefinite signature $(- + +)$.

In this article, we assume that the connection ∇ of any Weyl structrue $([g], \nabla)$ is torsion-free. Under this torsion-free assumption, ∇ is uniquely determined by a representative g of $[g]$ and the 1-form ω satisfying (1). So a Weyl structure is defined by the pair (g, ω). A Weyl structure $([g], \nabla)$ is called *Einstein-Weyl* if and only if $R_{(ij)} = \Lambda g_{ij}$ for some function Λ where $R_{(ij)} = \frac{1}{2}(R_{ij} + R_{ji})$ is the symmetrized Ricci tensor.

Null surfaces and the integrability. Let $([g], \nabla)$ be a $(- + +)$ Weyl structure on a 3-dimensional manifold M. A 2-dimensional subspace $V \subset T_xM$ $(x \in M)$ is called *null plane* if $[g]$ degenerates on V. A 2-dimensional submanifold $S \subset M$ is called *null surface* if the tangent space T_xS is null plane for each $x \in S$. The following theorem is known [4, 10].

Theorem 2.1. *For a 3-dimensional $(- + +)$ Weyl structure $(M, [g], \nabla)$, the following properties are equivalent*

- *$([g], \nabla)$ is Einstein-Weyl, and*
- *For each null plane $V \subset T_xM$, there exists a totally geodesic null surface $S \subset M$ such that $x \in S$ and $T_xS = V$.*

Holomorphic disks. Let $\mathbb{D} = \{\lambda \in \mathbb{C} \mid |\lambda| \leq 1\}$ be the complex unit disk. For a complex manifold Z and its totally real submanifold Y with $\dim_{\mathbb{R}} Y = \dim_{\mathbb{C}} Z$, a holomorphic disk D on (Z, Y) is the image of a continuous map $(\mathbb{D}, \partial\mathbb{D}) \to (Z, Y)$ which is holomorphic on the interior of $\mathbb{D}$. By this definition, a holomorphic disk can be the image of a constant map, that is, a point. In this article, however, we notice to holomorphic disks with *partial index* 2 which are not points.

The notion of the partial index is as follows. Let N be the complex normal bundle on D in Z, and $N_{\mathbb{R}}$ be the real normal bundle on ∂D in Y. Then we can define the "virtual normal bundle" on $\mathbb{C}P^1 = D \cup_{\partial D} D$ as the double $N \cup_{N_{\mathbb{R}} \otimes \mathbb{C}} N$. If Z is a complex surface, then the virtual normal bundle is a holomorphic line bundle, and its degree is called the partial index of D. See [6] for the detail of this construction.

The idea of the LeBrun-Mason correspondence. LeBrun-Mason type twistor space is a pair (Z, P) of complex manifold Z and totally real submanifold P. For example, in the case explained in the beginning of the introduction, the twistor space (Z, P) is given by $Z = \mathbb{C}P^1 \times \mathbb{C}P^1$ and $P = \{(\lambda, \psi(\lambda)) \mid \lambda \in \mathbb{C}P^1\}$.

Now suppose that Z is a complex surface, and let $\{D_x\}_{x \in M}$ be the space of holomorphic disks on (Z, P) of partial index 2. Then LeBrun-Mason twistor theory indicates that the parameter space M is a real 3-dimensional manifold and that M has a natural Einstein-Weyl structure. This Einstein-Weyl structure is characterized by the following condition: for each $z \in P$, the set

$$\mathfrak{S}_z = \{x \in M \mid z \in \partial D_x\}$$

is a totally geodesic null surface on M. See [9, 10] for the detail.

3. Standard model

To establish a new correspondence, we start from a model case. Let S be the complex surface defined by

$$S = \left\{ [z_0 : z_1 : z_2 : z_3] \in \mathbb{C}P^3 \;\middle|\; z_1^2 = z_0 z_2 \right\}, \tag{2}$$

and let $S_{\mathbb{R}}$ be the real submanifold

$$S_{\mathbb{R}} = \left\{ \left[e^{-i\theta} : 1 : e^{i\theta} : v\right] \in S \;\middle|\; \theta \in \mathbb{R}/2\pi\mathbb{Z}, v \in \mathbb{R} \cup \{\infty\} \right\}. \tag{3}$$

The complex surface S has a unique singularity at $[0 : 0 : 0 : 1]$, and is bimeromorphic to the degree 2 Hirzebruch surface Σ_2 via the normalization $\Sigma_2 \to S$. The surface S is identified with the weighted projective space $\mathbb{P}(1, 1, 2)$ in the following way. Let $[y_0 : y_1 ; v]$ be the weighted homogeneous coordinate of weight $(1, 1, 2)$, that is, it is an equivalent class on $\mathbb{C}^3 \setminus \{(0, 0, 0)\}$ with the equivalence relation

$$[y_0 : y_1 ; v] = [\mu y_0 : \mu y_1 ; \mu^2 v] \qquad \text{for} \quad \mu \in \mathbb{C}^*.$$

The weighted projective space $\mathbb{P}(1, 1, 2)$ is the set of all these classes, and is identified with S by the bijection $\mathbb{P}(1, 1, 2) \to S$ given by

$$[y_0 : y_1 ; v] \longmapsto [y_0^2 : y_0 y_1 : y_1^2 : v]. \tag{4}$$

The singular point is $[0 : 0 ; 1]$.

The surface S is also identified with the 1-point compactification of the total space of $\mathcal{O}(2)$, where $\mathcal{O}(2)$ is the degree 2 complex line bundle over $\mathbb{C}P^1$. The projection $S \supset \mathcal{O}(2) \to \mathbb{C}P^1$ is $[y_0 : y_1; v] \mapsto [y_0 : y_1]$. Similarly, the real manifold $S_{\mathbb{R}}$ is the 1-point compactification of a cylinder $S^1 \times \mathbb{R} = \{(\theta, v)\}$.

Next we give the twistor correspondence for the twistor space $(S, S_{\mathbb{R}})$. For each $(t, z) \in \mathbb{R} \times \mathbb{C}$, we define a holomorphic disk $D_{(t,z)}$ as the image of

$$(\mathbb{D}, \partial\mathbb{D}) \longrightarrow (S, S_{\mathbb{R}}) \ : \ \omega \longmapsto \left[\omega^{-1} : 1 : \omega : t + (z\omega + \bar{z}\omega^{-1})/2\right]. \tag{5}$$

Then we obtain a smooth family of holomorphic disks $\{D_{(t,z)}\}$ on $(S, S_{\mathbb{R}})$ parametrized by $(t, z) \in \mathbb{R} \times \mathbb{C}$.

Each holomorphic disk $D_{(t,z)}$ is the "half" of a holomorphic section of $\mathcal{O}(2)$. This observation implies that the partial index of $D_{(t,z)}$ is equal to 2. LeBrun-Mason theory indicates that there exists a natural indefinite Einstein-Weyl structure on the parameter space $\mathbb{R} \times \mathbb{C}$. This Einstein-Weyl structure is characterized by the following property: the set

$$\mathfrak{S}_u = \{(t, z) \in \mathbb{R} \times \mathbb{C} \mid u \in \partial D_{(t,z)}\} \qquad (u \in S_{\mathbb{R}})$$

is a totally geodesic null surface for each u. Actually, when $u = (\theta, v) \in S^1 \times \mathbb{R} \subset S_{\mathbb{R}}$, the set $\mathfrak{S}_u$ is the plane defined by $t + (ze^{i\theta} + \bar{z}e^{-i\theta})/2 = v$. Since the flat Einstein-Weyl structure

$$g_0 = -dt^2 + |dz|^2, \qquad \omega_0 = 0 \tag{6}$$

satisfies the required property, this is the wanted Einstein-Weyl structure.
S^1-action. We fix an S^1-action on the flat Einstein-Weyl space $\mathbb{R} \times \mathbb{C}$ by $\sigma \cdot (t, z) = (t, e^{i\sigma} z)$ where $(t, z) \in \mathbb{R} \times \mathbb{C}$ and $\sigma \in \mathbb{R}/2\pi\mathbb{Z} \cong S^1$. We put $M = \{(t, z) \in \mathbb{R} \times \mathbb{C} \mid z \neq 0\}$ on which the S^1-action is free. We also use the coordinate $(t, \rho, \psi) \in \mathbb{R} \times \mathbb{R}_+ \times S^1 \cong M$ defined by $z = \rho e^{i\psi}$, then the S^1-action is $\sigma \cdot (t, \rho, \psi) = (t, \rho, \psi + \sigma)$. Notice that (t, ρ) gives a coordinate on the orbit space $\mathcal{H} := M/S^1 = \mathbb{R} \times \mathbb{R}_+$. If we use the cylindrical coordinate (t, ρ, ψ), the flat Einstein-Weyl structure (g_0, ω_0) given by (6) is also written as

$$g = \frac{-dt^2 + d\rho^2}{\rho^2} + d\psi^2, \qquad \omega = \frac{d\rho}{\rho}. \tag{7}$$

Correspondingly, we introduce a $\mathbb{C}^*$-action on S by

$$\mu \cdot [y_0 : y_1; v] = [\mu^{-1/2} y_0 : \mu^{1/2} y_1; v] \qquad \mu \in \mathbb{C}^*. \tag{8}$$

Then the restricted action of $U(1) \subset \mathbb{C}^*$ preserves the real submanifold $S_{\mathbb{R}}$. Further, these S^1-actions are equivariant under the twistor correspondence, that is, we have

$$D_{\sigma\cdot(t,z)} = e^{-i\sigma} D_{(t,z)}.$$

4. S^1-invariant Einstein-Weyl structures

In this section, we give a family of S^1-invariant Einstein-Weyl structures by deforming the standard one.

Proposition 4.1. *Let V be an open set on $\mathcal{H} = \mathbb{R} \times \mathbb{R}_+$ and we put $U := V \times S^1 \subset M$. Let $A_i = A_i(t,\rho)$ $(i = 0, 1)$ be C^{k-1}-functions on V satisfying $A_1^2 - A_0^2 > 0$. Then the C^{k-1}-Weyl structure (g, ω) on U defined by*

$$g = \frac{A_1^2 - A_0^2}{\rho^2}(-dt^2 + d\rho^2) + d\psi^2, \qquad \omega = \frac{2A_0A_1dt + (A_0^2 + A_1^2)d\rho}{\rho(A_1^2 - A_0^2)} \tag{9}$$

is Einstein-Weyl if and only if the following equations hold:

$$(A_0)_\rho - (A_1)_t = \frac{A_0}{\rho}, \tag{10}$$

$$(A_0)_t - (A_1)_\rho = 0. \tag{11}$$

The proof is given by either a direct calculation or the following arguments. Actually we will establish a twistor correspondence for this Weyl structure, which means that this is Einstein-Weyl.

Remark 4.1. The Weyl structure (9) and the equations (10) and (11) are the indefinite analogues of the positive definite case investigated by Calderbank & Pedersen [2]. Their case (positive definite case) is related to Joyce's torus invariant self-dual metric [5]. By this reason, we call the system of equations (10) and (11) the *indefinite Joyce equation.* Notice that the constant solution $(A_0, A_1) = (0, 1)$ gives the standard Einstein-Weyl structure (7).

Wave equation. Now we explain the relation to the wave equation

$$\Box\tilde{u} := \left(-\frac{\partial^2}{\partial t^2} + \frac{\partial^2}{\partial x_1^2} + \frac{\partial^2}{\partial x_2^2}\right)\tilde{u} = 0. \tag{12}$$

If a C^k-solution $\tilde{u}$ of (12) has the cylindrical symmetry, that is, if there is a function $u = u(t,\rho)$ of $(t,\rho) \in \mathbb{R} \times \mathbb{R}_{\geq 0}$ satisfying $\tilde{u}(t,x) = u(t,|x|)$, then

u satisfies the equation

$$-\frac{\partial^2 u}{\partial t^2}+\frac{\partial^2 u}{\partial \rho^2}+\frac{1}{\rho}\frac{\partial u}{\partial \rho}=0. \tag{13}$$

We call the equation (13) the *cylindrical wave equation.*

The indefinite Joyce equation (10) and (11) is directly related to the cylindrical wave equation (13) as follows. If $u=u(t,\rho)$ is a C^k-solution of (13), then we obtain a C^{k-1}-solution (A_0, A_1) of (10) and (11) on $\mathcal{H}=\{(t,\rho)\}$ by putting

$$A_0=\rho\, u_t, \qquad A_1=1+\rho\, u_\rho. \tag{14}$$

Conversely, a C^{k-1}-solution (A_0,A_1) determines a C^k-solution u of (13) up to constant. Indeed, (10) is the integrability condition for u, and (11) is equivalent to the cylindrical wave equation (13). Notice that the standard case $(A_0,A_1)=(0,1)$ corresponds to the trivial solution $u=0$.

Further, the following fact is known (see [1, 3, 11]).

Theorem 4.1. *Suppose $u(t,\rho)$ is a C^k-solution to the cylindrical wave equation (13). Then there exists a unique C^k-function $h(s)$ on $\mathbb{R}$ such that*

$$u(t,\rho)=\frac{1}{2\pi}\int_0^{2\pi} h(t+\rho\cos\theta)d\theta \tag{15}$$

Though the correspondence $u\to h$ is given by using the (well-known) Poisson formula, we omit the detail.

5. Twistor space and the main theorem

For each function $h\in C^k(\mathbb{R})$, we define a real submanifold $P_h\subset\mathcal{O}(2)$ by

$$P_h=\left\{[1:e^{i\theta+h(v)};e^{i\theta+h(v)}v]\in\mathcal{O}(2)\;\middle|\; v\in\mathbb{R},\,\theta\in S^1\right\}. \tag{16}$$

Notice that P_h is preserved by the $U(1)$-action given by (8). The main theorem is the following.

Theorem 5.1. *The following objects correspond in the sence of LeBrun-Mason correspondence:*

- *the Einstein-Weyl space (g,ω) defined by a solution (A_0,A_1) to (10) and (11), and*
- *the pair $(\mathcal{O}(2),P_h)$.*

Here h and (A_0,A_1) are related by (14) and (15).

We give the proof of this theorem in the rest of this article.

6. Deformation of holomorphic disks

In this section, we construct the family of holomorphic disks on the twistor space $(\mathcal{O}(2), P_h)$ where P_h is the real submanifold defined by (16).

When $h = 0$ (i.e. the standard case), holomorphic disks $\{D_{(t,z)}\}$ are explicitly given by (5). If we use the weighted homogeneous coordinate on $\mathcal{O}(2)$, the map (5) is also written as

$$\mathbb{D} \longrightarrow \mathcal{O}(2) \ : \ \omega \longmapsto [1 : \omega \, ; \, \omega(t + (z\omega + \bar{z}\omega^{-1})/2)].$$

By the cylindrical coordinate $(t, z) = (t, \rho e^{i\psi})$ on $\mathbb{R} \times \mathbb{C}$, the boundary $\partial D_{(t,z)}$ is the image of

$$\partial\mathbb{D} \longrightarrow P_0 \ : \ e^{i\theta} \longmapsto [1 : e^{i\theta} \, ; \, e^{i\theta} \, \Xi(t, \rho \, ; \, e^{i(\theta+\psi)})] \tag{17}$$

$$\text{where} \qquad \Xi(t, \rho \, ; \, e^{i\theta}) = t + \rho\cos\theta.$$

Now we deform (17). Since the twistor space $(\mathcal{O}(2), P_h)$ has the S^1-symmetry, it is enough to consider the case when $\psi = 0$. Let us introduce an unknown function $\Theta(e^{i\theta})$ on S^1 satisfying

$$[1 : \Theta(e^{i\theta}) \, ; \, \Theta(e^{i\theta})\Xi(t, \rho; e^{i\theta})] \in P_h. \tag{18}$$

For this, we require $|\Theta(e^{i\theta})| = \exp(h(\Xi(t, \rho; e^{i\theta})))$, so $\Theta(e^{i\theta})$ can be written as

$$\Theta(e^{i\theta}) = \exp(i\theta + i\kappa(\theta) + h(\Xi(t, \rho; e^{i\theta}))) \tag{19}$$

with unknown real valued function $\kappa(\theta)$.

We want to determine Θ so that (18) extends holomorphically to the interior of $\mathbb{D} = \{\omega \subset \mathbb{C} \mid |\omega| \leq 1\}$ for $\omega - e^{i\theta}$. We notice to the Fourier expansion:

$$h(\Xi(t, \rho; e^{i\theta})) = \sum_{k=-\infty}^{\infty} H_k(t, \rho) e^{ik\theta}, \tag{20}$$

and let us put

$$u(t, \rho) = H_0(t, \rho), \quad H_\pm(t, \rho; e^{i\theta}) = \sum_{k=1}^{\infty} H_{\pm k}(t, \rho) e^{\pm ik\theta}.$$

Notice that $u(t, \rho)$ is given by

$$u(t, \rho) = \frac{1}{2\pi} \int_0^{2\pi} h(t + \rho\cos\theta) d\theta,$$

i.e. we recovered the equation (15).

For our purpose, the exponent of (19) cannot contain negative powers of $\omega = e^{i\theta}$, so the real valued function $\kappa(\theta)$ must be in the form

$$\kappa(\theta) = \psi - i(H_+(t, \rho; e^{i\theta}) - H_-(t, \rho; e^{i\theta}))$$

where $\psi \in \mathbb{R}/2\pi\mathbb{Z} \simeq S^1$ is an arbitrary constant. Hence $\Theta(\omega)$ is written as

$$\Theta(\omega) = \Theta(t, \rho, \psi\,;\, \omega) = \omega \exp\left(i\psi + u(t, \rho) + 2H_+(t, \rho\,;\, \omega)\right). \tag{21}$$

Notice that we have $\kappa = 0$ for the standard case. For each (t, ρ, ψ), the map

$$\omega \longmapsto [1 : \Theta(t, \rho, \psi\,;\, \omega)\,;\, \Theta(t, \rho, \psi\,;\, \omega)\Xi(t, \rho\,;\, \omega)]$$

can be extended to $|\omega| \leq 1$ holomorphically, and its image gives a holomorphic disk on $(\mathcal{O}(2), P_h)$. If we put

$$\mathfrak{f}(t, \rho, \psi\,;\, \omega) = [1 : \Theta\,;\, \Theta\Xi],$$

we have obtained the following.

Proposition 6.1. *Let* $M = \{(t, \rho, \psi) \in \mathbb{R} \times \mathbb{R}_+ \times S^1\}$ *and* $\mathcal{W} = M \times \mathbb{D}$, $\mathcal{W}_\mathbb{R} = M \times \partial\mathbb{D}$. *Then* $\{\mathfrak{f}(t, \rho, \psi\,;\, \cdot)\}_{(t,\rho,\psi)\in M}$ *gives a family of holomorphic disks on* $(\mathcal{O}(2), P_h)$.

We obtain the following double fibration:

$$\begin{array}{ccc} & (\mathcal{W}, \mathcal{W}_\mathbb{R}) & \\ {}^{p}\swarrow & & \searrow^{(\mathfrak{f}, \mathfrak{f}_\mathbb{R})} \\ M & & (\mathcal{O}(2), P_h) \end{array} \tag{22}$$

where $\mathfrak{f}_\mathbb{R}$ is the restriction of $\mathfrak{f}$ on $\mathcal{W}_\mathbb{R}$.

7. Distribution

For a given $h \in C^k(\mathbb{R})$, let us define $A_i = A_i(t, \rho)$ $(i = 1, 2)$ by (14) and (15). We define tangent vectors $\widetilde{\mathfrak{m}}_1$ and $\widetilde{\mathfrak{m}}_2$ on $\mathcal{W}_\mathbb{R} = \{(t, \rho, \psi, e^{i\theta})\}$ by

$$\begin{aligned} \widetilde{\mathfrak{m}}_1 &= \rho\left(-\frac{\partial}{\partial t} + \frac{\partial}{\partial \rho}\right) + \tan\frac{\theta}{2}\,(A_1 - A_0)\frac{\partial}{\partial \psi} - \tan\frac{\theta}{2}\frac{\partial}{\partial \theta}, \\ \widetilde{\mathfrak{m}}_2 &= \rho\left(\frac{\partial}{\partial t} + \frac{\partial}{\partial \rho}\right) - \cot\frac{\theta}{2}\,(A_1 + A_0)\frac{\partial}{\partial \psi} + \cot\frac{\theta}{2}\frac{\partial}{\partial \theta}. \end{aligned} \tag{23}$$

Lemma 7.1. $(\mathfrak{f}_\mathbb{R})_*(\widetilde{\mathfrak{m}}_1) = (\mathfrak{f}_\mathbb{R})_*(\widetilde{\mathfrak{m}}_2) = 0.$

Proof. Since $\Xi_*(\widetilde{\mathfrak{m}}_i) = 0$ is easy, we only need to check $\Theta_*(\widetilde{\mathfrak{m}}_i) = 0$. We put $\omega = e^{i\theta}$, then we obtain

$$\begin{aligned} \widetilde{\mathfrak{m}}_1 &= \rho\left(-\frac{\partial}{\partial t} + \frac{\partial}{\partial \rho}\right) + i\,\frac{1-\omega}{1+\omega}(A_1 - A_0)\frac{\partial}{\partial \psi} + \frac{1-\omega}{1+\omega}\,\omega\frac{\partial}{\partial \omega}, \\ \widetilde{\mathfrak{m}}_2 &= \rho\left(\frac{\partial}{\partial t} + \frac{\partial}{\partial \rho}\right) + i\,\frac{1+\omega}{1-\omega}(A_1 + A_0)\frac{\partial}{\partial \psi} + \frac{1+\omega}{1-\omega}\,\omega\frac{\partial}{\partial \omega}. \end{aligned} \tag{24}$$

We first calculate $\Theta_*(\widetilde{\mathfrak{m}}_1)$. Let us define $\mathfrak{a}_\pm$ respectively by

$$\mathfrak{a}_\pm = \rho\left(-\frac{\partial}{\partial t} + \frac{\partial}{\partial \rho}\right) \pm i(A_1 - A_0)\frac{\partial}{\partial \psi} \pm \omega\frac{\partial}{\partial \omega},$$

then we have $(1+\omega)\widetilde{\mathfrak{m}}_1 = \mathfrak{a}_+ + \omega\mathfrak{a}_-$. Notice that, as differential operators, $\mathfrak{a}_\pm$ preserve the degree of ω. By the expansion (20) and $\Xi_*((1+\omega)\widetilde{\mathfrak{m}}_1) = 0$, we obtain for every $k \in \mathbb{Z}$

$$\mathfrak{a}_+\left(H_k(t,\rho)\omega^k\right) + \omega\mathfrak{a}_-\left(H_{k-1}(t,\rho)\omega^{k-1}\right) = 0.$$

Hence we obtain

$$(1+\omega)\widetilde{\mathfrak{m}}_1\left(H_+(t,\rho,\omega)\right) = \mathfrak{a}_+(H_1(t,\rho)\omega) = -\omega\mathfrak{a}_-\left(u(t,\rho)\right) = -\omega(-\rho u_t + \rho u_\rho).$$

By the definition of Θ in (21), we obtain

$$\Theta_*(\widetilde{\mathfrak{m}}_1) = \frac{1-\omega}{1+\omega}\left(1 - A_1 + A_0 - \rho u_t + \rho u_\rho\right)\Theta. \tag{25}$$

In a similar calculation for $\widetilde{\mathfrak{m}}_2$, we obtain

$$\Theta_*(\widetilde{\mathfrak{m}}_2) = \frac{1+\omega}{1-\omega}\left(1 - A_1 - A_0 + \rho u_t + \rho u_\rho\right)\Theta. \tag{26}$$

By definition we have $(A_0, A_1) = (\rho u_t, 1 + \rho u_\rho)$, so $\Theta_*(\widetilde{\mathfrak{m}}_i) = 0$ hold. $\square$

Remark 7.1. We can extend the formula (24) to $|\omega| \leq 1$, so $\widetilde{\mathfrak{m}}_1$ and $\widetilde{\mathfrak{m}}_2$ can be extended on $\mathcal{W}$. Further, we can show $\ker \mathfrak{f}_*^{0,1} = \mathrm{Span}\langle \widetilde{\mathfrak{m}}_1, \widetilde{\mathfrak{m}}_2, \frac{\partial}{\partial \bar{\omega}}\rangle$, where $\mathfrak{f}_*^{0,1}$ is the composition of $\mathfrak{f}_* : T_{\mathbb{C}}\mathcal{W} \to T_{\mathbb{C}}(\mathcal{O}(2))$ and the projection to the anti-holomorphic tangent bundle : $T_{\mathbb{C}}(\mathcal{O}(2)) \to T^{0,1}(\mathcal{O}(2))$.

Remark 7.2. As already mentioned, (A_0, A_1) satisfies the indefinite Joyce equation (10) and (11). We can also check directly that the involutive condition for $\{\widetilde{\mathfrak{m}}_1, \widetilde{\mathfrak{m}}_2\}$ is equivalent to (10) and (11).

8. Recovering the Einstein-Weyl structure

Here we review some general theory (see [10]), apply it to our case, and recover the Einstein-Weyl structure.

General method. Let M be a 3-dimensional manifold and $\varpi : \mathcal{X} \to M$ be a $\mathbb{C}P^1$-bundle equipped with an involution ρ satisfying the following conditions:

- ρ is a fiberwise antiholomorphic map with fixed points,
- the fixed point set $\mathcal{X}_\rho$ is an S^1-bundle over M and $\mathcal{X}_\rho$ divides $\mathcal{X}$ into two components $\mathcal{X}_\pm$.

Proposition 8.1 ([10] Prop. 8.1). *Suppose that there is a complex 3-plane distribution $\varPi \subset T_{\mathbb{C}}\mathcal{X}$ satisfying following conditions:*

1. $\rho_* \varPi = \overline{\varPi}$,
2. *the restriction of $\varPi$ to $\mathcal{X}_+$ is C^k $(k \geq 1)$ and involutive,*
3. $\varPi + \overline{\varPi} = T_{\mathbb{C}}\mathcal{X}$ *on* $\mathcal{X} \setminus \mathcal{X}_\rho$,
4. $\varPi \cap \ker \varpi_*$ *is the $(0,1)$-tangent space of the $\mathbb{C}P^1$ fibers of ϖ,*
5. *the restriction of $\varPi$ to a fiber of $\mathcal{X}$ has $c_1 = -4$ with respect to the complex orientation, and*
6. *the map $\mathcal{X} \to \mathbb{P}(TM) : z \mapsto \varpi_*(\varPi \cap \overline{\varPi})_z$ is not constant along each fiber of ϖ.*

Then M admits a unique C^{k-1} indefinite Einstein-Weyl structure $([g], \nabla)$ such that the null surfaces are the projections via ϖ of the integral manifolds of real 2-plane distribution $\varPi \cap T\mathcal{X}_\rho$ on $\mathcal{X}_\rho$.

Notice that the conditions 1 to 6 in the above Proposition are local conditions on M. Hence the distribution $\varPi$ defines an Einstein-Weyl structure on an open set of M on which $\varPi$ satisfies all these conditions.

Application to our case. Now we apply Proposition 8.1 to our situation. Let $h \in C^k(\mathbb{R})$ be the given function and (A_0, A_1) be as in Section 7. We put $M = \{(t, \rho, \psi) \in \mathbb{R} \times \mathbb{R}_+ \times S^1\}$ and $\mathcal{X} = \mathcal{W} = M \times \mathbb{C}P^1 = \{(t, \rho, \psi, \omega)\}$. We introduce an involution ρ on $\mathcal{X}$ by $\omega \mapsto \bar{\omega}^{-1}$, and define a complex rank 3 distribution $\varPi$ on $\mathcal{X}$ by

$$\varPi = \operatorname{Span}\left\{ \widetilde{\mathfrak{m}}_1, \widetilde{\mathfrak{m}}_2, \frac{\partial}{\partial \bar{\omega}} \right\}$$

where $\widetilde{\mathfrak{m}}_i$ are defined by (24). Notice that $\varPi$ is involutive.

Lemma 8.1. *Let Д be the integrable distribution on $\mathcal{W}$ defined as above. Then the condition 6 in Proposition 8.1 does not hold if and only if $A_1^2 - A_0^2 = 0$.*

Proof. The real distribution L on $\mathcal{X}$ satisfying $L \otimes \mathbb{C} = \text{Д} \cap \overline{\text{Д}}$ on $\mathcal{X} \setminus \mathcal{X}_\rho$ is spanned by the real tangent vector $\widetilde{l}$ of the form

$$\widetilde{l} = (A_1 + A_0)|1 + \omega|^2 \widetilde{\mathfrak{m}}_1 - (A_1 - A_0)|1 - \omega|^2 \widetilde{\mathfrak{m}}_2 + \gamma \frac{\partial}{\partial \bar{\omega}} \tag{27}$$

where γ is determined so that $\widetilde{l}$ is real. Hence $\varpi_*(\text{Д} \cap \overline{\text{Д}})_\omega$ is spanned by $l = \varpi_*(\widetilde{l})$. If $A_1 - A_0 = 0$ (respectively $A_1 + A_0 = 0$), then l is proportional to

$$\varpi_*(\widetilde{\mathfrak{m}}_1)_\omega = \rho \left(-\frac{\partial}{\partial t} + \frac{\partial}{\partial \rho} \right)$$

(respectively $\varpi_*(\widetilde{\mathfrak{m}}_2)_\omega$). Similarly, we can check that $\varpi_*(\text{Д} \cap \overline{\text{Д}})_\omega$ is not constant if $A_1^2 - A_0^2 \neq 0$. Thus the condition 6 does not hold if and only if $A_1^2 - A_0^2 = 0$. □

Hence if we put $V = \{(t, \rho) \in \mathbb{R} \times \mathbb{R}_+ \mid A_1^2 - A_0^2 > 0\}$ and $U = V \times S^1$, then we obtain a unique Einstein-Weyl structure on U. Finally, to complete the proof of Theorem 5.1, we check the following.

Lemma 8.2. *The unique Einstein-Weyl structure induced on U is explicitly given by the formula* (9).

Proof. It is enough to check that $\varpi_*(\text{Д})_\omega$ is a null plane for each $\omega = e^{i\theta} \in \mathcal{X}_\rho$. The subspace $\varpi_*(\text{Д})_\omega$ is spanned by

$$\begin{aligned} \mathfrak{m}_1(\theta) &= \varpi_*(\widetilde{\mathfrak{m}}_1)_\omega = \rho \left(-\frac{\partial}{\partial t} + \frac{\partial}{\partial \rho} \right) + \tan\frac{\theta}{2}(A_1 - A_0)\frac{\partial}{\partial \psi}, \\ \mathfrak{m}_2(\theta) &= \varpi_*(\widetilde{\mathfrak{m}}_2)_\omega = \rho \left(\frac{\partial}{\partial t} + \frac{\partial}{\partial \rho} \right) - \cot\frac{\theta}{2}(A_1 + A_0)\frac{\partial}{\partial \psi}. \end{aligned} \tag{28}$$

For the indefinite metric g defined in (9), we can directly check

$$g(\mathfrak{m}_1(\theta), \mathfrak{m}_1(\theta)) = g(\mathfrak{m}_2(\theta), \mathfrak{m}_2(\theta)) = g(\mathfrak{m}_1(\theta), \mathfrak{m}_2(\theta)) = 0.$$

Hence $\varpi_*(\text{Д})_\omega$ is null. □

References

1. H. Bateman, The solution of partial differential equations by means of definite integrals, *Proc. Lond. Math. Soc.* (2) **1**, (1904), 451–458.
2. D. M. J. Calderbank, H. Pedersen, Selfdual Einstein metrics with torus symmetry, *J. Diff. Geom.* **60** (2002) 485–521.
3. R. Courant, D. Hilbert, *Methods of mathematical physics*, vol. 2, Intersciences, New York (1962).
4. N. J. Hitchin, Complex manifolds and Einstein's equations, *Twistor Geometry and Non-Linear Systems*, Lecture Notes in Mathematics vol. 970, Springer (1982), 73–99.
5. D. D. Joyce, Explicit construction of self-dual 4-manifolds, *Duke Math. J.* **77** (1995) 519–552.
6. C. LeBrun: Twistors, holomorphic disks, and Riemann surfaces with boundary, *Perspectives in Riemannian geometry*, CRM Proc. Lecture Notes, 40, Amer. Math. Soc. Providence, RI, (2006) 209-221.
7. C. LeBrun, L. J. Mason, Zoll manifolds and complex surfaces, *J. Diff. Geom.* **61**, (2002) 453-535.
8. C. LeBrun, L. J. Mason, Nonlinear gravitons, null geodesics, and holomorphic disks, *Duke Math. J.* **136** (2007) 205–273.
9. C. LeBrun, L. J. Mason, The Einstein-Weyl equations, scattering maps, and holomorphic disks, *Math. Res. Lett.* **16** (2009) 291–301.
10. F. Nakata, A construction of Einstein-Weyl spaces via LeBrun-Mason type twistor correspondence, *Comm. Math. Phys.* **289** (2009) 663–699.
11. N. M. J. Woodhouse, Cylindrical gravitational waves, *Class. Quantum Grav.* **6** (1989) 933–943.

Received December 19, 2014
Revised April 6, 2015

Current Developments in Differential Geometry
and its Related Fields pp. 57 – 75

A FAMILY OF SURFACES IN E^3 GIVEN BY AN OVER-DETERMINED SYSTEM

Naoya ANDO *

Graduate School of Science and Technology, Kumamoto University,
2-39-1 Kurokami, Kumamoto 860-8555 Japan
E-mail: ando@sci.kumamoto-u.ac.jp

In this paper, we find and study a family of surfaces in E^3 given by an over-determined system in the form of $F_u = \alpha + \beta e^F$, $F_v = \delta e^{-F}$. We will see that such a family of surfaces is characterized in terms of a 1-form and two principal distributions. We will obtain a characterization of a family of surfaces which has a minimal surface.

Keywords: Over-determined system, compatibility condition, principal distribution, canonical pre-divergence, minimal surface.

1. Introduction

1.1. *Over-determined systems on surfaces*

According to the fundamental theorem of the theory of surfaces, an immersion into the 3-dimensional Euclidean space E^3 is uniquely determined by the induced metric and the shape operator, up to a motion of E^3. The uniqueness does not necessarily hold by the induced metric and principal distributions. In relation to such a situation, we study over-determined systems of the following type:

$$F_u = \alpha + \beta e^F, \qquad F_v = \gamma + \delta e^{-F}, \tag{1}$$

where α, β, γ, δ are smooth functions of two variables u, v. Let M be a two-dimensional manifold and $\iota : M \longrightarrow E^3$ an immersion of M into E^3. Suppose that there exists no umbilical point of ι and that the curvature K of the induced metric g by ι is nowhere zero. Let $\mathcal{D}_1$, $\mathcal{D}_2$ be principal distributions of ι orthogonal to each other and (u, v) local coordinates satisfying $\partial/\partial u \in \mathcal{D}_1$ and $\partial/\partial v \in \mathcal{D}_2$. Then g is locally represented as $g = A^2du^2 + B^2dv^2$, where A, B are smooth, positive-valued functions. Let k_1, k_2 be principal curvatures of ι corresponding to $\mathcal{D}_1$, $\mathcal{D}_2$, respectively.

*The author is partially supported by Grant-in-Aid for Scientific Research (No. 21740054), Japan Society for the Promotion of Science.

Then we have the equations of Gauss and Codazzi-Mainardi:

$$\begin{aligned} &k_1k_2 = K, \\ &(k_1)_v = -(\log A)_v(k_1 - k_2), \quad (k_2)_u = (\log B)_u(k_1 - k_2). \end{aligned} \tag{2}$$

Applying $k_2 = K/k_1$ to the second and the third relations of (2), we obtain (1), where $F := \log k_1^2$ and

$$\begin{aligned} \alpha &:= (\log K^2B^2)_u, \quad & \beta &:= -\frac{(\log B^2)_u}{K}, \\ \gamma &:= -(\log A^2)_v, \quad & \delta &:= K(\log A^2)_v. \end{aligned} \tag{3}$$

The set of solutions of (1) with (3) is determined by the induced metric and two principal distributions, and if the system has plural solutions, then arbitrary distinct two solutions give immersions which share the induced metric and principal distributions but do not share principal curvatures.

In [7], we described the solutions of over-determined systems in the form of (1) with the compatibility condition

$$\beta_v + \beta\gamma = 0, \quad \alpha_v - \gamma_u + 2\beta\delta = 0, \quad -\delta_u + \alpha\delta = 0. \tag{4}$$

If (1) satisfies the compatibility condition (4), then for an arbitrarily given initial value, there exists a unique solution of (1) with the initial value at a point ([13, p. 393]). In [7], we saw that a molding surface S in E^3 is characterized by (4) for the corresponding over-determined system and that the system on a molding surface S satisfies not only (4) but also $\beta\delta \equiv 0$. The condition $\beta\delta \equiv 0$ for S is equivalent to the condition that a family of lines of curvature consists of geodesics. Originally, this remarkable property of molding surfaces was stated by Cartan ([10, pp. 152–153]) and provided with a careful proof by Bryant-Chern-Griffiths ([9, pp. 277–281]). Molding surfaces are closely related to parallel curved surfaces, which were studied by the author ([1], [2], [3]). By definition, for a parallel curved surface S in E^3, there exists a plane P such that at each point of S, at least one principal direction is parallel to P. A canonical parallel curved surface is represented as a disjoint union of plane curves which are congruent in E^3 with one another and tangent to principal directions. These curves are lines of curvature and geodesics. Referring to [3] and [7], we see that for a surface S in E^3 with no umbilical point and nowhere zero Gaussian curvature, S is molding if and only if a neighborhood of each point of S is a canonical parallel curved surface.

In [5], a surface in E^3 with a family of geodesics of curvature is characterized in terms of local representation of the first fundamental form. Let

(M, g) be a two-dimensional Riemannian manifold with nowhere zero curvature and $\mathcal{D}_1$, $\mathcal{D}_2$ one-dimensional distributions on M orthogonal to each other at any point with respect to g. Suppose that the integral curves of $\mathcal{D}_2$ are geodesics. Then M can be locally and isometrically immersed in E^3 so that $\mathcal{D}_1$, $\mathcal{D}_2$ give principal distributions if and only if we can find local coordinates (u, v) on a neighborhood of each point of M satisfying $\partial/\partial u \in \mathcal{D}_1$, $\partial/\partial v \in \mathcal{D}_2$ and $g = A^2du^2 + dv^2$, where A is represented as $A = A_1(u)A_2 + A_3(u)$, and A_2 is a smooth function of two variables u, v satisfying $(A_2)_v = \sin(\alpha_1(u) + \alpha_2(v))$ for smooth functions α_1, α_2 of one variable and A_1, A_3 are smooth functions of one variable ([5]). In addition, in the case where M can be immersed in the above manner, the surface is molding if and only if α_1 is constant ([5]).

Although an over-determined system (1) on the surface does not satisfy the compatibility condition (4), it is possible that the surface is not uniquely determined by the induced metric and principal distributions, up to a motion of the space. In [8], the author studied an over-determined system in the form of (1) which does not satisfy the compatibility condition but has plural solutions. Such a system has just two solutions. The author obtained a characterization of a surface on which (1) has just two solutions, in terms of the sinh-Gordon equation ([8]). An over-determined system (1) has just two solutions on a surface in E^3 with nonzero constant mean curvature which is not part of any surface of revolution. There exists a non-isothermic surface on which (1) has just two solutions ([8]).

1.2. *The purposes*

Suppose that $F := \log k_1^2$ is a solution of (1) with (3). Then $F' := \log k_1^2 A^2$ is a solution of the following over-determined system

$$F'_u = \alpha' + \beta' e^{F'}, \qquad F'_v = \delta' e^{-F'}, \tag{5}$$

where

$$\alpha' := (\log K^2 A^2 B^2)_u, \quad \beta' := -\frac{2B_u}{KA^2B}, \quad \delta' := 2KAA_v. \tag{6}$$

We set

$$I := \frac{B_u}{A}, \qquad J := \frac{A_v}{B}.$$

Then $K \neq 0$ implies $I_u + J_v \neq 0$ and α', β', δ' are represented as

$$\alpha' = (\log(I_u + J_v)^2)_u, \quad \beta' = \frac{2I}{I_u + J_v}, \quad \delta' = -2(I_u + J_v)J. \tag{7}$$

Let I, J be smooth functions of two variables u, v defined on a domain D of $\boldsymbol{R}^2$ satisfying $I_u + J_v \neq 0$. Let α', β', δ' be smooth functions represented as in (7). Suppose that there exists a solution F' of (5). Then from the fundamental theorem of the theory of surfaces, we see that for each pair (A, B) of smooth, positive-valued functions A, B satisfying $B_u/A = I$ and $A_v/B = J$, there exists an immersion of a neighborhood of each point of D into E^3, unique up to an isometry of E^3, satisfying

(i) the induced metric is given by $g = A^2du^2 + B^2dv^2$;
(ii) $\partial/\partial u$, $\partial/\partial v$ give principal distributions;
(iii) $k_1 := e^{F'/2}/A$, $k_2 := K/k_1$ are principal curvatures corresponding to $\partial/\partial u$, $\partial/\partial v$, respectively, where $K := -(I_u + J_v)/AB$.

For smooth functions I, J of two variables, there exist smooth, positive-valued functions A, B satisfying $B_u/A = I$ and $A_v/B = J$. Let f be a solution of the following semilinear hyperbolic differential equation of second order:

$$f_{uv} + Je^f f_u + Ie^{-f} f_v + J_u e^f - I_v e^{-f} = 0 \tag{8}$$

(see Appendix below for the existence and the uniqueness of a solution of a semilinear hyperbolic equation with a given initial data). Then there exists a smooth, positive-valued function A satisfying

$$(\log A)_u = Ie^{-f} - f_u, \quad (\log A)_v = Je^f, \tag{9}$$

since (8) is the compatibility condition of (9). If we set $B := e^f A$, then A, B satisfy $B_u/A = I$ and $A_v/B = J$. Hence in the case where an over-determined system (5) with α', β', δ' as in (7) has a solution, the system gives a family of surfaces in E^3. There exists a one-to-one correspondence between the family of surfaces given by each solution of (5) with (7) and the set of solutions of (8).

The set of solutions of (1) with (3) is determined by the metric g and distributions $\mathcal{D}_1$, $\mathcal{D}_2$, while the set of solutions of (5) with (6) depends on (u, v). However the family of surfaces given by an over-determined system (5) with (7) is independent of (u, v) and determined, up to a motion of E^3, by $\mathcal{D}_1$, $\mathcal{D}_2$ and a 1-form ω which is related to the canonical pre-divergence V_K of $(M, g, \mathcal{D}_1, \mathcal{D}_2)$.

Let M and ι be as in the beginning of Subsection 1.1. Let V_K be the canonical pre-divergence of $(M, g, \mathcal{D}_1, \mathcal{D}_2)$: V_K is a vector field on M given by the sum of the geodesic curvature vectors of the integral curves of $\mathcal{D}_1$, $\mathcal{D}_2$ through each point (see [6] for properties of V_K). Let θ_K be a 1-form on M defined by $\theta_K(w) := g(V_K, w)$ for each tangent vector w at each point of M. We see that θ_K is determined by g and $\mathcal{D}_1$, $\mathcal{D}_2$. Let M be oriented and $*$ Hodge's $*$-operator. If (u, v) are local coordinates which satisfy $\partial/\partial u \in \mathcal{D}_1$, $\partial/\partial v \in \mathcal{D}_2$ and give the positive orientation of M, then $\omega := - * \theta_K$ is represented as

$$\omega = -Jdu + Idv,$$

where $I := B_u/A$, $J := A_v/B$. If we give M the opposite orientation, then ω points to the opposite direction. By $K \neq 0$ together with

$$d\omega = (I_u + J_v)du \wedge dv,$$

we obtain $d\omega \neq 0$. In Section 2, referring to [8] in the case where the space is Euclidean, we will see that a family of surfaces as above is characterized in terms of a 1-form ω and two principal distributions.

We set

$$X := \beta'_v, \quad Y := \alpha'_v + 2\beta'\delta', \quad Z := -\delta'_u + \alpha'\delta'. \tag{10}$$

Let F' be a solution of an over-determined system (5). Then by $F'_{uv} = F'_{vu}$, we obtain

$$Xe^{2F'} + Ye^{F'} + Z = 0. \tag{11}$$

Therefore we see that if $X \neq 0$, then

$$e^{F'} \in \left\{ \frac{-Y + \sqrt{Y^2 - 4XZ}}{2X}, \frac{-Y - \sqrt{Y^2 - 4XZ}}{2X} \right\}.$$

The compatibility condition of (5) is given by $X \equiv 0$, $Y \equiv 0$ and $Z \equiv 0$. Referring to [7], we see that if (5) with (7) satisfies the compatibility condition, then $IJ \equiv 0$. Referring to [8], we see that (5) with (7) has just two solutions if and only if the family of surfaces given by the system has a surface in E^3 with nonzero constant mean curvature which is not part of any surface of revolution. In Section 3, we will see that (5) with (7) has a unique solution so that $Y^2 = 4XZ$ holds if and only if the family of surfaces given by the system has a minimal surface in E^3 which is not part of any catenoid.

2. A two-dimensional manifold equipped with a 1-form and two one-dimensional distributions

Let M be a two-dimensional manifold. Let $\mathcal{D}_1$, $\mathcal{D}_2$ be two smooth one-dimensional distributions on M satisfying $\mathcal{D}_1(p) \neq \mathcal{D}_2(p)$ for any $p \in M$.

Let ω be a 1-form on M satisfying $d\omega \neq 0$ at any point of M. Let (u, v) be local coordinates on a neighborhood of each point of M. Suppose $\partial/\partial u \in \mathcal{D}_1$ and $\partial/\partial v \in \mathcal{D}_2$. We represent ω as $\omega = -Jdu + Idv$. Since $d\omega = (I_u + J_v)du \wedge dv$, we obtain $I_u + J_v \neq 0$.

Let $(\tilde{u}, \tilde{v})$ be local coordinates on a neighborhood of each point of M satisfying $\partial/\partial\tilde{u} \in \mathcal{D}_1$ and $\partial/\partial\tilde{v} \in \mathcal{D}_2$. Let $\tilde{I}$, $\tilde{J}$ be smooth functions determined by ω and $(\tilde{u}, \tilde{v})$: $\omega = -\tilde{J}d\tilde{u} + \tilde{I}d\tilde{v}$. Then we obtain

$$I = \tilde{I}\frac{d\tilde{v}}{dv}, \quad J = \tilde{J}\frac{d\tilde{u}}{du}, \quad I_u + J_v = (\tilde{I}_{\tilde{u}} + \tilde{J}_{\tilde{v}})\frac{d\tilde{u}}{du}\frac{d\tilde{v}}{dv}. \tag{12}$$

Let F' be a solution of (5) with α', β', δ' as in (7) for local coordinates (u, v). Then for smooth functions $\tilde{\alpha}'$, $\tilde{\beta}'$, $\tilde{\delta}'$ represented by $\tilde{I}$, $\tilde{J}$ and $(\tilde{u}, \tilde{v})$ as in (7),

$$\tilde{F}' := F' - \log\left(\frac{d\tilde{u}}{du}\right)^2$$

is a solution of (5) with $u := \tilde{u}$, $v := \tilde{v}$, $\alpha' := \tilde{\alpha}'$, $\beta' := \tilde{\beta}'$, $\delta' := \tilde{\delta}'$. Therefore we see that whether (5) with (7) has a solution depends only on $(\omega, \mathcal{D}_1, \mathcal{D}_2)$ and does not depend on (u, v). In addition, the cardinality of the set of solutions of (5) with (7) depends only on $(\omega, \mathcal{D}_1, \mathcal{D}_2)$ and does not depend on (u, v).

Referring to [8], we prove the following in this section.

Theorem 2.1. *Let M be an orientable two-dimensional manifold. Let $\mathcal{D}_1$, $\mathcal{D}_2$ be two one-dimensional distributions on M satisfying $\mathcal{D}_1(p) \neq \mathcal{D}_2(p)$ for any $p \in M$. Let ω be a 1-form on M satisfying $d\omega \neq 0$ at any point of M. Then for each $p \in M$, the following are mutually equivalent:*

- (i) *There exists an immersion of a neighborhood of p into E^3 satisfying*
  - (a) *$\mathcal{D}_1$, $\mathcal{D}_2$ give principal distributions,*
  - (b) *$\omega = -*\theta_K$, where $*$ is Hodge's $*$-operator with respect to the induced metric g and some orientation of M, and θ_K is a 1-form defined by $\theta_K(\,\cdot\,) = g(V_K, \,\cdot\,)$ for the canonical pre-divergence V_K of $(M, g, \mathcal{D}_1, \mathcal{D}_2)$;*
- (ii) *An over-determined system (5) with (7) has a solution on a neighborhood of p;*
- (iii) *There exist local coordinates (s, t) on a neighborhood of p satisfying $\partial/\partial s \in \mathcal{D}_1$, $\partial/\partial t \in \mathcal{D}_2$ and*

$$(I_u + J_v)^2 = \left(-2\int_{u_0}^{u}(I_u + J_v)I du + \left(dt\left(\frac{\partial}{\partial v}\right)\right)^2\right) \times \left(-2\int_{v_0}^{v}(I_u + J_v)J dv + \left(ds\left(\frac{\partial}{\partial u}\right)\right)^2\right), \tag{13}$$

where (u_0, v_0) *corresponds to a point* p.

Remark 2.1. Whether there exist (s, t) as in (iii) of Theorem 2.1 depends only on a triplet $(\omega, \mathcal{D}_1, \mathcal{D}_2)$ and does not depend on (u, v).

Remark 2.2. Suppose that a triplet $(\omega, \mathcal{D}_1, \mathcal{D}_2)$ satisfies (13) on M and that the cardinality of the set of solutions of (5) with (7) is constant on M. Let F', $\tilde{F}'$ be as above. Since $e^{F'}du^2 = e^{\tilde{F}'}d\tilde{u}^2$, there exists a smooth tensor field Q_0 of type $(0, 2)$ on M or its universal covering space $\tilde{M}$ defined by $Q_0 := e^{F'}du^2$ on a neighborhood U of each point of M or $\tilde{M}$. There exists a Riemannian metric g on U satisfying $g(\partial/\partial u, \partial/\partial v) = 0$ and $\omega = -*\theta_K$, where $*$ is Hodge's $*$-operator with respect to g and some orientation of U, and θ_K is a 1-form determined by $(g, \mathcal{D}_1, \mathcal{D}_2)$ as in Subsection 1.2. We represent g as $g = A^2du^2 + B^2dv^2$. Let k_1 be a smooth function on U determined by (Q_0, g) and represented as $k_1 = e^{F'/2}/A$. Then we can find an immersion of U into E^3 satisfying (i), (ii), (iii) in Subsection 1.2. Such an immersion does not depend on (u, v).

We already know that (i) is equivalent to (ii) in Theorem 2.1. In order to complete a proof of Theorem 2.1, we will show that (ii) is equivalent to (iii) in Theorem 2.1. Then we need the following.

Proposition 2.1. *There exists a solution F' of (5) if and only if we can find smooth functions a', b', c', d' of two variables u, v satisfying $a' + b' + c' + d' = 0$ and*

$$a'_u = \alpha', \quad e^{-b'}b'_u = \beta' e^{a'}, \quad c'_v = 0, \quad e^{-d'}d'_v = -\delta' e^{c'}. \tag{14}$$

In addition, if there exists a solution F' of (5), then we can find a', b', c', d' satisfying (14) and $F' = a' + b' = -c' - d'$; if there exist a', b', c', d' satisfying $a' + b' + c' + d' = 0$ and (14), then $F' := a' + b' = -c' - d'$ is a solution of (5).

Referring to [8] if necessary, we can obtain Proposition 2.1.

Proof of Theorem 2.1. We see that

$$\begin{aligned} a' &:= \log(I_u + J_v)^2, \\ b' &:= -\log\left(-2\int_{u_0}^{u}(I_u + J_v)Idu + \psi(v)\right), \\ c' &:= 0, \\ d' &:= -\log\left(-2\int_{v_0}^{v}(I_u + J_v)Jdv + \phi(u)\right) \end{aligned} \tag{15}$$

satisfy (14) with (7), where ϕ, ψ are smooth, positive-valued functions of one variable. In addition, if (5) with (7) has a solution on a neighborhood of p, then a', b', c', d' in (15) with suitable ϕ, ψ satisfy $a' + b' + c' + d' = 0$. Therefore we obtain

$$\begin{aligned} (I_u + J_v)^2 = &\left(-2\int_{u_0}^{u}(I_u + J_v)Idu + \psi(v)\right) \\ &\times\left(-2\int_{v_0}^{v}(I_u + J_v)Jdv + \phi(u)\right). \end{aligned} \tag{16}$$

Let s be a smooth function of one variable u satisfying $ds/du = \sqrt{\phi(u)}$, and t a smooth function of one variable v satisfying $dt/dv = \sqrt{\psi(v)}$. Then (s, t) are local coordinates on a neighborhood of p satisfying $\partial/\partial s \in \mathcal{D}_1$, $\partial/\partial t \in \mathcal{D}_2$, and we see that (s, t) satisfy (13). Hence we obtain (iii) from (ii). Suppose (iii). Then a', b', c', d' in (15) with $\phi := (ds/du)^2$, $\psi := (dt/dv)^2$ satisfy $a'+b'+c'+d' = 0$ and (14) with (7). Therefore from Proposition 2.1, we obtain (ii). □

Hence we can assign a two-dimensional manifold M equipped with $(\omega, \mathcal{D}_1, \mathcal{D}_2)$ satisfying (13) to a family of surfaces in E^3 given by an over-determined system (5) with (7). In addition, a triplet $(\omega, \mathcal{D}_1, \mathcal{D}_2)$ satisfying (13) locally determines such a family of surfaces.

Example 2.1. Suppose $\omega|_{\mathcal{D}_2} \equiv 0$. Then $I \equiv 0$. Therefore (16) is represented as

$$J_v(u, v)^2 = \psi(v)(-J(u, v)^2 + J(u, v_0)^2 + \phi(u)). \tag{17}$$

We see that J satisfies (17) if and only if there exists a smooth function α_1 of one variable satisfying $J = A_1(u)\sin(\alpha_1(u) + \alpha_2(v))$, where $A_1(u) := \sqrt{J(u, v_0)^2 + \phi(u)}$ and α_2 is a smooth function of one variable satisfying $\alpha_2' = \sqrt{\psi}$. Let A_2 be a smooth function of two variables satisfying $(A_2)_v = \sin(\alpha_1(u) + \alpha_2(v))$. Let A_3 be a smooth function of one variable such that $A := A_1(u)A_2 + A_3(u)$ is positive-valued. Then for $B := 1$, we obtain

$B_u/A = I$ and $A_v/B = J$, and we see that $g := A^2du^2 + B^2dv^2$ is the induced metric by an immersion as in (i) of Theorem 2.1. As is explained in Subsection 1.1, such an immersion was studied in [5]. Each member of the family of surfaces given by $(\omega, \mathcal{D}_1, \mathcal{D}_2)$ is determined by A_3 and $f := -\log A$ satisfies (8).

Example 2.2. Suppose that ω is locally represented as $\omega = -w_v du + w_u dv$, where w is a smooth function on a neighborhood of each $p \in M$. Then (16) is represented as

$$\begin{aligned}(w_{uu} + w_{vv})^2 = &\left(-2\int_{u_0}^{u}(w_{uu} + w_{vv})w_u du + \psi(v)\right)\\ &\times\left(-2\int_{v_0}^{v}(w_{uu} + w_{vv})w_v dv + \phi(u)\right).\end{aligned} \tag{18}$$

In addition, if w satisfies the sinh-Gordon equation $w_{uu} + w_{vv} = -\sinh 2w$, then

$$\phi(u) := \cosh 2w(u, v_0) + 1, \quad \psi(v) := \cosh 2w(u_0, v) - 1$$

satisfy (18) and in the case where (5) with (7) does not satisfy the compatibility condition, (5) has just two solutions ([8]). If w satisfies $w_{uu} + w_{vv} = e^{-2w}$, then

$$\phi(u) := e^{-2w(u,v_0)}, \quad \psi(v) := e^{-2w(u_0,v)}$$

satisfy (18) and in Section 3, we will obtain a characterization of (5) with (7) which has a unique solution so that $Y^2 = 4XZ$ holds, in terms of the equation $w_{uu} + w_{vv} = e^{-2w}$. The equation (8) for $\omega = -w_v du + w_u dv$ is represented as

$$f_{uv} + w_v e^f f_u + w_u e^{-f} f_v + 2w_{uv}\sinh f = 0. \tag{19}$$

We see that (19) has a trivial solution $f \equiv 0$ and that in the case where (5) with (7) has a solution, the surface corresponding to $f \equiv 0$ is isothermic.

Let $\mathcal{D}_1$, $\mathcal{D}_2$ and ω be as in the beginning of the present section. Let (u, v) be local coordinates satisfying $\partial/\partial u \in \mathcal{D}_1$, $\partial/\partial v \in \mathcal{D}_2$. Let α', β', δ' be determined by I, J and (u, v) as in (7), and X, Y, Z as in (10). Let $(\tilde{u}, \tilde{v})$ be local coordinates satisfying $\partial/\partial\tilde{u} \in \mathcal{D}_1$, $\partial/\partial\tilde{v} \in \mathcal{D}_2$ and $\tilde{X}$, $\tilde{Y}$, $\tilde{Z}$ as in (10) for $(\tilde{u}, \tilde{v})$. Let Q be a smooth tensor field of type $(0, 2)$ on M which is locally represented as $Q = e^q du^2 = e^{\tilde{q}} d\tilde{u}^2$. Then by (12), we obtain

$$e^q X = e^{\tilde{q}}\tilde{X}\frac{d\tilde{u}}{du}\frac{d\tilde{v}}{dv}, \quad Y = \tilde{Y}\frac{d\tilde{u}}{du}\frac{d\tilde{v}}{dv}, \quad e^{-q}Z = e^{-\tilde{q}}\tilde{Z}\frac{d\tilde{u}}{du}\frac{d\tilde{v}}{dv}.$$

Therefore we see that there exist smooth tensor fields c_2, c_1, c_0 of type $(0,2)$ on M which are locally represented as

$$c_2 = e^q X dudv, \quad c_1 = Y dudv, \quad c_0 = e^{-q} Z dudv.$$

For each real number x, we set

$$P(x) := x^2 c_2 + x c_1 + c_0.$$

We see that P is determined by a triplet $(\omega, \mathcal{D}_1, \mathcal{D}_2)$ and Q. We call P the *Codazzi-Mainardi polynomial* of $(M, \omega, \mathcal{D}_1, \mathcal{D}_2)$ for a tensor field Q. If $(\omega, \mathcal{D}_1, \mathcal{D}_2)$ satisfies (13), then the Codazzi-Mainardi polynomial P for $Q = Q_0 = e^{F'} du^2$ satisfies $P(1) = 0$. The compatibility condition of (5) with α', β', δ' as in (7) is given by $P \equiv 0$. Suppose that $(\omega, \mathcal{D}_1, \mathcal{D}_2)$ satisfies (13) and $P \not\equiv 0$. In addition, suppose that there exist distinct two tensor fields Q_1, Q_2 on M which are locally represented as $Q_1 = e^{F_1'} du^2$, $Q_2 = e^{F_2'} du^2$, respectively so that F_1', F_2' are just two solutions of (5) with (7). Then the Codazzi-Mainardi polynomial P for $Q = Q_2$ satisfies $P(e^{F_1' - F_2'}) = 0$.

Remark 2.3. Let K, (u, v), A, B, k_1, k_2 be as in the beginning of Subsection 1.1 and F' as in the beginning of Subsection 1.2. Then by direct computations, we obtain

$$\frac{K}{2AB}(e^{F'} X + Y + e^{-F'} Z) = c_{20} k_1^2 + c_{11} k_1 k_2 + c_{02} k_2^2, \tag{20}$$

where

$$\begin{aligned} c_{20} &:= \frac{1}{AB}\{(\log |K| A^2)_v (\log B)_u - (\log B)_{uv}\}, \\ c_{11} &:= \frac{1}{AB}\{(\log |K| AB)_{uv} - 4(\log A)_v (\log B)_u\}, \\ c_{02} &:= \frac{1}{AB}\{(\log |K| B^2)_u (\log A)_v - (\log A)_{uv}\}. \end{aligned}$$

The right side of (20) vanishes ([3], [4], [6]) and the left side of (20) also vanishes, because of (11). If M is oriented, then c_{20}, c_{11}, c_{02} are determined by the induced metric g and two principal distributions $\mathcal{D}_1$, $\mathcal{D}_2$, and do not depend on local coordinates (u, v) which satisfy $\partial/\partial u \in \mathcal{D}_1$, $\partial/\partial v \in \mathcal{D}_2$ and give the positive orientation of M. The Codazzi-Mainardi polynomial was originally defined by c_{20}, c_{11}, c_{02} ([3], [4]). In [6], c_{20}, c_{11}, c_{02} were represented by the geodesic curvatures of lines of curvature.

3. The case where $Y^2 = 4XZ$

Let X, Y, Z be as in (10) and $\tilde{X}$, $\tilde{Y}$, $\tilde{Z}$ as in (10) for local coordinates $(\tilde{u}, \tilde{v})$ satisfying $\partial/\partial\tilde{u} \in \mathcal{D}_1$, $\partial/\partial\tilde{v} \in \mathcal{D}_2$. Then $Y^2 = 4XZ$ is equivalent to $\tilde{Y}^2 = 4\tilde{X}\tilde{Z}$ and therefore whether $Y^2 = 4XZ$ holds depends only on $(\omega, \mathcal{D}_1, \mathcal{D}_2)$. We will prove

Theorem 3.1. *Let M be a two-dimensional manifold. Let $\mathcal{D}_1$, $\mathcal{D}_2$ be two smooth one-dimensional distributions on M satisfying $\mathcal{D}_1(p) \neq \mathcal{D}_2(p)$ for any $p \in M$. Let ω be a 1-form on M satisfying $d\omega \neq 0$ at any point of M. Then a triplet $(\omega, \mathcal{D}_1, \mathcal{D}_2)$ satisfies (13), $P \not\equiv 0$ and $Y^2 = 4XZ$ on M if and only if on a neighborhood of each point of M, there exist local coordinates (u, v) satisfying $\partial/\partial u \in \mathcal{D}_1$, $\partial/\partial v \in \mathcal{D}_2$ and a smooth function w satisfying*

$$w_u = I, \quad w_v = J, \quad w_{uv} + 2w_u w_v \neq 0, \quad w_{uu} + w_{vv} = e^{-2w}.$$

Remark 3.1. The condition $w_{uv} + 2w_u w_v \neq 0$ corresponds to $P \not\equiv 0$.

Remark 3.2. If $(\omega, \mathcal{D}_1, \mathcal{D}_2)$ satisfies (13), $P \not\equiv 0$ and $Y^2 = 4XZ$, then we obtain $X \neq 0$ and $-Y/2X > 0$, and we see that (5) with (7) has a unique solution $F' = \log(-Y/2X)$.

Remark 3.3. For an oriented minimal surface in E^3 with no umbilical point, $\omega := - * \theta_K$ and two principal distributions $\mathcal{D}_1$, $\mathcal{D}_2$ satisfy (13) and $Y^2 = 4XZ$ (refer to [3]). We know that a minimal surface in E^3 with no umbilical point and $P \not\equiv 0$ is characterized by the equation $w_{uu} + w_{vv} = e^{-2w}$. Theorem 3.1 says that a triplet $(\omega, \mathcal{D}_1, \mathcal{D}_2)$ satisfies (13), $P \not\equiv 0$ and $Y^2 = 4XZ$ if and only if the family of surfaces given by (5) with (7) has a minimal surface in E^3 which is not part of any catenoid.

Remark 3.4. In [12], we can find a representation of solutions of the equation $w_{uu} + w_{vv} = e^{-2w}$.

Remark 3.5. Referring to [8], we see that on a neighborhood of each point of M, an over-determined system (5) with (7) for $(\omega, \mathcal{D}_1, \mathcal{D}_2)$ has just two solutions if and only if there exist local coordinates (u, v) satisfying $\partial/\partial u \in \mathcal{D}_1$, $\partial/\partial v \in \mathcal{D}_2$ and a smooth nowhere zero function w satisfying

$$w_u = I, \quad w_v = J, \quad (\log|\tanh w|)_{uv} \neq 0, \quad w_{uu} + w_{vv} = -\sinh 2w.$$

In order to prove Theorem 3.1, we need

Proposition 3.1. *Let α', β', δ' be smooth functions of two variables. Then an over-determined system* (5) *has a unique solution so that $Y^2 = 4XZ$ holds if and only if there exist a smooth function f of one variable u and a smooth, positive-valued function Δ of two variables u, v satisfying*

$$\alpha' = -f(u) + 2(\log\Delta)_u, \quad \beta' = \frac{f(u)}{\Delta} + \left(\frac{1}{\Delta}\right)_u, \quad \delta' = \Delta_v \tag{21}$$

and

$$(\log\Delta)_{uv} \neq ((\log\Delta)_u - f(u))(\log\Delta)_v. \tag{22}$$

Proof. Suppose that (5) has a unique solution F' so that $Y^2 = 4XZ$ holds. Then all of X, Y, Z are nonzero and $\Delta := -Y/2X$ is positive-valued so that $e^{F'} = \Delta$ holds. From the second relation in (5), we obtain $\Delta_v = \delta'$. By this together with $2X\Delta + Y = 0$, we obtain $(\alpha' + 2\beta'\Delta)_v = 0$. Therefore there exists a smooth function f of one variable u satisfying $\alpha' = f(u) - 2\beta'\Delta$. By this together with the first relation in (5), we obtain $\beta' = f(u)/\Delta + (1/\Delta)_u$. This implies $\alpha' = -f(u) + 2(\log\Delta)_u$. Suppose that α', β', δ' are represented as in (21). Then $F' := \log\Delta$ is a solution of (5). Since

$$-\Delta X = \frac{Y}{2} = -\frac{Z}{\Delta} = F'_{uv} - (F'_u - f(u))F'_v,$$

we obtain $Y^2 = 4XZ$, and we see that F' is a unique solution if and only if $F'_{uv} \neq (F'_u - f(u))F'_v$ holds. □

Remark 3.6. Let h be a smooth function of one variable u satisfying $dh/du = f$. Then

$$a' := -h(u) + 2\log\Delta, \quad b' := h(u) - \log\Delta, \quad c' := 0, \quad d' := -\log\Delta$$

satisfy $a' + b' + c' + d' = 0$ and (14) with (21).

Proof of Theorem 3.1. Suppose that $(\omega, \mathcal{D}_1, \mathcal{D}_2)$ satisfies (13), $P \not\equiv 0$ and $Y^2 = 4XZ$ on M. Then noticing that (5) with (7) has a unique solution, we see by Proposition 3.1 that there exist a smooth function f of one variable u and a smooth, positive-valued function Δ of two variables u, v satisfying

$$(\log(I_u + J_v)^2)_u = -f(u) + 2(\log\Delta)_u, \tag{23}$$

$$\frac{2I}{I_u + J_v} = \frac{f(u)}{\Delta} + \left(\frac{1}{\Delta}\right)_u, \tag{24}$$

$$-2(I_u + J_v)J = \Delta_v. \tag{25}$$

Let ϕ be a smooth, positive-valued function of one variable u satisfying $d(\log\phi^2)/du = f$. Then from (23), we see that there exists a smooth, positive-valued function ψ of one variable v satisfying

$$\Delta^2 = \phi(u)^2\psi(v)^2(I_u + J_v)^2.$$

Therefore we obtain

$$\Delta = \varepsilon\phi(u)\psi(v)(I_u + J_v), \tag{26}$$

where $\varepsilon := +1$ or -1. By (24), (25) and (26), we obtain

$$\begin{aligned}(\log\Delta/\phi(u)^2)_u &= -2\varepsilon I\phi(u)\psi(v),\\ (\log\Delta/\phi(u)^2)_v &= -\frac{2\varepsilon J}{\phi(u)\psi(v)}.\end{aligned} \tag{27}$$

We set $\tilde\Delta := \Delta/\phi(u)^2$. Let $\tilde u$ be a smooth function of one variable u satisfying $d\tilde u/du = \phi$ and $\tilde v$ a smooth function of one variable v satisfying $d\tilde v/dv = \varepsilon/\psi$. Then $(\tilde u, \tilde v)$ are local coordinates satisfying $\partial/\partial\tilde u \in \mathcal{D}_1$, $\partial/\partial\tilde v \in \mathcal{D}_2$. By (12), we rewrite (26) and (27) into

$$\tilde\Delta = \tilde I_{\tilde u} + \tilde J_{\tilde v}, \quad (\log\tilde\Delta)_{\tilde u} = -2\tilde I, \quad (\log\tilde\Delta)_{\tilde v} = -2\tilde J. \tag{28}$$

If we set $w := -(1/2)\log\tilde\Delta$, then we can rewrite (28) into

$$w_{\tilde u} = \tilde I, \qquad w_{\tilde v} = \tilde J, \qquad w_{\tilde u\tilde u} + w_{\tilde v\tilde v} = e^{-2w}.$$

We see that (22) is equivalent to $w_{\tilde u\tilde v} + 2w_{\tilde u}w_{\tilde v} \neq 0$. If there exist local coordinates (u, v) satisfying $\partial/\partial u \in \mathcal{D}_1$, $\partial/\partial v \in \mathcal{D}_2$ and a smooth function w satisfying $w_u = I$, $w_v = J$ and $w_{uu} + w_{vv} = e^{-2w}$, then $\Delta := e^{-2w}$ and $f := 0$ satisfy (21), where α', β', δ' are as in (7). Hence we have proved Theorem 3.1. □

Example 3.1. Let x, y, z be functions of two variables u, v defined by

$$\begin{aligned}x(u,v) &:= -\frac{1}{3}u^3 + uv^2 + \frac{1}{4}u,\\ y(u,v) &:= -u^2v + \frac{1}{3}v^3 - \frac{1}{4}v,\\ z(u,v) &:= \frac{1}{2}u^2 - \frac{1}{2}v^2.\end{aligned}$$

We define a minimal immersion ι by $\iota(u,v) := (x(u,v), y(u,v), z(u,v))$. The image by ι is Enneper's surface. We obtain

$$\begin{aligned}\iota_u &= \left(-u^2 + v^2 + \frac{1}{4},\ -2uv,\ u\right),\\ \iota_v &= \left(2uv,\ -u^2 + v^2 - \frac{1}{4},\ -v\right).\end{aligned}$$

Therefore the first fundamental form of ι is given by

$$d\iota \cdot d\iota = \left(u^2 + v^2 + \frac{1}{4}\right)^2 (du^2 + dv^2).$$

Let ν be a unit normal vector field of ι. Then we can suppose

$$\nu = \frac{1}{u^2 + v^2 + 1/4}\left(u,\ v,\ u^2 + v^2 - \frac{1}{4}\right).$$

Therefore the second fundamental form of ι is given by

$$-d\iota \cdot d\nu = -du^2 + dv^2.$$

We see that $\partial/\partial u$ and $\partial/\partial v$ give principal directions at any point. If we set $A := u^2 + v^2 + 1/4$, then the Gaussian curvature K of ι is given by $K = -1/A^4$. By (6) with $B := A$, we obtain

$$\alpha' = -(\log A^4)_u, \quad \beta' = (A^2)_u, \quad \delta' = \left(\frac{1}{A^2}\right)_v. \tag{29}$$

Therefore by (10), we obtain

$$X = (A^2)_{uv}, \quad Y = -2\frac{(A^2)_{uv}}{A^2}, \quad Z = \frac{(A^2)_{uv}}{A^4}. \tag{30}$$

In particular, we obtain $Y^2 = 4XZ$. By $A = u^2 + v^2 + 1/4$, we see that if $uv \neq 0$, then all of X, Y, Z are nonzero. We set $w = \log A$. Then we immediately see $w_u = I$, $w_v = J$. From $K = -1/A^4$, we obtain $w_{uu} + w_{vv} = e^{-2w}$. Since

$$w_{uv} + 2w_u w_v = \frac{(A^2)_{uv}}{2A^2} = \frac{4uv}{(u^2 + v^2 + 1/4)^2},$$

we see that if $uv \neq 0$, then $w_{uv} + 2w_u w_v \neq 0$.

Example 3.2. Let x, y, z be functions of two variables u, v defined by

$$x(u, v) := \sinh u \cos v, \quad y(u, v) := \sinh u \sin v, \quad z(u, v) := v.$$

We define a minimal immersion ι by $\iota(u, v) := (x(u, v), y(u, v), z(u, v))$. The image by ι is the helicoid. The first fundamental form of ι is given by $d\iota \cdot d\iota = \cosh^2 u\,(du^2 + dv^2)$. Let ν be a unit normal vector field of ι. Then the second fundamental form of ι is given by $-d\iota \cdot d\nu = -2du dv$. We set $\tilde{u} := (1/\sqrt{2})(u + v)$, $\tilde{v} := (1/\sqrt{2})(u - v)$. Then $(\tilde{u}, \tilde{v})$ are local coordinates satisfying

$$d\iota \cdot d\iota = \left(\cosh^2 \frac{1}{\sqrt{2}}(\tilde{u} + \tilde{v})\right)(d\tilde{u}^2 + d\tilde{v}^2),$$
$$-d\iota \cdot d\nu = -d\tilde{u}^2 + d\tilde{v}^2.$$

We see that $\partial/\partial\tilde{u}$ and $\partial/\partial\tilde{v}$ give principal directions at any point. If we set $A := \cosh(1/\sqrt{2})(\tilde{u}+\tilde{v})$, then the Gaussian curvature K of ι is given by $K = -1/A^4$. By (6) with $B := A$, we obtain (29) and (30) with $(u,v) := (\tilde{u},\tilde{v})$, and therefore we obtain $Y^2 = 4XZ$. By $A = \cosh(1/\sqrt{2})(\tilde{u}+\tilde{v})$, we see that all of X, Y, Z are nonzero. We set $w = \log A$. Then $w_{\tilde{u}} = I$, $w_{\tilde{v}} = J$ with $(u,v) := (\tilde{u},\tilde{v})$. From $K = -1/A^4$, we obtain $w_{\tilde{u}\tilde{u}} + w_{\tilde{v}\tilde{v}} = e^{-2w}$. Since

$$w_{\tilde{u}\tilde{v}} + 2w_{\tilde{u}}w_{\tilde{v}} = \frac{(A^2)_{\tilde{u}\tilde{v}}}{2A^2} = \frac{\cosh\sqrt{2}(\tilde{u}+\tilde{v})}{2\cosh^2(1/\sqrt{2})(\tilde{u}+\tilde{v})},$$

we obtain $w_{\tilde{u}\tilde{v}} + 2w_{\tilde{u}}w_{\tilde{v}} \neq 0$.

Example 3.3. Let x, y, z be functions of two variables u, v defined by

$$x(u,v) := \arg\frac{\xi+\sqrt{-1}}{\xi-\sqrt{-1}}, \quad y(u,v) := \arg\frac{\xi+1}{\xi-1}, \quad z(u,v) := \log\left|\frac{\xi^2+1}{\xi^2-1}\right|,$$

where $\xi := u+\sqrt{-1}v$ satisfies $|\xi| < 1$ and we suppose $x, y \in (-3\pi/2, -\pi/2)$. We define a minimal immersion ι by $\iota(u,v) := (x(u,v), y(u,v), z(u,v))$. The image by ι is a fundamental piece of Scherk's surface. Since

$$\frac{\partial x}{\partial\xi} = -\frac{1}{\xi^2+1}, \quad \frac{\partial y}{\partial\xi} = \frac{\sqrt{-1}}{\xi^2-1}, \quad \frac{\partial z}{\partial\xi} = -\frac{2\xi}{\xi^4-1},$$

the first fundamental form of ι is given by

$$d\iota\cdot d\iota = dx^2+dy^2+dz^2 = 4\frac{(|\xi|^2+1)^2}{|\xi^4-1|^2}(du^2+dv^2). \tag{31}$$

We set

$$f := 2\left(\frac{\partial x}{\partial\xi} - \sqrt{-1}\frac{\partial y}{\partial\xi}\right) = \frac{4}{\xi^4-1},$$
$$g := \frac{\partial z}{\partial\xi}\Big/\left(\frac{\partial x}{\partial\xi} - \sqrt{-1}\frac{\partial y}{\partial\xi}\right) = -\xi.$$

Then the second fundamental form of ι is given by

$$-\mathrm{Re}\left(f\frac{dg}{d\xi}d\xi^2\right) = 4\mathrm{Re}\left(\frac{1}{\xi^4-1}d\xi^2\right). \tag{32}$$

Let $\tilde{\xi}$ be a holomorphic function of ξ satisfying $(d\tilde{\xi}/d\xi)^2 = 4/(1-\xi^4)$ and $\tilde{u}$, $\tilde{v}$ the real and the imaginary parts of $\tilde{\xi}$, respectively. Then from (32), we see that the second fundamental form of ι is given by $-d\tilde{u}^2+d\tilde{v}^2$, and from (31), we obtain $d\iota\cdot d\iota = A^2(d\tilde{u}^2+d\tilde{v}^2)$, where $A := (1+|\xi|^2)/|1-\xi^4|^{1/2}$. We see that $\partial/\partial\tilde{u}$ and $\partial/\partial\tilde{v}$ give principal directions at any point. The Gaussian curvature K of ι is given by $K = -1/A^4$. By (6) with $B := A$, we obtain (29) and (30) with $(u,v) := (\tilde{u},\tilde{v})$, and therefore we obtain $Y^2 = 4XZ$. By $A = (1+|\xi|^2)/|1-\xi^4|^{1/2}$, we obtain

$$(A^2)_\xi = 2A^2\left(\frac{\bar{\xi}}{1+|\xi|^2} + \frac{\xi^3}{1-\xi^4}\right).$$

By this together with $(d\tilde{\xi}/d\xi)^2 = 4/(1-\xi^4)$, we obtain

$$\begin{aligned}(A^2)_{\tilde{\xi}\tilde{\xi}} &= (A^2)_\xi \frac{d^2\xi}{d\tilde{\xi}^2} + (A^2)_{\xi\xi}\left(\frac{d\xi}{d\tilde{\xi}}\right)^2 \\ &= \frac{A^2}{2}\frac{3\xi^2 + \bar{\xi}^2 + 8\xi^2|\xi|^2 + \xi^6 + 3\xi^2|\xi|^4}{(1+|\xi|^2)^2(1-\xi^4)}.\end{aligned}$$

Therefore we see that the imaginary part of $(A^2)_{\tilde{\xi}\tilde{\xi}}$ is nonzero on an open dense set O of the domain of ι. This means that $(A^2)_{\tilde{u}\tilde{v}}$ is nonzero on O and therefore we see that all of X, Y, Z are nonzero on O. We set $w = \log A$. Then $w_{\tilde{u}} = I$, $w_{\tilde{v}} = J$ with $(u,v) := (\tilde{u},\tilde{v})$ and $w_{\tilde{u}\tilde{u}} + w_{\tilde{v}\tilde{v}} = e^{-2w}$. By

$$2w_{\tilde{\xi}\tilde{\xi}} + 4w_{\tilde{\xi}}^2 = \frac{(A^2)_{\tilde{\xi}\tilde{\xi}}}{A^2},$$

we obtain $w_{\tilde{u}\tilde{v}} + 2w_{\tilde{u}}w_{\tilde{v}} \neq 0$ on O.

Remark 3.7. Suppose that the family of surfaces given by (5) with (7) has a surface which is part of a catenoid. Then there exist local coordinates (u,v) satisfying $\partial/\partial u \in \mathcal{D}_1$, $\partial/\partial v \in \mathcal{D}_2$, $I \equiv 0$ and $J = \tanh v$. Then a pair (A,B) of positive-valued functions A, B satisfies $B_u/A = I$ and $A_v/B = J$ if and only if B is of one variable v so that $A_v = B(v)\tanh v$. This condition means that A is represented as $A = A_2(v) + A_3(u)$, where A_2 and A_3 are smooth functions of one variable. Each member of the family of surfaces is determined by A_3.

Remark 3.8. For $(\omega, \mathcal{D}_1, \mathcal{D}_2)$ satisfying (13), $P \not\equiv 0$ and $Y^2 = 4XZ$ on M, there exists no umbilical point of an immersion as in (i) of Theorem 2.1. Noticing (20), we see that if a point p of M is umbilical with respect to the immersion, then we have

$$(\log K^2)_v(\log B)_u + (\log K^2)_{uv} + (\log K^2)_u(\log A)_v = 0 \tag{33}$$

at p. Applying $w_u = B_u/A$, $w_v = A_v/B$ and $w_{uu} + w_{vv} = e^{-2w}$ to (33), we obtain

$$(w_{uv} + 2w_u w_v)\left(2 + \frac{B}{A} + \frac{A}{B}\right) = 0 \tag{34}$$

at p. However, since A, B are positive-valued and since $w_{uv} + 2w_u w_v \neq 0$ by Theorem 3.1, we see that (34) does not hold.

Appendix. Semilinear hyperbolic differential equations

In order to understand a family of surfaces given by each solution of (5) with (7), we need to understand the existence and the uniqueness of a solution of a semilinear hyperbolic differential equation with a given initial data, because of a one-to-one correspondence between the above family of surfaces and the set of solutions of (8).

If we set $t := u + v$, $x := u - v$, then (8) can be rewritten into

$$f_{tt} - f_{xx} = \Phi(t, x, f, f_t, f_x), \tag{A.1}$$

where

$$\begin{aligned}\Phi(t, x, f, p, q) := &-(Je^f + Ie^{-f})p - (Je^f - Ie^{-f})q \\ &-(J_t e^f + J_x e^f - I_t e^{-f} + I_x e^{-f}).\end{aligned}$$

If we set $F := {}^t(f_t, f_x, f)$, then (A.1) can be rewritten into

$$F_t + AF_x + B = \begin{pmatrix} 0 \\ 0 \\ 0 \end{pmatrix}, \tag{A.2}$$

where

$$A := \begin{pmatrix} 0 & -1 & 0 \\ -1 & 0 & 0 \\ 0 & 0 & 0 \end{pmatrix}, \quad B := \begin{pmatrix} -\Phi(t, x, f, f_t, f_x) \\ 0 \\ -f_t \end{pmatrix}. \tag{A.3}$$

We see that A is diagonalizable and has distinct three eigenvalues 1, -1, 0. Therefore there exist three eigenvectors of A which form an element of $SL(3, \boldsymbol{R})$.

For $k \in \boldsymbol{N}$, let A be a square matrix of degree k such that the components are smooth functions on an open set O of the tx-plane and B a smooth $\boldsymbol{R}^k$-valued function on an open set of $\boldsymbol{R}^{k+2}$. Let F be a smooth $\boldsymbol{R}^k$-valued function on O satisfying

$$F_t(t, x) + A(t, x)F_x(t, x) + B(t, x, F(t, x)) = \begin{pmatrix} 0 \\ 0 \\ \vdots \\ 0 \end{pmatrix} \tag{A.4}$$

for any $(t, x) \in O$. The equation (A.4) is said to be *hyperbolic* if A satisfies the following:

(i) for any $(t,x) \in O$, $A(t,x)$ is diagonalizable so that all the eigenvalues are real;

(ii) there exist k smooth $\boldsymbol{R}^k$-valued functions on O which give eigenvectors of $A(t,x)$ and form an element of $SL(k,\boldsymbol{R})$ at each $(t,x) \in O$.

Therefore the equation (A.2) with (A.3) is hyperbolic.

Suppose that (A.4) is hyperbolic. Noticing (A.3), we suppose that A in (A.4) is constant. Then there exists an element S of $SL(k,\boldsymbol{R})$ satisfying $SA = \Lambda S$, where Λ is a diagonal matrix of degree k. Let λ^i be the (i,i)-component of Λ and V^i a vector field on the tx-plane defined by $V^i := \partial/\partial t + \lambda^i \partial/\partial x$. We set $\tilde{F} := SF$. We denote by $\tilde{F}^i$ the i-th component of $\tilde{F}$ and by $\beta^i(t,x,\tilde{F}(t,x))$ the i-th component of $SB(t,x,F(t,x))$. Then we obtain

$$(V^i\tilde{F}^i)(t,x) + \beta^i(t,x,\tilde{F}(t,x)) = 0 \tag{A.5}$$

($i = 1,2,\ldots,k$). Therefore integrating both sides of (A.5) along each integral curve of V^i, we obtain

$$\begin{aligned} &\tilde{F}^i(\tau,\xi) - \psi^i(\xi - \lambda^i\tau) \\ &= \int_0^{-\tau} \beta^i(t+\tau, \lambda^i t + \xi, \tilde{F}(t+\tau, \lambda^i t + \xi))dt, \end{aligned} \tag{A.6}$$

where $\psi^i(\xi - \lambda^i\tau) := \tilde{F}^i(0, \xi - \lambda^i\tau)$.

Let $\beta^1,\ldots,\beta^k$ be smooth functions on an open set of $\boldsymbol{R}^{k+2}$. Let $\psi^1,\ldots,\psi^k$ be smooth functions of one variable x on a neighborhood I of $x_0 \in \boldsymbol{R}$. Let $\tilde{F}$ be a smooth $\boldsymbol{R}^k$-valued function on a neighborhood $\tilde{O}$ of $(0,x_0)$ in $\boldsymbol{R}^2$ satisfying $\tilde{F}^i(0,x) = \psi^i(x)$ for any $x \in I$ satisfying $(0,x) \in \tilde{O}$. Let $T(\tilde{F})$ be a smooth $\boldsymbol{R}^k$-valued function on a neighborhood of $(0,x_0)$ in $\boldsymbol{R}^2$ defined by

$$\begin{aligned} &(T(\tilde{F}))^i(\tau,\xi) \\ &:= \psi^i(\xi - \lambda^i\tau) + \int_0^{-\tau} \beta^i(t+\tau, \lambda^i t + \xi, \tilde{F}(t+\tau, \lambda^i t + \xi))dt. \end{aligned}$$

Then we obtain $T(\tilde{F})(0,x) = \tilde{F}(0,x)$. For each $n \in \boldsymbol{N}$, we set $T^{n+1}(\tilde{F}) := T(T^n(\tilde{F}))$ by induction. Then referring to [11, pp. 467–468], we see that on a suitable neighborhood O of $(0,x_0)$, $\{T^n(\tilde{F})\}$ uniformly converges to a continuous function $\tilde{F}_0$. We see that $\tilde{F}_0$ is of class C^1: $\{\partial T^n(\tilde{F})/\partial t\}$ and $\{\partial T^n(\tilde{F})/\partial x\}$ uniformly converge. Similarly, we see that $\tilde{F}_0$ is of class C^∞. A smooth function $\tilde{F} := \tilde{F}_0$ satisfies (A.6) and $\tilde{F}_0^i(0,x) = \psi^i(x)$. Therefore if we set $F_0 := S^{-1}\tilde{F}_0$, then $F := F_0$ is a solution of (A.4). Referring to

[11, pp. 448–449], we see that F_0 is uniquely determined by the values of F_0 on the x-axis.

References

1. N. Ando, A class of real-analytic surfaces in the 3-Euclidean space, *Tsukuba J. Math.* **26**, 251–267 (2002).
2. N. Ando, Parallel curved surfaces, *Tsukuba J. Math.* **28**, 223–243 (2004).
3. N. Ando, A two-dimensional Riemannian manifold with two one-dimensional distributions, *Kyushu J. Math.* **59**, 285–299 (2005).
4. N. Ando, Semisurfaces and the equations of Codazzi-Mainardi, *Tsukuba J. Math.* **30**, 1–30 (2006).
5. N. Ando, A surface which has a family of geodesics of curvature, *Beiträge zur Algebra und Geometrie* **48**, 237–250 (2007).
6. N. Ando, The geodesic curvatures of lines of curvature, preprint.
7. N. Ando, Molding surfaces and Liouville's equation, preprint.
8. N. Ando, Over-determined systems in relation to principal curvatures, preprint.
9. R.L. Bryant, S.S. Chern and P.A. Griffiths, Exterior differential systems, *Proceedings of the 1980 Beijing Symposium on Differential Geometry and Differential Equations*, Vol. 1 (1980) 219–338, Science Press, Beijing, 1982.
10. E. Cartan, *Les systèmes différentiels extérieurs et leurs applications géométriques*, 2nd ed., Hermann, 1971.
11. R. Courant and D. Hilbert, *Methods of mathematical physics*, Vol. II, John Wiley & Sons, 1962.
12. H. Furuhata and T. Kurose, Self-dual centroaffine surfaces of codimension two with constant affine mean curvature, *Bull. Belg. Math. Soc.* **9**, 1–15 (2002).
13. J.J. Stoker, *Differential geometry*, John Wiley & Sons, 1969.

Received December 9, 2014
Revised March 3, 2015

Current Developments in Differential Geometry
and its Related Fields pp. 77 – 92

SOME REMARKS ON NONCOMMUTATIVE INSTANTONS

Nikolay A. IVANOV *

Department of Mathematics, Veliko Tarnovo University, Bulgaria
E-mail: Nikolay.Antonov.Ivanov@gmail.com

We make some comments on noncommutative $U(N)$-instantons on $\mathbb{R}^4_\theta$. We elaborate on the equations for the ASD-connection for free modules. Further we make some remarks on the computation of the topological index of ADHM instantons.

Keywords: Noncommutative $\mathbb{R}^4$, ADHM-construction, $U(N)$-connection, topological index.

1. Introduction

The idea of noncommutative field theories is old (see [20]). Much later there had been significant interest motivated by operator algebra theory (from mathematics side) and by string theory (from physics side). In [5] and [7], Connes and Rieffel studied the space of minima of a version of Yang-Mills functional on noncommutative two-tori. This YM-functional is defined by considering a connection on a projective module, by taking the corresponding curvature two-form, by collapsing it into itself and by taking trace. The usual Yang-Mills functional is constructed by integrating the wedge product of a curvature forms with itself over the underlying four dimensional space. In their paper [6] Connes, Douglas and Schwarz considered matrix theory, compactified on tori. In [18] Seiberg and Witten considered a B-field in the presence of a Dp-brane (which prevents the B-field to be gauged out). Noncommutative field theory has become a separate theory, having its mathematical ([4, 5]) and physical ([9, 22]) inclinations.

Conventional instantons are topological field configurations of gauge fields. They have been introduced by Belavin, Polyakov, Schwarz and Tyupkin [3] and studied extensively further. Noncommutative instantons are an analog pertaining to the noncommutative field theory. They have been introduced by Nekrasov and Schwatz [14] and developed further in [12, 13, 16, 19, 23]. There are more recent discussions on the topic (see [11, 17]).

*Partially supported by "St. Cyril and St. Methodius University" grant RD-09-422-13/09.04.2014

2. Noncommutative spaces, algebras of operators and the Fock-Hilbert space

Consider the space $\mathbb{R}^4_\theta$ with four coordinate operators $\hat{x}_1, \hat{x}_2, \hat{x}_3, \hat{x}_4$ satisfying the relations $[\hat{x}_i, \hat{x}_j] = i\theta_{ij}$, where θ_{ij} is a real nondegenerate antisymmetric matrix. After SO(4) rotation of coordinates, θ can be brought to the form:

$$\theta_{ij} = \begin{pmatrix} 0 & \theta_{12} & 0 & 0 \\ -\theta_{12} & 0 & 0 & 0 \\ 0 & 0 & 0 & \theta_{34} \\ 0 & 0 & -\theta_{34} & 0 \end{pmatrix},$$

where $\theta_{12} > 0,\ \theta_{12} + \theta_{34} \geq 0$.

A theory built up by such operators is nonlocal: $\sigma_{\hat{x}_1} \cdot \sigma_{\hat{x}_2} \geq \frac{1}{2}|\theta_{12}|$; $\sigma_{\hat{x}_3} \cdot \sigma_{\hat{x}_4} \geq \frac{1}{2}|\theta_{34}|$. Here $\sigma_{\hat{x}_i} = \sqrt{\langle \hat{x}_i^2 \rangle - \langle \hat{x}_i \rangle^2}$.

Since the pairs of operators $(\hat{x}_1, \hat{x}_2)$ and $(\hat{x}_3, \hat{x}_4)$ satisfy the canonical commutation relations, by the Stone-von Neumann theorem they can be analytically represented as the position and momentum operators on $L^2(\mathbb{R}^2)$:

$$\begin{aligned} \hat{x}_1\psi(x_1, x_3) &= x_1\psi(x_1, x_3), \quad \hat{x}_2\psi(x_1, x_3) = -i\theta_{12}\frac{d\psi(x_1, x_3)}{dx_1} \\ \hat{x}_3\psi(x_1, x_3) &= x_3\psi(x_1, x_3), \quad \hat{x}_4\psi(x_1, x_3) = -i\theta_{34}\frac{d\psi(x_1, x_3)}{dx_3}. \end{aligned} \tag{1}$$

One can consider appropriate algebras $\mathcal{A}$ on $\mathbb{R}^4_\theta$. To start, consider these examples [12, 19]:

- $\mathcal{S}(\mathbb{R}^4) = \{f \in C^\infty(\mathbb{R}^4) \mid \sup_{x\in\mathbb{R}^4} |x^\alpha D^\beta f(x)| < \infty,\ \forall \alpha, \beta \in \mathbb{N}_0^4\}$;
- $\Gamma^m_\rho(\mathbb{R}^4) = \{f \in C^\infty(\mathbb{R}^4) \mid |\partial_\alpha f(x)| \leq C_\alpha(1+|x|^2)^{\frac{1}{2}(m-\rho|\alpha|)},\ \forall \alpha \in \mathbb{N}_0^4\}$, where $m < 0,\ 0 < \rho \leq 1$, and $|\alpha| = \alpha_1 + \alpha_2 + \alpha_3 + \alpha_4$;
- $\mathcal{K}^\infty(\mathbb{R}^4) = \{f \in C^\infty(\mathbb{R}^4) \mid \lim_{|x|\to\infty} |\partial_\alpha f(x)| = 0,\ \forall \alpha \in \mathbb{N}_0^4\}$.

Next we use the Weyl formula which assigns to a function on $\mathbb{R}^4$ an operator ([22]) on $\mathbb{R}^4_\theta$:

$$f(x) \mapsto \hat{\mathcal{W}}[f(x)] = \int d^4x\ f(x)\hat{\Delta}(x),\ \hat{\Delta}(x) = \frac{1}{(2\pi)^4}\int d^4k\ \mathrm{e}^{ik^j\hat{x}_j}\mathrm{e}^{-ik^j x_j}.$$

We have $\hat{\mathcal{W}}[\mathrm{e}^{ik^j x_j}] = \mathrm{e}^{ik^j\hat{x}_j}$ and the reverse formula is

$$\mathrm{Tr}(\hat{\mathcal{W}}[f]\hat{\Delta}(x)) = f(x).$$

We set $\mathcal{S} = \hat{\mathcal{W}}[\mathcal{S}(\mathbb{R}^4)],\ \Gamma^m_\rho = \hat{\mathcal{W}}[\Gamma^m_\rho(\mathbb{R}^4)],\ \mathcal{K}^\infty = \hat{\mathcal{W}}[\mathcal{K}^\infty(\mathbb{R}^4)]$. We note the following:

Lemma 2.1. *These $\mathcal{S}$, Γ^m_ρ $(m < 0,\ 0 < \rho \leq 1)$ and $\mathcal{K}^\infty$ are norm-dense subalgebras of the algebra of compact operators $\mathcal{K}$ on $L^2(\mathbb{R}^2)$.*

Proof. It is well known that $\mathcal{S}$ is a norm dense subalgebra of $\mathcal{K}$ and it is known that all these sets are algebras, containing $\mathcal{S}$ (see [12, 19]). It is enough to show that they are all subalgebras of $\mathcal{K}$.

According to the Baker-Campbell-Hausdorff Formula ([10]) we have $[\mathrm{e}^{ik^j\hat{x}_j}\psi](x_1,x_3) = \mathrm{e}^{i(k^1k^2\theta_{12}+k^3k^4\theta_{34})/2}\mathrm{e}^{i(k^1x_1+k^3x_3)}\psi(x_1+\theta_{12}k^2, x_3+\theta_{34}k^4)$. Suppose that $f(x_1,x_3) \in C_0(\mathbb{R}^2)$ is a function from one of the above algebras (in particular it goes to zero at infinity). Let $\{\psi_n(x_1,x_3)\}_{n=1}^\infty$ be a sequence from the unit ball of $L^2(\mathbb{R}^2)$. We have

$$\begin{aligned}
\hat{\mathcal{W}}[f]\psi_n(x_1,x_3) &= \int d^4y\ f(y)\frac{1}{(2\pi)^4}\int d^4k\ \mathrm{e}^{ik^j\hat{x}_j}\mathrm{e}^{-ik^jy_j}\psi_n(x_1,x_3)\\
&= \frac{1}{(2\pi)^4}\int\int d^4yd^4kf(y)\mathrm{e}^{-ik^jy_j}\mathrm{e}^{i(k^1k^2\theta_{12}+k^3k^4\theta_{34})/2}\mathrm{e}^{i(k^1x_1+k^3x_3)}.\\
&\qquad .\psi_n(x_1+\theta_{12}k^2, x_3+\theta_{34}k^4)\\
&= \frac{1}{(2\pi)^4}\int\int d^4ydk^2dk^4f(y)\delta(k^2\theta_{12}+x_1-y_1)\delta(k^4\theta_{34}+x_3-y_3).\\
&\qquad .\mathrm{e}^{\ i(k^2y_2+k^4y_4)}\psi_n(x_1+\theta_{12}k^2, x_3+\theta_{34}k^4)\\
&= \frac{1}{(2\pi)^4}\int\int dy_2dy_4dk^2dk^4f(x_1+\theta_{12}k^2, y_2, x_3+\theta_{34}k^4, y_4).\\
&\qquad .\mathrm{e}^{-i(k^2y_2+k^4y_4)}\psi_n(x_1+\theta_{12}k^2, x_3+\theta_{34}k^4)\\
&= \frac{1}{(2\pi)^4\theta_{12}\theta_{34}}\int\int d^4y\mathrm{e}^{-i[y_2(y_1-x_1)/\theta_{12}+y_4(y_3-x_3)/\theta_{34}]}\\
&\qquad f(y_1,y_2,y_3,y_4)\psi_n(y_1,y_3).
\end{aligned}$$

For y_1-x_1 and y_3-x_3 both away from zero there is dampening coming from f and the rapidly oscillating phase. Thus the volume is concentrated around the origin. Compactness argument shows that we can find a fundamental subsequence, hence the operator is compact. □

To an algebra $\mathcal{A}$ of the above form we can adjoint a unit. The algebra we obtain in this way we will denote by $\tilde{\mathcal{A}}$. It follows from the above lemma that the K-theories of all these algebras coinside: If $\mathcal{A} = \mathcal{K}^\infty$, Γ^m_ρ, or $\mathcal{S}$, then

$$\begin{aligned}
&K_0(\mathcal{A}) = K_0(\mathcal{K}) = \mathbb{Z},\ \ K_1(\mathcal{A}) = K_1(\mathcal{K}) = \{0\},\\
&K_0(\tilde{\mathcal{A}}) = K_0(\tilde{\mathcal{K}}) = \mathbb{Z}^2 \text{ and } K_1(\tilde{\mathcal{A}}) = K_1(\tilde{\mathcal{K}}) = \{0\}.
\end{aligned} \tag{2}$$

$\mathcal{A}$ is closed with respect to taking derivatives, which are defined in the following way:

$$\partial_1(\hat{f}) \equiv \frac{i}{\theta_{12}}[\hat{x}_2, \hat{f}], \quad \partial_2(\hat{f}) \equiv \frac{-i}{\theta_{12}}[\hat{x}_1, \hat{f}],$$

$$\partial_3(\hat{f}) \equiv \frac{i}{\theta_{34}}[\hat{x}_4, \hat{f}], \quad \partial_4(\hat{f}) \equiv \frac{-i}{\theta_{34}}[\hat{x}_3, \hat{f}].$$

Note that these are all inner derivations and we have $\partial_1(\hat{x}_1^n) = n\hat{x}_1^{n-1}$, etc.

The Lie algebra $\mathbb{R}^4$ acts on $\mathcal{A}$ via the derivatives. The derivatives commute, due to the Bianchi identity.

We have two commuting sets of operators, satisfying the canonical commutation relations:

$$c_1 \equiv \frac{1}{\sqrt{2\theta_{12}}}\hat{z_1} = \frac{1}{\sqrt{2\theta_{12}}}(\hat{x_1} + i\hat{x_2}), \quad c_1^* \equiv \frac{1}{\sqrt{2\theta_{12}}}\hat{z_1}^* = \frac{1}{\sqrt{2\theta_{12}}}(\hat{x_1} - i\hat{x_2}),$$

$$c_2 \equiv \frac{1}{\sqrt{2\theta_{34}}}\hat{z_2} = \frac{1}{\sqrt{2\theta_{34}}}(\hat{x_3} + i\hat{x_4}), \quad c_2^* \equiv \frac{1}{\sqrt{2\theta_{34}}}\hat{z_2}^* = \frac{1}{\sqrt{2\theta_{34}}}(\hat{x_3} - i\hat{x_4}).$$

The commutation relations are:

$$[c_1, c_2] = [c_1, c_2^*] = [c_1^*, c_2] = [c_1^*, c_2^*] = 0, \ [c_1, c_1^*] = [c_2, c_2^*] = I.$$

The Hilbert space on which these operators are represented is $\mathcal{F} = \mathcal{F}_1 \otimes \mathcal{F}_2$, which is the tensor product of two copies of the Fock space. There is a standard orthonormal basis of $\mathcal{F}$ given by the set $\{\ |m,n\rangle \mid m,n \in \mathbb{Z},\ m,n \geq 0\}$. On vectors of this set c_i's act as follows:

$$c_1(|m,n\rangle) = \sqrt{m+1}|m+1,n\rangle; \quad c_2(|m,n\rangle) = \sqrt{n+1}|m,n+1\rangle;$$
$$c_1^*(|m,n\rangle) = \sqrt{m}|m-1,n\rangle, \text{ if } m \geq 1; \quad c_1^*(|0,n\rangle) = 0;$$
$$c_2^*(|m,n\rangle) = \sqrt{n}|m,n-1\rangle, \quad \text{if } n \geq 1; \quad c_2^*(|m,0\rangle) = 0.$$

When one uses the representation of equations (1), one takes $\mathcal{F}_1 \cong \mathcal{F}_2 \subset L^2(\mathbb{R})$ and thus $\mathcal{F} \subset L^2(\mathbb{R}^2)$. The vacuum state is given by $|0,0\rangle = \dfrac{1}{2\pi}\mathrm{e}^{-(x_1^2+x_3^2)/2}$. The excited states of c_i's are given by the Hermite polynomials of x_1 and x_3, multiplied by $\frac{1}{2\pi}\mathrm{e}^{-(x_1^2+x_3^2)/2}$. In fact, it turns out that $\mathcal{F}_1 \cong \mathcal{F}_2 \cong \mathcal{S}(\mathbb{R})$ and $\mathcal{F} \cong \mathcal{S}(\mathbb{R}^2)$.

Instead, we can take as creation operators

$$z_1 = x_1 + ix_2 \text{ and } z_2 = x_3 + ix_4,$$

and as annihilation operators

$$\frac{\partial}{\partial z_1} \equiv \frac{1}{2}(\frac{\partial}{\partial x_1} - i\frac{\partial}{\partial x_2}) \text{ and } \frac{\partial}{\partial z_2} \equiv \frac{1}{2}(\frac{\partial}{\partial x_3} - i\frac{\partial}{\partial x_4}).$$

These operators can be implemented on the Segal-Bargman space $\mathcal{H}L^2(\mathbb{C}^2, \mu_1)$ of all holomorphic functions on $\mathbb{C}^2$, f, for which $\|f\|_1 < \infty$. The scalar product (and the norm) is given by

$$\langle f, g\rangle_1 \equiv \frac{1}{\pi^2}\int_{\mathbb{C}^2} \bar{f} g e^{-|z_1|^2-|z_2|^2} dz_1 dz_2.$$

Then in the above notations one has $|m,n\rangle \equiv z_1^m z_2^n/\sqrt{m!n!}$.

3. Vector bundles, connections and curvatures

Let F be a (principal or complex vector) bundle with structure group G over a manifold M and let $\pi : F \to M$ be a bundle projection. The bundle can be given by a set of local trivializations and transformations between them which are elements of G. The set of sections $\Gamma(M, F)$ is the set of all (continuous, differentiable, etc.) maps $\gamma : M \to F$ such that $\gamma(x) \in \pi^{-1}(x)$. It follows from the Serre-Swan theorem ([21]) that every vector bundle on a locally compact manifold is projective (i.e. a summand in a free bundle). It is important to note that $\Gamma(M, F)$ is a module over $C^\infty(M, \mathbb{C})$.

In the noncommutative case one does not have "points" because of nonlocality. Therefore in order to define "bundles", one needs to consider (right) modules.

Let A be a C^*-algebra ([4]) and let Λ be a Lie group, acting on A via derivations (LieΛ is embedded in the algebra of derivations of A). Let A^∞ be a dense $*$-subalgebra of smooth (with respect to LieΛ) elements. To each right A-module Ξ corresponds a unique (up to isomorphism) right A^∞-module Ξ^∞ for which $\Xi = \Xi^\infty \otimes_{A^\infty} A$. A right A^∞-module Ξ^∞ is projective if there exists a right A^∞-module $\tilde{\Xi}^\infty$ such that $\Xi^\infty \oplus \tilde{\Xi}^\infty \cong \mathbb{C}^n \otimes A^\infty$ for some $n \in \mathbb{N}$.

A connection on Ξ^∞ is a $\mathbb{C}$-linear map $\nabla : \Xi^\infty \to \Xi^\infty \otimes (\mathrm{Lie}\Lambda)^*$ satisfying the condition $\nabla_X(\xi \cdot x) = \nabla(\xi) \cdot x + \xi \cdot \delta_X(x)$, where $X \in \mathrm{Lie}\Lambda$, δ_X is the corresponding element from the algebra of derivations, $x \in A^\infty$, and $\xi \in \Xi^\infty$. The curvature $\Theta \in \mathrm{End}_{A^\infty}(\Xi^\infty) \otimes \wedge^2(\mathrm{Lie}\Lambda)^*$ of ∇ is given by

$$\Theta(X, Y) = \nabla_X\nabla_Y - \nabla_Y\nabla_X - \nabla_{[X,Y]} \in \mathrm{End}_{A^\infty}(\Xi^\infty), \quad X, Y \in \mathrm{Lie}\Lambda.$$

In the $\mathbb{R}^4_\theta$ case we have that $\Lambda = \mathbb{R}^4$ is commutative. If we denote $F_{jk} = \Theta(\partial_j, \partial_k)$ then we will have $F_{jk} = \nabla_j\nabla_k - \nabla_k\nabla_j$. Also in our case $A = \mathcal{A} = \mathcal{K}$ and A^∞ is one of the algebras from section 2. Below we drop the ∞ signs for shortness.

Every connection on $\mathbb{R}^4_\theta$ is of the form $\nabla_j = \partial_j + A_j$, where $A_j \in \mathrm{End}_A(\Xi)$. Its curvature is

$$F_{mn} = \partial_m A_n - \partial_n A_m + [A_m, A_n].$$

We now consider a connection of the form $P\partial_j$, where P is the projection onto Ξ in $\Xi \oplus \tilde{\Xi} \cong \mathbb{C}^n \otimes A$. In the commutative case ([1]) one can find local linear maps u for which $P = uu^*$ and $u^*u = 1$. In the noncommutative case this is possible only if:

Lemma 3.1. *Assume that $U \in M_{n\times k}(\tilde{A})$ is such that $UU^* = P$ and $U^*U = 1_k$. Then the module $PM_{n\times n}(\tilde{A})$ is free.*

Proof. If we denote by $u_1, u_2, \ldots, u_k$ the columns of U, then we will have the following isomorphism $\iota : \mathrm{Span}_{\tilde{A}}\{u_1, \ldots, u_k\} \to \mathbb{C}^k \otimes \tilde{A}$ given by

$$\iota : u_1 \cdot a_1 + \ldots u_k \cdot a_k \mapsto (a_1, \ldots, a_k).$$

The coefficients can be uniquely recovered by multiplying $u_1 \cdot a_1 + \ldots u_k \cdot a_k = U(a_1, \ldots, a_k)^T$ on the left by U^*.
Next we have

$$\begin{aligned} PM_{n\times n}(\tilde{A}) &= UU^*M_{n\times n}(\tilde{A}) = UU^*UM_{k\times k}(\tilde{A}) \\ &= U1_kM_{k\times k}(\tilde{A}) = UM_{k\times k}(\tilde{A}) = \mathrm{Span}_{\tilde{A}}\{u_1, \ldots, u_k\}^k, \end{aligned}$$

the range projection of U^* being 1_k. □

In this case the connection is

$$A = U^*dU, \text{ or } A_j = U^*\partial_j U \text{ and satisfies } A = -(dU^*)U, \qquad (3)$$

because of the relation $U^*U = 1_k$. Its curvature is

$$\begin{aligned} F_{mn} &= (\partial_m U^*)\partial_n U + U^*\partial_m\partial_n U - (\partial_n U^*)\partial_m U - U^*\partial_n\partial_m U \\ &\quad - (\partial_m U^*)UU^*\partial_n U + (\partial_n U^*)UU^*\partial_m U \\ &= (\partial_m U^*)\partial_n U - (\partial_n U^*)\partial_m U + (\partial_n U^*)UU^*\partial_m U - (\partial_m U^*)UU^*\partial_n U \\ &= (\partial_m U^*)(1 - UU^*)\partial_n U - (\partial_n U^*)(1 - UU^*)\partial_m U, \end{aligned}$$

or

$$F_{mn} = (\partial_m U^*)(1 - UU^*)\partial_n U - (\partial_n U^*)(1 - UU^*)\partial_m U.$$

We note that $U^*(1 - UU^*) = (1 - UU^*)U = 0$, and therefore if we set $Q = (1 - UU^*)$ we would have

$$Q^2 = (1-UU^*)(1-UU^*) = 1-UU^*+(1-UU^*)UU^* = 1-UU^* = Q. \quad (4)$$

We have

$$\begin{aligned}(\partial_1 U^*)(1-UU^*)\partial_2 U &= \frac{i}{\theta_{12}}[\hat{x}_2, U^*](1-UU^*)\frac{-i}{\theta_{12}}\,[\hat{x}_1, U] \\ &= \frac{1}{\theta_{12}^2}(\hat{x}_2 U^* - U^*\hat{x}_2)(1-UU^*)(\hat{x}_1 U - U\hat{x}_1) \\ &= \frac{-1}{\theta_{12}^2}U^*\hat{x}_2(1-UU^*)\hat{x}_1 U.\end{aligned}$$

Analogously

$$(\partial_2 U^*)(1-UU^*)\partial_1 U = \frac{-1}{\theta_{12}^2}U^*\hat{x}_1(1-UU^*)\hat{x}_2 U$$

and therefore

$$F_{12} = \frac{1}{\theta_{12}^2}U^*(\hat{x}_1(1-UU^*)\hat{x}_2 - \hat{x}_2(1-UU^*)\hat{x}_1)U. \tag{5}$$

Further

$$F_{13} = \frac{1}{\theta_{12}\theta_{34}}U^*(\hat{x}_2(1-UU^*)\hat{x}_4 - \hat{x}_4(1-UU^*)\hat{x}_2)U, \tag{6}$$

$$F_{14} = \frac{1}{\theta_{12}\theta_{34}}U^*(\hat{x}_3(1-UU^*)\hat{x}_2 - \hat{x}_2(1-UU^*)\hat{x}_3)U, \tag{7}$$

$$F_{23} = \frac{1}{\theta_{12}\theta_{34}}U^*(\hat{x}_4(1-UU^*)\hat{x}_1 - \hat{x}_1(1-UU^*)\hat{x}_4)U, \tag{8}$$

$$F_{24} = \frac{1}{\theta_{12}\theta_{34}}U^*(\hat{x}_1(1-UU^*)\hat{x}_3 - \hat{x}_3(1-UU^*)\hat{x}_1)U, \tag{9}$$

$$F_{34} = \frac{1}{\theta_{34}^2}U^*(\hat{x}_3(1-UU^*)\hat{x}_4 - \hat{x}_4(1-UU^*)\hat{x}_3)U. \tag{10}$$

All projective modules over $\tilde{\mathcal{A}}$ are of the form $\mathcal{E}_{kn} \equiv \mathcal{F}^{*k} \oplus \tilde{\mathcal{A}}^n$, where $n, k \in \mathbb{N}$ ([12, 16]). This follows essentially from equations (2). Every connection on $\mathcal{E}_{kn}$ is of the form

$$\nabla_j = \begin{pmatrix} i\theta_{jl}^{-1}\hat{x}_l & 0 \\ 0 & \partial_j \end{pmatrix} + \begin{pmatrix} C_j & |B_j\rangle \\ \langle E_j| & \hat{D}_j \end{pmatrix},$$

where $C_j \in M_{kk}(\mathbb{C})$, $|B_j\rangle, |E_j\rangle \in M_{kn}(\mathcal{F})$, $\hat{D}_j \in M_{nn}(\tilde{\mathcal{A}})$, and $\langle E_j| = |E_j\rangle^*$. We also note that the curvature of

$$\bar{\nabla}_j \equiv \begin{pmatrix} i\theta_{jl}^{-1}\hat{x}_l & 0 \\ 0 & \partial_j \end{pmatrix}$$

is $F_{jl} = -i\theta_{jl}^{-1}$.

The operator trace of endomorphisms of such modules is given ([12]) by

$$\mathrm{Tr}\begin{pmatrix} C & |B\rangle \\ \langle E| & \hat{D} \end{pmatrix} = (2\pi)^2(\theta_1\theta_2)\sum_{l=1}^{k} C_{ll} + \sum_{l=1}^{n}\int D_{ll}(x), \tag{11}$$

where $D_{ab}(x)$ is the corresponding function, obtained by the reverse of the Weyl transform.

Every projective $\mathcal{A}$-module is of the form $P\mathcal{A}^N$, for some projection $P \in M_N(\mathcal{A})$. For unital algebras we have $\mathrm{End}(P\tilde{\mathcal{A}}^N) = PM_N(\tilde{\mathcal{A}})P$ (for nonunital algebras this is not true). The operator trace on $M_N(\mathcal{A})$ is ([23])

$$\begin{aligned} \mathrm{Tr}(\star) &\equiv \int \mathrm{Tr}_N(\star)d^4x = \pi^2|\theta_{12}\theta_{34}|\mathrm{Tr}_{\mathcal{F}}(\mathrm{Tr}_N(\star)) \\ &= \pi^2|\theta_{12}\theta_{34}| \sum_{n_1,n_2=0}^{\infty} \langle n_1, n_2|\mathrm{Tr}_N(\star)|n_1, n_2\rangle. \end{aligned} \tag{12}$$

4. Fields, gauges, instantons, ASD equations

The Yang-Mills action is given by

$$S = \frac{1}{g^2}\mathrm{Tr}\frac{1}{4}(F_{ij}F_{ij})d^4x, \tag{13}$$

where Tr is given by equation (11) or equation (12), respectively. Minimizing the action yields in a standard way we have the ASD equations:

$$F_{12} = -F_{34}, \quad F_{13} = F_{24}, \quad F_{14} = -F_{23}. \tag{14}$$

Let $\mathcal{A}$ be an algebra from Section 2. Usually gauge theories over submodules of $\tilde{\mathcal{A}}^N$ are considered as $U(N)$ gauge theories. For our purposes we define '$U(N)$ gauge group' as

$$U_0^N(\tilde{\mathcal{A}}) \equiv \{\exp(iH_1)\cdots\exp(iH_n) \mid n \in \mathbb{N},\ H_j \in M_N(\tilde{\mathcal{A}}),\ H_j^* = H_j,\ \forall j\}.$$

Since it is usually supposed that a connection takes values in the Lie algebra of the gauge group, here we will suppose that a connection takes values in the set of all antihermitean operators of $M_N(\tilde{\mathcal{A}})$. Note that this choice is consistent with the special case of equation (3).

It is clear that ∂_j is a derivation for all j (actually it is an inner derivation). Taking a $U \in U_0^N(\tilde{\mathcal{A}})$ and applying ∂_j to $UU^* = 1$ gives

$$\partial_j(1) = \partial_j(UU^*) = \partial_j(U)U^* + U\partial_j(U^*).$$

Thus $U\partial_j(U^*) = -\partial_j(U)U^*$. Multiplying by U^* to the left gives

$$\partial_j(U^*) = -U^*\partial_j(U)U^*.$$

Under the action of U a connection transforms as

$$\nabla_j = \partial_j + A_j \mapsto U^*(\partial_j + A_j)U = \partial_j + U^* A_j U + U^* \partial_j U.$$

A standard computation yields the transformation

$$F_{mn} \mapsto U^* F_{mn} U.$$

In some sense the unitary group $U_0^N(\mathcal{A})$ is the largest possible:

Lemma 4.1. *$U_0^N(\tilde{\mathcal{A}})$ is norm dense in the group $U^N(\tilde{\mathcal{K}})$ of all unitary operators from $M_N(\tilde{\mathcal{K}})$.*

Proof. $U_0^N(\tilde{\mathcal{A}})$ is norm-dense in $U_0^N(\tilde{\mathcal{K}})$. From Proposition 2.1.6 of [15] it follows that $U_0^N(\tilde{\mathcal{K}})$ is the connected component of the identity operator. But $M_N(\tilde{\mathcal{K}})$ is stable, because $M_N(\tilde{\mathcal{K}}) \otimes \tilde{\mathcal{K}} \cong M_N(\tilde{\mathcal{K}} \otimes \tilde{\mathcal{K}}) \cong M_N(\tilde{\mathcal{K}})$ (the C^*-norm on $\tilde{\mathcal{K}} \otimes \tilde{\mathcal{K}}$ is unique). Exercise 8.17 of [15] gives

$$U^N(\tilde{\mathcal{K}})/U_0^N(\tilde{\mathcal{K}}) \cong U(M_N(\tilde{\mathcal{K}}))/U_0(M_N(\tilde{\mathcal{K}})) \cong K_1(M_N(\tilde{\mathcal{K}})) \cong K_1(\tilde{\mathcal{K}}) \cong \{0\}.$$

The last equality is from equation (2). □

A submodule of $\mathcal{A}^N$ is sent to an isomorphic submodule. If we consider a theory on such a submodule as a Yang-Mills theory on $\mathcal{A}^N$, then the action (13) will be preserved when acting with elements of $U_0^N(\tilde{\mathcal{A}})$.

It is clear that an ASD-connection (i.e. an instanton) is mapped to an ASD-connection. This follows from the fact that the transformed connection will still satisfy equations (14). Moreover it is mapped to a connection of the same topological number. The topological number is defined as ([23])

$$Q \equiv \frac{1}{16\pi^2} \mathrm{Tr}(F_{mn} F_{mn}) = \frac{g^2}{4\pi^2} S.$$

In analogy we can consider antihermitean connections on the modules $\mathcal{E}_{kn}$:

$$\nabla_j = \begin{pmatrix} \tilde{\partial}_j & 0 \\ 0 & \partial_j \end{pmatrix} + i \begin{pmatrix} C_j & |B_j\rangle \\ \langle B_j| & \hat{D}_j \end{pmatrix}, \tag{15}$$

where $C_j^* = C_j$, $\hat{D}_j^* = \hat{D}_j$, $\langle B_j| = |B_j\rangle^*$ and $\tilde{\partial}_j \equiv i\theta_{jl}^{-1}\hat{x}_l$. In [12, 19] a different gauge condition has been considered. Applying ASD equations (14) to (15) twelve equations are obtained for the twelve unknown operators of the A_j's.

Lemma 4.2. *The curvature of* (15) *is*

$$(F_{mn})_{11} = i\tilde{\partial_m}C_n - i\tilde{\partial_n}C_m + C_nC_m - C_mC_n + |B_n\rangle\langle B_m| - |B_m\rangle\langle B_n|,$$
$$(F_{mn})_{12} = i\tilde{\partial_m}|B_n\rangle - i\tilde{\partial_n}|B_m\rangle + C_n|B_m\rangle - C_m|B_n\rangle + |B_n\rangle\hat{D_m} - |B_m\rangle\hat{D_n},$$
$$(F_{mn})_{21} = i\partial_m\langle B_n| - i\partial_n\langle B_m| + \langle B_n|C_m - \langle B_m|C_n + \hat{D_n}\langle B_m| - \hat{D_m}\langle B_n|,$$
$$(F_{mn})_{22} = i\partial_m\hat{D_n} - i\partial_n\hat{D_m} + \langle B_n|B_m\rangle - \langle B_m|B_n\rangle + \hat{D_n}\hat{D_m} - \hat{D_m}\hat{D_n}.$$

When we work with projective modules which are not free, we must use a more general connection of the form $A = \Pi d\Pi$, where Π is a projection on a free module.

5. ASD equations on free modules

In this section, for the sake of simplicity, we will assume that $\theta_{12} = \theta_{34} = 1$.

Plugging in equations (5) - (10) into equation (14) we get

$$\begin{aligned} &U^*(\hat{x}_1(1-UU^*)\hat{x}_2 - \hat{x}_2(1-UU^*)\hat{x}_1)U \\ &\qquad = U^*(\hat{x}_4(1-UU^*)\hat{x}_3 - \hat{x}_3(1-UU^*)\hat{x}_4)U, \end{aligned} \tag{16}$$

$$\begin{aligned} &U^*(\hat{x}_2(1-UU^*)\hat{x}_4 - \hat{x}_4(1-UU^*)\hat{x}_2)U \\ &\qquad = U^*(\hat{x}_1(1-UU^*)\hat{x}_3 - \hat{x}_3(1-UU^*)\hat{x}_1)U, \end{aligned} \tag{17}$$

$$\begin{aligned} &U^*(\hat{x}_3(1-UU^*)\hat{x}_2 - \hat{x}_2(1-UU^*)\hat{x}_3)U \\ &\qquad = U^*(\hat{x}_1(1-UU^*)\hat{x}_4 - \hat{x}_4(1-UU^*)\hat{x}_1)U. \end{aligned} \tag{18}$$

Setting $1 - UU^* = Q$ and making use of equation (4), we see that equations (16) - (18) are equivalent to

$$U^* \begin{pmatrix}\sigma_1x_1 & \sigma_2x_2 & \sigma_3x_3 & x_4\end{pmatrix} Q \begin{pmatrix} 0 & 1 & 0 & 0 \\ 1 & 0 & 0 & 0 \\ 0 & 0 & 0 & i \\ 0 & 0 & -i & 0 \end{pmatrix} Q \begin{pmatrix}\sigma_1x_1 \\ \sigma_2x_2 \\ \sigma_3x_3 \\ x_4\end{pmatrix} U = 0, \tag{19}$$

$$U^* \begin{pmatrix}\sigma_1x_1 & \sigma_2x_2 & \sigma_3x_3 & x_4\end{pmatrix} Q \begin{pmatrix} 0 & 0 & 1 & 0 \\ 0 & 0 & 0 & i \\ 1 & 0 & 0 & 0 \\ 0 & -i & 0 & 0 \end{pmatrix} Q \begin{pmatrix}\sigma_1x_1 \\ \sigma_2x_2 \\ \sigma_3x_3 \\ x_4\end{pmatrix} U = 0, \tag{20}$$

$$U^* \begin{pmatrix}\sigma_1x_1 & \sigma_2x_2 & \sigma_3x_3 & x_4\end{pmatrix} Q \begin{pmatrix} 0 & 0 & 0 & -i \\ 0 & 0 & -1 & 0 \\ 0 & -1 & 0 & 0 \\ i & 0 & 0 & 0 \end{pmatrix} Q \begin{pmatrix}\sigma_1x_1 \\ \sigma_2x_2 \\ \sigma_3x_3 \\ x_4\end{pmatrix} U = 0, \tag{21}$$

where for shortness we have written $U = U \otimes \mathrm{Id}_2$, $Q = Q \otimes \mathrm{Id}_2$ and $U^* = U^* \otimes \mathrm{Id}_2$. Here $\sigma_1, \sigma_2, \sigma_3$ are the Pauli matrices, they satisfy $\sigma_1\sigma_2 = i\sigma_3$, $\sigma_2\sigma_3 = i\sigma_1$, $\sigma_3\sigma_1 = i\sigma_2$, and anticommute: $\sigma_a\sigma_b = 2\delta_{ab}\mathrm{Id}$. Summing these three equations (19), (20) and (21) we get that (14) is equivalent to the equation

$$U^* \begin{pmatrix} \sigma_1 x_1 & \sigma_2 x_2 & \sigma_3 x_3 & x_4 \end{pmatrix} Q \begin{pmatrix} 0 & 1 & 1 & -i \\ 1 & 0 & -1 & i \\ 1 & -1 & 0 & i \\ i & -i & -i & 0 \end{pmatrix} Q \begin{pmatrix} \sigma_1 x_1 \\ \sigma_2 x_2 \\ \sigma_3 x_3 \\ x_4 \end{pmatrix} U = 0.$$

If we denoted

$$\Xi = \begin{pmatrix} 0 & 1 & 1 & -i \\ 1 & 0 & -1 & i \\ 1 & -1 & 0 & i \\ i & -i & -i & 0 \end{pmatrix} \quad \text{and} \quad \Omega = \begin{pmatrix} \frac{1}{2} & -\frac{1}{2} & -\frac{1}{2} & \frac{i}{2} \\ \frac{1}{\sqrt{2}} & 0 & 0 & -\frac{i}{\sqrt{2}} \\ \frac{1}{\sqrt{6}} & \sqrt{\frac{2}{3}} & 0 & \frac{i}{\sqrt{6}} \\ \frac{1}{2\sqrt{3}} & -\frac{1}{2\sqrt{3}} & \frac{\sqrt{3}}{2} & \frac{i}{2\sqrt{3}} \end{pmatrix}$$

we would have

$$\Xi = \Omega^* \begin{pmatrix} -3 & 0 & 0 & 0 \\ 0 & 1 & 0 & 0 \\ 0 & 0 & 1 & 0 \\ 0 & 0 & 0 & 1 \end{pmatrix} \Omega.$$

Thus we can put

$$p_0 = \frac{\sigma_1 x_1}{2} - \frac{\sigma_2 x_2}{2} - \frac{\sigma_3 x_3}{2} + \frac{i x_4}{2}, \qquad p_1 = \frac{\sigma_1 x_1}{\sqrt{2}} - \frac{i x_4}{\sqrt{2}},$$
$$p_2 = \frac{\sigma_1 x_1}{\sqrt{6}} + \sqrt{\frac{2}{3}}\sigma_2 x_2 + \frac{i x_4}{\sqrt{6}}, \quad p_3 = \frac{\sigma_1 x_1}{2\sqrt{3}} - \frac{\sigma_2 x_2}{2\sqrt{3}} + \frac{\sqrt{3}\sigma_3 x_3}{2} + \frac{i x_4}{2\sqrt{3}}$$

and obtain

Proposition 5.1. *The ASD equations* (14) *are equivalent to the equation*

$$U^*[-3p_0^* Q p_0 + p_1^* Q p_1 + p_2^* Q p_2 + p_3^* Q p_3]U = 0. \tag{22}$$

Note that (22) is a Schrödinger-like equation, due to the fact that the presence of x_i's in the positive operators $U^* p_0^* Q p_0 U$, $U^* p_1^* Q p_1 U$, $U^* p_2^* Q p_2 U$, and $U^* p_3^* Q p_3 U$ corresponds to taking derivatives.

6. ADHM construction and topological charge

We now recall the ADHM-construction (ADHM stands for Atiyah, Drinfeld, Hitchin and Manin). In the conventional case it has been introduced in [2]. The analogue in the noncommutative case has been done in [14].

Consider complex vector spaces V and W, $\dim_{\mathbb{C}} V = k$, $\dim_{\mathbb{C}} W = n$ ([23]). Let $B_1, B_2 \in M_{kk}(\mathbb{C})$, $I \in M_{kn}(\mathbb{C})$, $J \in M_{nk}(\mathbb{C})$ be matrices, satisfying

$$[B_1, B_1^*] + [B_2, B_2^*] + II^* - J^*J = 2(\theta_{12} + \theta_{34})\mathrm{Id}_n \tag{23}$$

and

$$[B_1, B_2] + IJ = 0. \tag{24}$$

Next consider the operator

$$\Delta^* = \begin{pmatrix} I & B_2 + \hat{z}_2 & B_1 + \hat{z}_1 \\ J^* & -B_1^* - \hat{z}_1^* & B_2^* + \hat{z}_2^* \end{pmatrix},$$

where $\hat{z}_1 = \hat{x_2} + i\hat{x_1}$ and $\hat{z}_2 = \hat{x_4} + i\hat{x_3}$. Conditions (23) and (24) are equivalent to

$$\Delta^*\Delta = \begin{pmatrix} \Gamma & 0 \\ 0 & \Gamma \end{pmatrix},$$

where $\Gamma \in \mathbb{A}$ is some operator, where $\mathbb{A} \equiv \mathrm{Alg}(1, \hat{x_1}, \hat{x_2}, \hat{x_3}, \hat{x_4})$. It turns out that due to equations (23) and (24) $\Delta^*\Delta$ has no kernel ([13, 19]). Therefore it is formally invertible. If we denoted by Υ the right $\mathbb{A}$-module generated by $(\Delta^*\Delta)^{-1/2}\Delta^*$, i.e. $\Upsilon = (\Delta^{-1}\Delta)^{-1/2}\mathbb{A}$, then it would turn out that $\Pi' = \Delta(\Delta^*\Delta)^{-1}\Delta^*$ would be the projection onto Υ. Therefore $\Pi = 1 - \Pi'$ would be the projection onto the orthogonal complement Ψ of Υ in the module $(V \oplus V \oplus W) \otimes \mathbb{A}$ with respect to the inner product $\langle (a_1, \dots, a_{2k+n}),\ (b_1, \dots, b_{2k+n}) \rangle = a_1^* b_1 + \cdots + a_{2k+n}^* b_{2k+n}$.

From now on we assume that $\theta_{12} = \theta_{34} = 1$.

The solutions of $\Delta^* U = 0$ can be combined into a rank n free module part V, consisting of n columns, and a part K, consisting of k vectors from $\mathcal{K}$. The zero modes of

$$\begin{pmatrix} B_2 + \hat{z}_2 & B_1 + \hat{z}_1 \\ -B_1^* - \hat{z}_1^* & B_2^* + \hat{z}_2^* \end{pmatrix} \tag{25}$$

comprise of a k dimensional space, because (25) is in fact an elliptic pseudodifferential operator of order k, and therefore has index k ([19]). If $|v_1\rangle, \dots, |v_k\rangle$ is an orthonormal family of zero modes of (25) we can set

$$K_{1,(n+1)} = |v_1\rangle\langle 0, 0| + \dots |v_k\rangle\langle 0, k-1|$$

and all the other entries of $K \in M_{(n+2k),(n+2k)}$ equal to zero. Then we can write $U = V + K$ and have:

$$U^*U = 1_n \oplus \mathbb{I}_k, \quad UU^* = \Pi, \quad V^*V = 1_n, \quad K^*K = \mathbb{I}_k,$$

where $\mathbb{I}_k$ is the projection of $M_{(n+2k),(n+2k)}$ with all entries equal to zero, except the $(n+1)\times(n+1)$-st entry, which is $|0,0\rangle\langle 0,0|+\ldots|0,k-1\rangle\langle 0,k-1|$.

We now want to find the connection $\nabla = d + A$ of the module Ψ. We have ([1])

$$\begin{aligned}\nabla(U\psi) &= \Pi d(U\psi) = UU^*d(U\psi) = U[U^*(dU)\psi + U^*Ud\psi] \\ &= U[U^*(dU)\psi + (1_n \oplus \mathbb{I}_k)(d\psi)].\end{aligned}$$

Therefore $A = U^*dU$.

The curvature is

$$\begin{aligned}F_{pq} &= \partial_p U^*\partial_q U - \partial_q U^*\partial_p U + (U^*\partial_p U)(U^*\partial_q U) - (U^*\partial_q U)(U^*\partial_p U) \\ &= \partial_p U^*(1 - UU^*)\partial_q U - \partial_q U^*(1 - UU^*)\partial_p U \\ &\quad + (\partial_p \mathbb{I}_k)U^*\partial_q U - (\partial_q \mathbb{I}_k)U^*\partial_p U.\end{aligned}$$

The computation of the topological index of the ADHM connection in [23] uses the assumption that $U^*U = 1_n$, which is incorrect. Thus in the curvature the term $(\partial_p\mathbb{I}_k)U^*\partial_q U - (\partial_q\mathbb{I}_k)U^*\partial_p U$ has been omitted. The computation uses the Corrigan's identity

$$\mathrm{Tr}_{2k+n}(F_{pq}F_{pq}) = \frac{1}{2}\partial_q\partial_q \mathrm{Tr}[\sigma_p b^*(2 - \Delta\Gamma^{-1}\Delta^*)b\bar{\sigma}_p\Gamma^{-1}],$$

where

$$b = \begin{pmatrix} 0 & 0 \\ 1 & 0 \\ 0 & 1 \end{pmatrix}.$$

Its derivation does not depend on the assumption $U^*U = 1_n$ ([8]), and therefore a correction is due only to the analysis of the missing term in the curvature. In [16] the analysis of the topological index does not use the Corrigan's identiry and is done more carefully.

Further we obtain

$$\begin{aligned}F_{pq}F_{pq} = {} & [\partial_p U^*(1 - UU^*)\partial_q U - \partial_q U^*(1 - UU^*)\partial_p U]^2 \\ & + [\partial_p U^*(1 - UU^*)\partial_q U - \partial_q U^*(1 - UU^*)\partial_p U] \\ & \qquad [(\partial_p\mathbb{I}_k)U^*\partial_q U - (\partial_q\mathbb{I}_k)U^*\partial_p U] \\ & + [(\partial_p\mathbb{I}_k)U^*\partial_q U - (\partial_q\mathbb{I}_k)U^*\partial_p U] \\ & \qquad [\partial_p U^*(1 - UU^*)\partial_q U - \partial_q U^*(1 - UU^*)\partial_p U] \\ & + [(\partial_p\mathbb{I}_k)U^*\partial_q U - (\partial_q\mathbb{I}_k)U^*\partial_p U]^2.\end{aligned} \tag{26}$$

The first term is the one that contributes to the index. We will show that the other terms give zero in $\mathrm{Tr}_{\mathcal{H}}(\mathrm{Tr}_{2k+n}(\cdot))$.

For those terms we need the following formulae:

$$\partial_1(\mathbb{I}_k) = \partial_2(\mathbb{I}_k) = 0,$$

$$\partial_3(\mathbb{I}_k) = \frac{k}{2i}(|0,k\rangle\langle 0,k-1| + |0,k-1\rangle\langle 0,k|,$$

$$\partial_4(\mathbb{I}_k) = \frac{k}{2}(|0,k\rangle\langle 0,k-1| - |0,k-1\rangle\langle 0,k|,$$

$$\partial_3 K = \sum_{j=0}^{k-1} \Big\{ \frac{z_2 - z_2^*}{2i}|v_j\rangle\langle j| - \frac{j}{2i}|v_j\rangle\langle j-1| + \frac{j+1}{2i}|v_j\rangle\langle j+1| \Big\},$$

$$\partial_4 K = \sum_{j=0}^{k-1} \Big\{ \frac{z_2 + z_2^*}{2}|v_j\rangle\langle j| - \frac{j}{2}|v_j\rangle\langle j-1| - \frac{j+1}{2}|v_j\rangle\langle j+1| \Big\}.$$

The eventual nonzero contributions can arize only for $p = 3,\ q = 4$. The case $p = 4,\ q = 3$ is symmetric. The second and the third terms of (26) are equal by the trace property. Clearly the only nonzero element of $\partial_3 U(\partial_3\mathbb{I}_k)U^*$ is $\frac{k}{2i}\frac{k}{2i}|0,k\rangle\langle v_k|$ and in $\partial_4 U(\partial_4\mathbb{I}_k)U^*$ it is $\frac{-k}{2}\frac{k}{2i}|0,k\rangle\langle v_k|$. Therefore $(1 - UU^*)\partial_3 U(\partial_3\mathbb{I}_k)U^* = 0$ and the second and third terms of (26) are zero. The last term of (26) is (note that we can write K instead of U in this)

$$\begin{aligned}
&[(\partial_3\mathbb{I}_k)U^*\partial_4 U - (\partial_4\mathbb{I}_k)U^*\partial_3 U]^2 \\
&\quad = (\partial_3\mathbb{I}_k)K^*\partial_4 K(\partial_3\mathbb{I}_k)K^*\partial_4 K - (\partial_3\mathbb{I}_k)K^*\partial_4 K(\partial_4\mathbb{I}_k)K^*\partial_3 K \\
&\qquad - (\partial_4\mathbb{I}_k)K^*\partial_3 K(\partial_3\mathbb{I}_k)K^*\partial_4 K + (\partial_4\mathbb{I}_k)K^*\partial_3 K(\partial_4\mathbb{I}_k)K^*\partial_3 K.
\end{aligned}$$

The second and the third term have the same contribution by the trace property. From the above observations it is easy to see that the nonzero contibution to the trace in the first term is

$$\frac{k}{2i}|0,k\rangle\langle 0,k-1||0,k-1\rangle\langle v_k|\frac{-k}{2}|v_k\rangle\langle 0,k|\frac{k}{2i}|0,k\rangle\langle 0,k-1||0,k-1\rangle.$$

$$\langle v_k|\frac{-k}{2}|v_k\rangle\langle 0,k| = \frac{-k^4}{16}|0,k\rangle\langle 0,k|.$$

In completely analogous way we obtain that the only nonzero contribution to the trace in the second term is

$$\frac{k^4}{16}|0,k\rangle\langle 0,k|,$$

and in the fourth term is

$$\frac{-k^4}{16}|0,k\rangle\langle 0,k|.$$

Therefore

$$\mathrm{Tr}_{\mathcal{H}}(\mathrm{Tr}_{2k+n}([(\partial_p \mathbb{I}_k)U^*\partial_q U - (\partial_q \mathbb{I}_k)U^*\partial_p U]^2)) = -\frac{k^4}{16} + \frac{k^4}{16} + \frac{k^4}{16} - \frac{k^4}{16} = 0.$$

Thus we conclude that the topological index of an ADHM connection is equal to k, as proposed.

References

1. M. Atiyah, *Geometry of Yang-Mills Fields*, Lezioni Fermaine, (1979).
2. M. Atiyah, V. Drinfeld, N. Hitchin, Yu. Manin, *Construction of Instantons*, *Phys. Lett.* **65A**(3), (1978), 185–187.
3. A. Belavin, A. Polyakov, A. Schwarz, Yu. Tyupkin *Pseudoparticle Solutions of the Yang-Mills Equations*, *Phys. Lett.* **59B**(1), (1975), 85–87.
4. A. Connes, C^**-algebres et géométrie différentielle*, *CR Acad. Sci. Paris* Sér. AB **290**, (1980); english translation: arXiv:hep-th/0101093.
5. A. Connes, *Noncommutative Geometry*, Academic Press, (1994).
6. A. Connes, M. Douglas, A. Schwarz, Noncommutative geometry and Matrix theory, *J. High Energy Phys.* **02** (1998), 003, hep-th/9711162.
7. A. Connes, M. Rieffel, Yang-Mills for Non-commutative Two-tori, *Cont. Math.* **62** (1987), 237–266.
8. N. Dorey, V. Khoze, M. Mattis, Multi-instanton Calculus in N=2 Supersymmetric Gauge Theory, *Phys. Rev.* D 54 (1996), 2921–2943; hep-th/9603136.
9. M. Douglas, N. Nekrasov, Noncommutative Field Theory, *Rev. Mod. Phys.* **73** (2001), 977–1029; hep-th/0106048.
10. B. Hall, *Quantum Theory for Mathematicians*, Springer, (2013).
11. M. Hamanaka, T. Nakatsu, *Noncommutative Instantons Revisited*, J. Phys.: Conf. Ser. **411**, 012016, (2013).
12. A. Konechny, A. Schwarz, Introduction to M(atrix) Theory and Noncommutative Geometry I, II, *Phys. Rep.* **360**, 5-6, (2002), 353–465; hep-th/0012145, hep-th/0107251.
13. N. Nekrasov, Noncommutative Instantons Revisited, *Comm. Math. Phys.* **241**(1), (2003), 143–160; hep-th/0010017.
14. N. Nekrasov, A. Schwarz, Instantons on Noncommutative $\mathbb{R}_4$, and $(2,0)$ Superconformal Six Dimensional Theory, *Comm. Math. Phys.* **198**(3), (1998), 689–703; hep-th/9802068.

15. M. Rϕrdam, F. Larsen, N. Laustsen, *An Introduction to K-Theory for C^*-Algebras*, London Math. Soc. Student Text **49**, Cambridge University Press, 2000.
16. A. Sako, *Instanton number of noncommutative U(n) gauge theory*, J. High Energy Phys. **04**,(2003), 023, hep-th/0209139.
17. A. Sako, Recent Developments in Instantons in Noncommutative $\mathbb{R}^4$, *Adv. Math Phys.* **2010** (2010).
18. N. Seiberg, E. Witten, *String Theory and Noncommutative Geometry, J. High Energy Phys.* **09** (1999), 032, hep-th/9908142.
19. A. Schwarz, Noncommutative Instantons: a New Approach, *Comm. Math. Phys.* **221**(2), (2001), 433–450; hep-th/0102182.
20. H. Snyder, Quantized Spacetime, *Phys. Rev.* **71**, 38 (1947), 38–41.
21. R. Swan, Vector Bundles and Projective Modules, *Trans. Amer. Math. Soc.* **105**(2), (1962), 264-277 .
22. R. Szabo, Quantum Field Theory on Noncommutative Spaces, *Phys. Rep.* **378**, 4, (2003), 207–299; hep-th/0109162.
23. Y. Tian, C. Zhu, X. Song, Topological Charge of Noncommutative ADHM Instanton, *Mod. Phys. Lett. A* **18** (2003), 1691–1703; hep-th/0211225.

Received February 1, 2015
Revised March 5, 2015

Current Developments in Differential Geometry
and its Related Fields pp. 93 – 114

ALMOST CR STRUCTURE ON THE TWISTOR SPACE OF A QUATERNIONIC CR MANIFOLD

Hiroyuki KAMADA *
Miyagi University of Education,
149 Aramaki-aoba, Aoba-ku, Sendai, 980-0845 Japan
E-mail: hkamada@staff.miyakyo-u.ac.jp

Shin NAYATANI †
Graduate School of Mathematics, Nagoya University,
Chikusa-ku, Nagoya, 464-8602 Japan
E-mail: nayatani@math.nagoya-u.ac.jp

We define an almost CR structure on the twistor space of a quaternionic CR manifold. It is done by choosing a compatible quaternionic pseudohermitian structure, and the proof that the almost CR structure so defined is independent of the choice constitutes the core of this paper. For this, we should suppose that the quaternionic CR structure satisfies a stronger integrability condition, which we newly introduce.

Keywords: Quaternionic CR structure, twistor space.

1. Introduction

A quaternionic CR manifold is a manifold equipped with a quaternionic CR structure introduced in [2] as a new geometric structure modeled on a real hypersurface in a quaternionic manifold. At each point of such a manifold, there is a family of CR hyperplanes in the tangent space parametrized by a unit sphere S^2. Collecting all such CR hyperplanes over all points of the manifold, we obtain a canonical S^2-bundle over the quaternionic CR manifold, called the *twistor space*.

In this paper we define an almost CR structure on the twistor space of a quaternionic CR manifold which is ultra pseudoconvex. To do so we follow the construction in the case of a quaternionic manifold: each point of the twistor space corresponds tautologically to a complex structure on a hyperplane of the tangent space at the base point. So we lift it to the

*The first author is partially supported by Grant-in-Aid for Scientific Research (No. 24540062), Japan Society for the Promotion of Science.
†The second author is partially supported by Grant-in-Aid for Scientific Research (No. 25610009), Japan Society for the Promotion of Science.

horizontal subspace of the tangent space of the original point, and coupling with the canonical complex structure of the vertical subspace, which is the tangent space to the fiber $S^2 \approx \mathbb{C}P^1$, we obtain a CR hyperplane in the tangent space at the original point.

Here, the horizontality is defined by fixing a pseudohermitian structure on the base manifold compatible with its quaternionic CR structure and using the corresponding canonical connection, which is the quaternionic CR analogue of the Tanaka-Webster connection in CR geometry. It, therefore, remains to verify that the above definition of almost CR structure is independent of the choice of pseudohermitian structure used, and this is the main task of the present paper.

In the course of verification, we realized that the integrability condition which we formerly imposed on the quaternionic CR structure seemed insufficient to imply the well-definedness. So we introduce the new notion of strong integrability of the quaternionic CR structure, and see that if it is imposed on the structure, the well-definedness does follow. It remains to see whether the almost CR structure is indeed integrable, and we will address this issue in a forthcoming work.

This paper is organized as follows. After a brief review of CR structure in §2, in §3 we review hyper CR and pseudohermitian structures and related geometric notions, like Levi form and (ultra) pseudoconvexity. Then we review quaternionic CR structure, which is defined as a covering of a manifold by open sets equipped with hyper CR structures. Therefore, for local study of the quaternionic CR structure, we may regard it as a hyper CR structure. In §4, we review the canonical connection associated with a given quaternionic pseudohermitian structure. It is a fundamental tool to study our structure geometrically, and is a quaternionic analogue of Tanaka-Webster connection in CR geometry. For use in later sections, we present the transfomation law of the connection under a conformal change of pseudohermitian structure. In §5, we introduce the new concept of strong integrability of a quaternionic CR structure. Then we introduce the twistor space of a quaternionic CR manifold and define an almost CR structure on it. It is done by choosing a compatible quaternionic pseudohermitian structure and using the horizontality defined by means of the associated canonical connection. Therefore, we must verify that the almost CR structure so defined is independent of the choice of pseudohermitian structure. This proof is the main accomplishment of this paper, and the strong integrability is necessary here. In §6, we review the comparison of quaternionic CR structure and Biquard's quaternionic contact structure. We newly observe that the quaternionic CR structure canonically extending a quaternionic contact structure satisfies the strong integrability condition. In §7, we give proofs of two lemmas stated and used in §5.

Throughout this paper, all manifolds are assumed to be connected, orientable and of class C^∞.

2. CR structure

In this section we briefly review the usual CR structure.

Definition 2.1 (CR structure). Let M be a (real) manifold of dimension $2n+1$. An *almost CR structure* on M is a pair (Q, J) of a corank 1 subbundle Q of the tangent bundle TM of M and a complex structure J of Q (i.e., an endomorphism J of Q satisfying $J^2 = -\mathrm{id}_Q$). Set $Q^{1,0} = \{Z \in Q \otimes \mathbb{C} \mid JZ = \sqrt{-1}Z\}$. An almost CR structure (Q, J) is said to be *partially integrable* if it satisfies

$$[\Gamma(Q^{1,0}), \Gamma(Q^{1,0})] \subset \Gamma(Q \otimes \mathbb{C}),$$

or equivalently,

$$[JX, Y] + [X, JY] \in \Gamma(Q), \;\; X, Y \in \Gamma(Q). \tag{1}$$

If (Q, J) satisfies the full Nijenhuis condition

$$[\Gamma(Q^{1,0}), \Gamma(Q^{1,0})] \subset \Gamma(Q^{1,0}),$$

which is equivalent to

$$[JX, JY] - [X, Y] - J([JX, Y] + [X, JY]) = 0, \;\; X, Y \in \Gamma(Q)$$

together with (1), then (Q, J) is said to be *integrable* and is called a *CR structure*.

Let (Q, J) be any partially integrable almost CR structure on M. Then there exists a 1-form θ on M, called a *compatible* 1-form, such that $\operatorname{Ker}\theta = Q$. The *Levi form* L_θ for (Q, J) is a J-invariant symmetric bilinear form on Q, defined by

$$L_\theta(X, Y) = d\theta(X, JY), \quad X, Y \in Q.$$

If L_θ is positive-definite for some θ, then (Q, J) is said to be *strongly pseudoconvex*, and such a θ is called a *pseudohermitian structure*.

Note that a compatible 1-form θ for (Q, J) is unique up to the multiplication by a nowhere-vanishing smooth function on M. When θ is replaced with $f\theta$ for a function $f \neq 0$, the corresponding Levi forms $L_{f\theta}$ and L_θ

are related by $L_{f\theta} = fL_\theta$ on Q. If (Q, J) is strongly pseudoconvex, then, for each choice of a pseudohermitian structure θ compatible with (Q, J), there exists a distinguished vector field T on M satisfying $\theta(T) = 1$ and $d\theta(T, \cdot) = 0$, that is, T is the Reeb field for the contact form θ.

3. Quaternionic CR structure

We next review the basic definitions and fundamental facts on quaternionic CR structures and quaternionic pseudohermitian structures. For details, see [2]. In what follows, let M be a manifold of dimension $4n+3$. Unless otherwise stated, (a, b, c) denotes a cyclic permutation of $(1, 2, 3)$.

Definition 3.1 (Hyper CR structure). A *hyper CR structure* on M is a triple of almost CR structures $(Q_a, I_a), a = 1, 2, 3$, satisfying

(i) Q_a and Q_b are transverse to each other;
(ii) $I_a(Q_a \cap Q_b) = Q_a \cap Q_c$ and $I_a(Q_a \cap Q_c) = Q_a \cap Q_b$;
(iii) $I_a I_b = I_c$ on $Q_b \cap Q_c$ and $I_b I_a = -I_c$ on $Q_a \cap Q_c$;
(iv) $[I_a X, Y] + [X, I_a Y] \in \Gamma(Q_a)$, $X, Y \in \Gamma(Q)$;
(v) $[I_a X, I_a Y] - [X, Y] - I_a([I_a X, Y] + [X, I_a Y]) \in \Gamma(Q)$, $X, Y \in \Gamma(Q)$.

Here, $Q := Q_1 \cap Q_2 \cap Q_3$. In particular, $I_a I_b = -I_b I_a = I_c$ holds on Q.

Proposition 3.1 ([2], Definitions 1.2, 1.3). *Let $\{(Q_a, I_a)\}_{a=1}^3$ be a hyper CR structure on M with corank 3 bundle $Q = Q_1 \cap Q_2 \cap Q_3$. Then*

- *There exists a triple of vector fields T_a, $a = 1, 2, 3$, on M transverse to Q such that*

$$T_a \in \Gamma(Q_b \cap Q_c), \quad I_a T_b = T_c = -I_b T_a.$$

Hence $Q_a = Q \oplus \mathbb{R}T_b \oplus \mathbb{R}T_c$, $TM = Q_a \oplus \mathbb{R}T_a$. (Such a triple (T_1, T_2, T_3) is called an admissible triple *for $\{(Q_a, I_a)\}$.)*
- *There exists an $\mathbb{R}^3$-valued 1-form $\theta = (\theta_1, \theta_2, \theta_3)$ on M such that*

$$Q_a = \operatorname{Ker} \theta_a, \quad \theta_a \circ I_b = \theta_c \text{ on } Q_b, \quad -\theta_b \circ I_a = \theta_c \text{ on } Q_a.$$

(Such a θ is called a compatible 1-form *for $\{(Q_a, I_a)\}$.)*
- *For an admissible triple (T_1, T_2, T_3) and a* compatible 1-form *$\theta = (\theta_1, \theta_2, \theta_3)$ for $\{(Q_a, I_a)\}$, the following holds:*

$$\theta_1(T_1) = \theta_2(T_2) = \theta_3(T_3) \neq 0, \quad \theta_a(T_b) = \theta_b(T_a) = 0.$$

Note. A compatible 1-form θ for a hyper CR structure $\{(Q_a, I_a)\}$ on M is unique up to conformal change $\theta \mapsto f\theta$ by a nowhere-vanishing smooth function f on M. For two admissible triples (T_1, T_2, T_3) and $(\hat{T}_1, \hat{T}_2, \hat{T}_3)$ satisfying $\theta_a(T_a) = \theta_a(\hat{T}_a) = 1$, there exists a section W of Q such that $\hat{T}_a = T_a + 2I_a W$.

For a nonzero vector $\mathbf{v} = (v_a) \in \mathbb{R}^3$, we define a 1-form $\theta_{\mathbf{v}}$ and a vector field $T_{\mathbf{v}}$ on M by $\theta_{\mathbf{v}} = \sum_{a=1}^3 v_a \theta_a$ and $T_{\mathbf{v}} = \sum_{a=1}^3 v_a T_a$, respectively. If $\mathbf{v} \in S^2$, where S^2 is the unit sphere in $\mathbb{R}^3$, then one can define a complex structure $I_{\mathbf{v}}$ on $Q_{\mathbf{v}} := \operatorname{Ker} \theta_{\mathbf{v}}$ by

$$I_{\mathbf{v}} = \sum_{a=1}^{3} v_a \widetilde{I}_a|_{Q_{\mathbf{v}}},$$

where $\widetilde{I}_a$ is an extension of I_a to an endomorphism of TM defined by $\widetilde{I}_a T_a = 0$. Note that $I_{\mathbf{v}}$ is independent of the particular choice of admissible triple (T_1, T_2, T_3).

Proposition 3.2 (Levi form. [2], p. 9). *Let $\{(Q_a, I_a)\}$ be a hyper CR structure on M with a compatible 1-form $\theta = (\theta_a)$ and set*

$$L_\theta(X, Y) = \frac{1}{2} \{d\theta_a(X, I_a Y) + d\theta_a(I_b X, I_c Y)\}, \quad X, Y \in Q. \tag{2}$$

Then the right-hand side of (2) is independent of the choice of (a, b, c), and L_θ is symmetric and invariant by $I_{\mathbf{v}}$ for any $\mathbf{v} \in S^2$. (We call L_θ the Levi form *of θ.)*

The strong pseudoconvexity and pseudohermitian structure for a hyper CR structure $\{(Q_a, I_a)\}$ are defined by using the Levi form L_θ in a way similar to those for a CR structure.

We next introduce another convexity condition, called the ultra pseudoconvexity, stronger than the strong pseudoconvexity:

Definition 3.2 (Ultra pseudoconvexity). A hyper CR structure $\{(Q_a, I_a)\}$ is said to be *ultra pseudoconvex* if the symmetric bilinear form h_θ on Q defined by

$$h_\theta(X, Y) = (2n + 4)L_\theta(X, Y) - \sum_{a=1}^{3} d\theta_a(X, I_a Y), \quad X, Y \in Q,$$

is positive-definite for some compatible $\mathbb{R}^3$-valued 1-form θ.

Note. • If a hyper CR structure is ultra pseudoconvex, then it is also strongly pseudoconvex, but the converse is not true in general.

- The standard hyper CR structure on a strictly convex real hypersurface in the quaternionic affine space $\mathbb{H}^{n+1}$ is ultra pseudoconvex.
- There exist ultra pseudoconvex hyper CR structures $\{(Q_a, I_a)\}$ such that some of $d\theta_a(\cdot, I_a\cdot)|_{Q\times Q}$, $a = 1, 2, 3$, are indefinite or degenerate. (See Examples in [2].)

Definition 3.3 (Quaternionic CR structure). A *quaternionic CR structure* on M is a covering of M by open sets equipped with hyper CR structures which satisfy the following condition: if $\{(Q_a, I_a)\}$ and $\{(Q'_a, I'_a)\}$ are two such hyper CR structures defined on open subsets U and U' respectively and $U \cap U' \neq \emptyset$, there exists a function $S = S_{UU'}\colon U \cap U' \to SO(3)$ such that

$$Q'_{\mathbf{v}} = Q_{S\mathbf{v}}, \quad I'_{\mathbf{v}} = I_{S\mathbf{v}}, \quad \mathbf{v} \in S^2. \tag{3}$$

A quaternionic CR structure is said to be *strongly pseudoconvex* (resp. *ultra pseudoconvex*) if all the local hyper CR structures are strongly pseudoconvex (resp. ultra pseudoconvex).

Given a quaternionic CR structure on M, there exist a rank $4n$ subbundle $Q \subset TM$ and an S^2-bundle $\mathbb{I} \subset \mathrm{End}(Q)$ over M consisting of complex structures on Q such that

$$Q_q = (Q_1)_q \cap (Q_2)_q \cap (Q_3)_q, \quad \mathbb{I}_q = \{(I_{\mathbf{v}})|_{Q_q} \mid \mathbf{v} \in S^2\}, \quad q \in U$$

hold for each local hyper CR structure $\{(Q_a, I_a)\}$ defined on $U \subset M$ constituting the quaternionic CR structure. We call such a pair $(Q, \mathbb{I})$ the *underlying $GL(n,\mathbb{H})\cdot Sp(1)$-structure.* Here, a $GL(n,\mathbb{H})\cdot Sp(1)$-structure $\mathbb{I}$ on a rank $4n$ vector bundle Q is an S^2-bundle $\mathbb{I}$ of complex structures on Q such that $\mathbb{I}$ is expressed locally as $\mathbb{I}_q = \{(I_{\mathbf{v}})_q = \sum_{a=1}^{3} v_a (I_a)_q \mid \mathbf{v} \in S^2\}$ for some local sections I_a, $a = 1, 2, 3$, satisfying the quaternion relations $I_a I_b = -I_b I_a = I_c$.

We next define the Levi form for a quaternionic CR structure.

Definition 3.4 (Levi form). Let $\Theta = \{\theta_U\}$ be a collection of local $\mathbb{R}^3$-valued 1-forms compatible with the local hyper CR structures constituting a quaternionic CR structure on M and $(Q, \mathbb{I})$ the underlying $GL(n,\mathbb{H})\cdot Sp(1)$-structure. Suppose that Θ further satisfies the following condition: For $\theta_U = (\theta_a)$ and $\theta_{U'} = (\theta'_a)$ in Θ compatible with the local hyper CR structures $\{(Q_a, I_a)\}$ and $\{(Q'_a, I'_a)\}$, respectively, the condition

$$\theta'_{\mathbf{v}} = \theta_{S\mathbf{v}}, \quad \mathbf{v} \in S^2, \tag{4}$$

holds for the same function $S = S_{UU'} : U \cap U' \to SO(3)$ as in (3). Then there exists a globally-defined symmetric bilinear form L_Θ on Q satisfying $L_\Theta|_U = L_{\theta_U}$ on each U. We call L_Θ the *Levi form* of Θ. If L_Θ is positive-definite, then we call Θ a *quaternionic pseudohermitian structure* and (M, Θ) a *quaternionic pseudohermitian manifold.*

Note.
- By standard argument using a partition of unity, one can show that any strongly pseudoconvex quaternionic CR manifold admits a quaternionic pseudohermitian structure compatible with the quaternionic CR structure.
- For Θ satisfying (4), there also exists a globally-defined bilinear form h_Θ on Q such that $h_\Theta|_U = h_{\theta_U}$. Thus, a quaternionic CR structure is ultra pseudoconvex (resp. strongly pseudoconvex) if and only if h_Θ (resp. L_Θ) is positive-definite on Q for some quaternionic pseudohermitian structure Θ compatible with the quaternionic CR structure.

A quaternionic pseudohermitian manifold (M, Θ) determines, by definition, a natural $Sp(n)\cdot Sp(1)$-structure $(L_\Theta, \mathbb{I})$ on $Q = \operatorname{Ker} \Theta$ and a natural $SO(3)$-structure $g_{TM/Q}$ on TM/Q. Here, an $Sp(n)\cdot Sp(1)$*-structure* on a rank $4n$ vector bundle Q is a pair $(g, \mathbb{I})$ of a $GL(n, \mathbb{H})\cdot Sp(1)$-structure $\mathbb{I}$ and a fiber-metric g on Q satisfying $g(IX, IY) = g(X, Y)$ for all $I \in \mathbb{I}$ and $X, Y \in Q$. The natural $SO(3)$-structure $g_{TM/Q}$ on TM/Q is defined by

$$g_{TM/Q} = |\theta_U|^2 = \sum_{a=1}^{3} (\theta_a)^2.$$

Note that $g_{TM/Q}$ is well-defined, since $|\theta_U|^2 = |\theta_{U'}|^2$ on $U \cap U' \neq \emptyset$.

For each choice of a rank 3 subbundle $Q^\perp \subset TM$ satisfying $TM = Q \oplus Q^\perp$ (such a bundle $Q^\perp$ is called a 3-plane field transverse to Q), one can introduce a fiber-metric $g^\perp$ on $Q^\perp$ such that $(Q^\perp, g^\perp)$ is identified with $(TM/Q, g_{TM/Q})$.

4. Canonical connection

In this section, we review the canonical connection for an ultra pseudoconvex quaternionic pseudohermitian manifold. To begin with, we recall the following (cf. Lemmas II.1.1 and II.2.1 in [1]):

Proposition 4.1 ([2], Proposition 3.1). *Let (M, Θ) be a quaternionic pseudohermitian manifold with $Q = \operatorname{Ker} \Theta$ and Levi form $g = L_\Theta$. Then for each choice of a 3-plane field $Q^\perp$ transverse to Q with metric $g^\perp$,* *there exists a unique affine connection* ∇ *on* M *that satisfies the following conditions*:

- ∇ *induces a metric connection on* (Q, g) *and that on* $(Q^\perp, g^\perp)$.
- *For* $X, Y \in Q$ *and* $U, V \in Q^\perp$, *the torsion* Tor^∇ *of* ∇ *satisfies*
  $\cdot\ \mathrm{Tor}^\nabla(X, Y)_Q = \mathrm{Tor}^\nabla(U, V)_{Q^\perp} = 0$;
  $\cdot\ X \mapsto \mathrm{Tor}^\nabla(U, X)_Q \in \mathrm{End}(Q)$ *is* g*-symmetric*;
  $\cdot\ U \mapsto \mathrm{Tor}^\nabla(U, X)_{Q^\perp} \in \mathrm{End}(Q^\perp)$ *is* $g^\perp$*-symmetric.*
  Here, E_Q *and* $E_{Q^\perp}$ *denote respectively the* Q*- and* $Q^\perp$*-components of a tangent vector* $E \in TM$ *with respect to* $TM = Q \oplus Q^\perp$.

Let (M, Θ) be a quaternionic pseudohermitian manifold of dimension $4n+3$ (> 7) with $Q = \mathrm{Ker}\,\Theta$. For a 3-plane field $Q^\perp$ transverse to Q, let ∇ be the affine connection on M in Proposition 4.1 associated with $Q^\perp$ and denote by $\nabla^Q : \Gamma(Q) \to \Gamma(Q^* \otimes Q)$ the Q-partial connection on Q induced by ∇. Then the obstruction ω^{obs} for ∇^Q to be an $Sp(n)\cdot Sp(1)$-Q-partial connection lives in $Q^* \otimes (sp(n) \oplus sp(1))^\perp (\subset Q^* \otimes so(4n))$. Now we have the following irreducible decomposition:

$$(Q_q^* \otimes (sp(n) \oplus sp(1))^\perp) \otimes \mathbb{C} \cong KS^3H \oplus \Lambda_0^3 ES^3H \oplus ES^3H \oplus KH \oplus \Lambda_0^3 EH \oplus EH, \qquad (5)$$

where E and H denote the standard complex representations $\mathbb{C}^{2n}$ of $Sp(n)(\subset SU(2n))$ and $\mathbb{C}^2$ of $Sp(1)(= SU(2))$, respectively, and K denotes a certain irreducible complex $Sp(n)$-module. We denote by ω^{EH} the EH-component of ω^{obs} with respect to (5).

Theorem 4.1 ([2], Theorem 3.5). *Let* (M, Θ) *be a quaternionic pseudohermitian manifold of dimension* > 7 *with* $Q = \mathrm{Ker}\,\Theta$. *If* (M, Θ) *is ultra pseudoconvex, then there exists a unique* 3*-plane field* $Q^\perp$ *transverse to* Q *such that the affine connection on* M *in Proposition 4.1 associated with* $Q^\perp$ *satisfies* $\omega^{EH} = 0$.

Definition 4.1 (Canonical connection). Let (M, Θ) be an ultra pseudoconvex quaternionic pseudohermitian manifold of dimension > 7. We call $Q^\perp$ of Theorem 4.1 the *canonical* 3*-plane field* for (M, Θ), and the associated connection, denoted by D, the *canonical connection.*

Later we will need the transformation law for the canonical connection under a conformal change of pseudohermitian structure. Since we need the formulas locally, we may work on a hyper CR manifold. We first recall the following:

Lemma 4.1 ([2], Proposition 3.7). *Let* (M,θ) *be an ultra pseudoconvex hyper CR manifold of dimension* $4n+3$ (>7) *with compatible* $\mathbb{R}^3$*-valued 1-form* θ. *Let* $\theta' = e^{2f}\theta$, *and* (T_1,T_2,T_3) *(resp.* (T'_1,T'_2,T'_3)*) the admissible triple spanning the canonical* 3*-plane field satisfying* $\theta_a(T_a)=1$ *(resp.* $\theta'_a(T'_a)=1$*), called* the canonical triple. *Then we have*

$$T'_a = e^{-2f}(T_a + 2I_aW),$$

where $W \in \Gamma(Q)$ *is uniquely determined by*

$$h_\theta(W,X) = -(2n+1)d_bf(X), \quad X \in Q.$$

By using this lemma, we can prove the following formulas:

Proposition 4.2. *Let* (M,θ) *and* $\theta' = e^{2f}\theta$ *be as in Lemma 4.1, and let* D *and* D' *be the canonical connections corresponding to* θ *and* θ', *respectively. Let* $W \in \Gamma(Q)$ *be as in Lemma 4.1. Then we have*

$$\begin{aligned} g(D'_XY,Z) = g(D_XY,Z)& \\ &+d_bf(X)g(Y,Z) + d_bf(Y)g(X,Z) - d_bf(Z)g(X,Y)\\ &+\sum_{a=1}^{3} d\theta_a(X,Y)g(I_aW,Z) - \sum_{a=1}^{3} d\theta_a(X,Z)g(I_aW,Y)\\ &-\sum_{a=1}^{3} d\theta_a(Y,Z)g(I_aW,X) \end{aligned} \tag{6}$$

and

$$\begin{aligned} &g(D'_{T_a}X,Y) = g(D_{T_a}X,Y) + df(T_a)g(X,Y)\\ &+\sum_{p=1}^{3} d\theta_p(T_a,X)g(I_pW,Y) - \sum_{p=1}^{3} d\theta_p(T_a,Y)g(I_pW,X)\\ &-2d_bf(X)g(I_aW,Y) + 2d_bf(Y)g(I_aW,X)\\ &-g(D_X(I_aW),Y) + g(D_Y(I_aW),X) + 2d\theta_a(X,Y)g(W,W) \end{aligned} \tag{7}$$

for $X,Y,Z \in \Gamma(Q)$.

Proof. The formula (6) follows from (3.6) in [2] for D:

$$\begin{aligned} 2g(D_XY,Z) = Xg(Y,Z) + Yg(X,Z) - Zg(X,Y) + g([X,Y]_Q,Z)&\\ -g([X,Z]_Q,Y) - g([Y,Z]_Q,X),& \end{aligned} \tag{8}$$

and that for D', together with

$$[E,X]'_Q = [E,X]_Q + 2\sum_{a=1}^{3}(X\theta_a(E) + d\theta_a(E,X))I_aW, \tag{9}$$

where $X, Y, Z \in \Gamma(Q)$ and $E \in \Gamma(TM)$. The derivation of (7) is slightly more involved. Recall that D also satisfies

$$g(D_U X, Y) = \frac{1}{2}(Ug(X,Y) + g([U,X]_Q, Y) - g(X, [U,Y]_Q)) \qquad (10)$$

for $X, Y \in \Gamma(Q)$, $U \in \Gamma(Q^\perp)$ (cf. (3.8) in [2]). Since $T_a = e^{2f}T'_a - 2I_aW$, we have

$$D'_{T_a}X = e^{2f}D'_{T'_a}X - 2D'_{I_aW}X.$$

Computing with (8), (10) for D and D' and (9), and then using $D_XY - D_YX = [X,Y]_Q$, we obtain (7). □

5. Strong integrability and almost CR structure on the twistor space

In this section, motivated by the inspection of a real hypersurface in a quaternionic manifold, we introduce a stronger notion of integrability condition of a quaternionic CR structure, called the *strong integrability*. We then show that an almost CR structure may be defined on the twistor space of a strongly integrable quaternionic CR manifold.

Let $\mathcal{N}$ be a real $(4n+4)$-dimensional quaternionic manifold, that is, a manifold which is equipped with a $GL(n+1,\mathbb{H})\cdot Sp(1)$-structure $\mathcal{I}$ on its tangent bundle and admits a compatible torsion-free affine connection $\mathcal{D}$. On a neighborhood $\mathcal{U}$ of each point of $\mathcal{N}$, there exist local almost complex structures $\mathcal{I}_a$, $a = 1,2,3$, satisfying $\mathcal{I}_a\mathcal{I}_b = \mathcal{I}_c = -\mathcal{I}_b\mathcal{I}_a$ and

$$\mathcal{I}|_{\mathcal{U}} = \left\{v_1\mathcal{I}_1 + v_2\mathcal{I}_2 + v_3\mathcal{I}_3 \mid {v_1}^2 + {v_2}^2 + {v_3}^2 = 1\right\} \subset \mathrm{End}(T\mathcal{U}).$$

A torsion-free affine connection $\mathcal{D}$ on $\mathcal{N}$ is said to be *compatible* with $\mathcal{I}$, if on any such $\mathcal{U}$ there exists an $so(3)$-valued 1-form (γ_{ab}) satisfying

$$\mathcal{D}\mathcal{I}_a = \gamma_{ab} \otimes \mathcal{I}_b + \gamma_{ac} \otimes \mathcal{I}_c. \qquad (11)$$

Let $M = \rho^{-1}(0) \subset \mathcal{N}$ be a real hypersurface with defining function $\rho : \mathcal{N} \to \mathbb{R}$. Set $U = M \cap \mathcal{U}$ for a neighborhood $\mathcal{U}$ as above and define

$$Q_a = TU \cap \mathcal{I}_a TU, \quad I_a = \mathcal{I}_a|_{Q_a}.$$

Then $\{(Q_a, I_a)\}$ becomes a hyper CR structure on U. Indeed, it suffices to verify the integrability conditions (iv) and (v) in Definition 3.1 and to do so, we compute (cf. p. 12 [2]): Let $X, Y \in \Gamma(Q|_U)$ and take $\mathcal{X}, \mathcal{Y} \in \Gamma(T\mathcal{U})$

satisfying $X = \mathcal{X}|_U, Y = \mathcal{Y}|_U$. By the torsion-freeness and compatibility condition (11) of $\mathcal{D}$, we have

$$\begin{aligned}
&[\mathcal{I}_a\mathcal{X}, \mathcal{I}_a\mathcal{Y}] - [\mathcal{X}, \mathcal{Y}] - \mathcal{I}_a([\mathcal{I}_a\mathcal{X}, \mathcal{Y}] + [\mathcal{X}, \mathcal{I}_a\mathcal{Y}]) \\
&\quad = (\mathcal{D}_{\mathcal{I}_a\mathcal{X}}\mathcal{I}_a)\mathcal{Y} - (\mathcal{D}_{\mathcal{I}_a\mathcal{Y}}\mathcal{I}_a)\mathcal{X} - \mathcal{I}_a(\mathcal{D}_{\mathcal{X}}\mathcal{I}_a)\mathcal{Y} + \mathcal{I}_a(\mathcal{D}_{\mathcal{Y}}\mathcal{I}_a)\mathcal{X} \\
&\quad = \sum_{p=1}^{3} \gamma_{ap}(\mathcal{I}_a\mathcal{X})\mathcal{I}_p\mathcal{Y} - \sum_{p=1}^{3} \gamma_{ap}(\mathcal{I}_a\mathcal{Y})\mathcal{I}_p\mathcal{X} \\
&\qquad\quad - \sum_{p=1}^{3} \gamma_{ap}(\mathcal{X})\mathcal{I}_a\mathcal{I}_p\mathcal{Y} + \sum_{p=1}^{3} \gamma_{ap}(\mathcal{Y})\mathcal{I}_a\mathcal{I}_p\mathcal{X},
\end{aligned}$$

whose restriction to U gives

$$\begin{aligned}
&[I_aX, I_aY] - [X, Y] - \mathcal{I}_a([I_aX, Y] + [X, I_aY]) \\
&\qquad\qquad = \sum_{p=1}^{3} \gamma_{ap}(I_aX)I_pY - \sum_{p=1}^{3} \gamma_{ap}(I_aY)I_pX \\
&\qquad\qquad\qquad - \sum_{p=1}^{3} \gamma_{ap}(X)I_aI_pY + \sum_{p=1}^{3} \gamma_{ap}(Y)I_aI_pX. \qquad (12)
\end{aligned}$$

Since the right-hand side lies in $\Gamma(Q|_U)$, we obtain $\mathcal{I}_a([I_aX, Y] + [X, I_aY]) \in \Gamma(TU)$, and this implies $[I_aX, Y] + [X, I_aY] \subset \Gamma(TU \cap \mathcal{I}_aTU)$. Thus (iv) holds and (v) also follows from (12).

The computation above motivates us to introduce the following

Definition 5.1 (Strong integrability). A quaternionic CR structure on a manifold M is said to be *strongly integrable* if the following conditions hold: For any local hyper CR structure $\{(Q_a, I_a)\}$ on $U \subset M$ constituting the quaternionic CR structure, there exists an $so(3)$-valued 1-form (γ_{ab}) on U such that

$$\begin{aligned}
&[I_aX, I_aY] - [X, Y] - I_a([I_aX, Y] + [X, I_aY]) \\
&= -\sum_{p=1}^{3} \gamma_{ap}(X)I_aI_pY + \sum_{p=1}^{3} \gamma_{ap}(Y)I_aI_pX \\
&\quad + \sum_{p=1}^{3} \gamma_{ap}(I_aX)I_pY - \sum_{p=1}^{3} \gamma_{ap}(I_aY)I_pX, \quad X, Y \in \Gamma(Q|_U). \quad (13)
\end{aligned}$$

Let M be a strongly integrable quaternionic CR manifold. We may regard its underlying $GL(n, \mathbb{H})\cdot Sp(1)$-structure $\mathbb{I}$ on Q as an S^2-bundle over M, which we denote by $\pi\colon \mathcal{Z} \to M$. Let $q \in M$ and take a local hyper CR structure $\{(Q_a, I_a)\}$ defined around q, compatible with the quaternionic

CR structure. Then $\mathcal{Z}_q$ may be written as $\mathcal{Z}_q = \{(I_{\mathbf{v}})_q \mid \mathbf{v} \in S^2\}$. Recall that $(I_{\mathbf{v}})_q$ is defined on the hyperplane $(Q_{\mathbf{v}})_q \subset T_qM$. In what follows we will often omit the subscript q.

We now assume that the quaternionic CR structure of M is ultra pseudoconvex. Fix a compatible quaternionic pseudohermitian structure Θ, and let D be the corresponding canonical connection. Let $\mathcal{W}$ be the rank 3 vector subbundle of End(Q) whose associated S^2-bundle is $\mathcal{Z}$. Note that $\mathcal{W}_q = \{(I_{\mathbf{v}})_q \mid \mathbf{v} \in \mathbb{R}^3\}$. Now D defines a connection $\nabla^{\mathcal{W}}$ on $\mathcal{W}$ by $\nabla^{\mathcal{W}}_X I = (D_X I)^{\mathcal{W}}$ for any section $I \in \Gamma(\mathcal{W})$, where $(\ \cdot\ )^{\mathcal{W}}$ denotes the orthogonal projection $\mathrm{End}(Q) \to \mathcal{W}$. Let $I = I_{\mathbf{v}} \in \mathcal{Z}$ and $q = \pi(I)$. Using $\nabla^{\mathcal{W}}$, we decompose the tangent space $T_I\mathcal{Z}$ of $\mathcal{Z}$ at I as $T_I\mathcal{Z} = \mathcal{V}_I \oplus \mathcal{H}_I$, where $\mathcal{V}_I = T_I\pi^{-1}(q)$ is the vertical subspace and $\mathcal{H}_I$ is the horizontal subspace with respect to $\nabla^{\mathcal{W}}$. Note that $\pi_*|_{\mathcal{H}_I} : \mathcal{H}_I \to T_qM$ is an isomorphism. We define a complex structure $\mathcal{J}$ on $\mathcal{P}_I = \mathcal{V}_I \oplus \widetilde{Q_{\mathbf{v}}}$, where $\widetilde{Q_{\mathbf{v}}} = (\pi_*|_{\mathcal{H}_I})^{-1}(Q_{\mathbf{v}})$, so that

(i) $\mathcal{J}|_{\mathcal{V}_I}$ is the natural complex structure J_0 of $\pi^{-1}(q) = S^2$.
(ii) $\mathcal{J}|_{\widetilde{Q_{\mathbf{v}}}}$ is the lift of $I_{\mathbf{v}} : Q_{\mathbf{v}} \to Q_{\mathbf{v}}$ via $\pi_*|_{\mathcal{H}_I} : \mathcal{H}_I \to T_qM$.

We are now in a position to state our main result:

Theorem 5.1. *Let M be a strongly integrable ultra pseudoconvex quaternionic CR manifold. Then $\mathcal{J} : \mathcal{P}_I \to \mathcal{P}_I$ is independent of the choice of a compatible quaternionic pseudohermitian structure Θ. Therefore, $(\mathcal{P}, \mathcal{J})$ is a well-defined almost CR structure on $\mathcal{Z}$.*

We call the S^2-bundle $\mathcal{Z}$ in Theorem 5.1 the *twistor space* of M.

Proof. Let Θ and Θ' be two compatible pseudohermitian structures, and let ∇ and ∇' be the connections on $\mathcal{W}$ defined as above with respect to Θ and Θ', respectively. It is sufficient to show that for $X \in Q_{\mathbf{v}}$,

$$\mathcal{J}\widetilde{X}^{\nabla} = J_0(\widetilde{X}^{\nabla} - \widetilde{X}^{\nabla'}) + \mathcal{J}'\widetilde{X}^{\nabla'}, \tag{14}$$

where $\widetilde{X}^{\nabla}$ (resp. $\widetilde{X}^{\nabla'}$) is the horizontal lift of X with respect to ∇ (resp. ∇') and $\mathcal{J}' : \widetilde{Q_{\mathbf{v}}}' \to \widetilde{Q_{\mathbf{v}}}'$ is defined as in (ii) above with respect to ∇'. Note that $\mathcal{J}\widetilde{X}^{\nabla}$ (resp. $\mathcal{J}'\widetilde{X}^{\nabla'}$) is the horizontal lift of IX with respect to ∇ (resp. ∇'). Now set $\alpha = \nabla' - \nabla$ and identify $\mathcal{V}_I \subset T_I\mathcal{Z}$ with a subspace in End(Q_q), where $q = \pi(I)$. Then it is easy to verify that $\widetilde{X}^{\nabla'} - \widetilde{X}^{\nabla} = -\alpha_X I$, and therefore $\mathcal{J}'\widetilde{X}^{\nabla'} - \mathcal{J}\widetilde{X}^{\nabla} = -\alpha_{IX} I$. Hence, (14) is equivalent to

$$\alpha_{IX} I = J_0(\alpha_X I), \tag{15}$$

that is, that the mapping $\varphi \colon X \in Q_{\mathbf{v}} \mapsto \alpha_X I \in \mathcal{V}_I$ satisfies $\varphi \circ I = J_0 \circ \varphi$.

Lemma 5.1. *Let I be a local section of $\mathcal{Z}$ and Φ be a section of $\mathcal{W}$ which is pointwise orthogonal to I. (This is equivalent to saying that Φ is anti-commuting with I, and means that $\Phi(q)$ is tangent to $\mathcal{V}_{I(q)}$ for each point q.) Then for $X \in Q$ we have*

$$g((\nabla' - \nabla)_X I, \Phi) = -4d_b f(I\Phi X) \\ -4\sum_{a=1}^{3} d\theta_a(X, I\Phi I_a W) - 8n\, g(I\Phi W, X). \quad (16)$$

We postpone the proof of this lemma to the final section.

We now return and verify (15), which is equivalent to

$$g((\nabla' - \nabla)_{IX} I, \Phi) + g((\nabla' - \nabla)_X I, I\Phi) = 0, \;\; X \in Q_{\mathbf{v}}. \quad (17)$$

(Note that $J_0(\Phi) = I\Phi$.) We may assume that $I = I_1$. Then $Q_{\mathbf{v}} = Q \oplus \mathbb{R}T_2 \oplus \mathbb{R}T_3$ and Φ is linear combination of J and K. Here and in the following, we also use the notation I, J, K for I_1, I_2, I_3, respectively.

To show (17) for $X \in Q$, we first recall

$$d\theta_a(X, Y) - d\theta_a(IX, IY) = \begin{cases} 0 & \text{if } a = 1, \\ -2g(X, I_a Y) & \text{if } a = 2, 3 \end{cases} \quad (18)$$

for $X, Y \in Q$. By (16), if $X \in Q$, then the left-hand side of (17) is equal to

$$4\sum_{a=1}^{3}(d\theta_a(X, \Phi I_a W) - d\theta_a(IX, I\Phi I_a W)) = -8\sum_{a=2}^{3} g(X, I_a \Phi I_a W). \quad (19)$$

Here we have used (18) for $Y = \Phi I_a W \in Q$. The right-hand side of (19) then vanishes, since $K\Phi K W = -J\Phi J W$, which follows from $I\Phi = -\Phi I$. Thus (17) is verified for $X \in Q$.

It remains to show (17) for $X \in \mathbb{R}T_2 \oplus \mathbb{R}T_3$. In fact, it suffices to do this for $X = T_2$, namely, to show that

$$g((\nabla' - \nabla)_{T_3} I, \Phi) + g((\nabla' - \nabla)_{T_2} I, I\Phi) = 0. \quad (20)$$

So we compute $g((\nabla' - \nabla)_{T_a} I, \Phi)$: Let $\{\varepsilon_1, \ldots, \varepsilon_{4n}\}$ be a local orthonormal frame field for Q with respect to g. By definition, we have

$$g((\nabla' - \nabla)_{T_a} I, \Phi) = \sum_{i=1}^{4n} g(((\nabla' - \nabla)_{T_a} I)\varepsilon_i, \Phi\varepsilon_i)$$

$$= \sum_{i=1}^{4n}\{g((D' - D)_{T_a}(I\varepsilon_i), \Phi\varepsilon_i) + g((D' - D)_{T_a}\varepsilon_i, I\Phi\varepsilon_i)\}$$

$$= 2\sum_{i=1}^{4n} g((D' - D)_{T_a}(I\varepsilon_i), \Phi\varepsilon_i),$$

where we have replaced ε_i by $I\varepsilon_i$ and used $I\Phi = -\Phi I$ for the last equality. Using (7) and $I\Phi = -\Phi I$, and rearranging, we obtain

$$g((\nabla' - \nabla)_{T_a} I, \Phi) = -4\sum_{p=1}^{3} d\theta_p(T_a, I\Phi I_p W) + 8d_b f(I\Phi I_a W)$$
$$+4\sum_{i=1}^{4n}\Big\{g(D_{\Phi\varepsilon_i}(I_a W), I\varepsilon_i) + d\theta_a(I\varepsilon_i, \Phi\varepsilon_i)g(W,W)\Big\}.$$

Note that we have also used

$$\sum_{i=1}^{4n} g(D_{I\varepsilon_i}(I_a W), \Phi\varepsilon_i) = -\sum_{i=1}^{4n} g(D_{\Phi\varepsilon_i}(I_a W), I\varepsilon_i),$$

which can be verified by using the fact that Φ is a linear combination of J, K. Therefore, the left-hand side of (20) is rewritten as

$$g((\nabla' - \nabla)_{T_3} I, \Phi) + g((\nabla' - \nabla)_{T_2} I, I\Phi)$$
$$= 4\sum_{p=1}^{3}(-d\theta_p(T_3, I\Phi I_p W) + d\theta_p(T_2, \Phi I_p W)) + 4\sum_{i=1}^{4n} g((D_{\Phi\varepsilon_i} I)(JW), I\varepsilon_i)$$
$$+4\sum_{i=1}^{4n}(d\theta_3(I\varepsilon_i, \Phi\varepsilon_i) + d\theta_2(I\varepsilon_i, I\Phi\varepsilon_i))\, g(W,W), \tag{21}$$

where we have used

$$\sum_{i=1}^{4n} g(D_{I\Phi\varepsilon_i}(JW), I\varepsilon_i) = -\sum_{i=1}^{4n} g(D_{\Phi I\varepsilon_i}(JW), I\varepsilon_i)$$
$$= -\sum_{i=1}^{4n} g(D_{\Phi\varepsilon_i}(JW), \varepsilon_i) = -\sum_{i=1}^{4n} g(ID_{\Phi\varepsilon_i}(JW), I\varepsilon_i).$$

In the last sum of (21), by replacing ε_i with $I\varepsilon_i$ and using $\Phi I = -I\Phi$ as in the previous computation and also using (18), we obtain

$$\sum_{i=1}^{4n} d\theta_2(I\varepsilon_i, I\Phi\varepsilon_i) = -\frac{1}{2}\sum_{i=1}^{4n}(d\theta_2(\varepsilon_i, \Phi\varepsilon_i) - d\theta_2(I\varepsilon_i, I\Phi\varepsilon_i))$$
$$= \sum_{i=1}^{4n} g(\varepsilon_i, J\Phi\varepsilon_i) = -4nu$$

if we write $\Phi = uJ + vK$. Likewise,

$$\sum_{i=1}^{4n} d\theta_3(I\varepsilon_i, \Phi\varepsilon_i) = -\sum_{i=1}^{4n} g(\varepsilon_i, J\Phi\varepsilon_i) = 4nu.$$

Thus we see that the last sum in the right-hand side of (21) vanishes.

Substituting $\Phi = uJ + vK$ and rearranging, we obtain

$$
\begin{aligned}
&g((\nabla'-\nabla)_{T_3}I,\Phi) + g((\nabla'-\nabla)_{T_2}I, I\Phi) = \\
&\quad 4u\Big[\sum_{i=1}^{4n} g((D_{J\varepsilon_i}I)(JW), I\varepsilon_i) - \big\{(d\theta_1(T_2,KW) + d\theta_1(T_3,JW)) \\
&\qquad +(d\theta_2(T_2,W)-d\theta_3(T_3,W)-d\theta_2(T_3,IW)-d\theta_3(T_2,IW))\big\}\Big] \\
&\quad +4v\Big[\sum_{i=1}^{4n} g((D_{K\varepsilon_i}I)(JW), I\varepsilon_i) - \big\{(-d\theta_1(T_2,JW)+d\theta_1(T_3,KW)) \\
&\qquad +(d\theta_2(T_2,IW)-d\theta_3(T_3,IW)+d\theta_2(T_3,W)+d\theta_3(T_2,W))\big\}\Big]. \qquad (22)
\end{aligned}
$$

To proceed further, we need the following

Lemma 5.2. *Let (M,Θ) be a quaternionic pseudohermitian manifold which is strongly integrable ultra pseudoconvex of dimension $4n+3$ (>7). Let $\{(Q_a,I_a)\}$ be a local hyper CR structure constituting the quaternionic CR structure, θ a compatible $\mathbb{R}^3$-valued 1-form, and (T_1,T_2,T_3) the canonical triple satisfying $\theta_a(T_a)=1$. Then*

$$(d\theta_2 - \sqrt{-1}d\theta_3)(T_2 - \sqrt{-1}T_3, Z) = 0, \qquad (23)$$

$$\sum_{k=1}^{n} g([e_{2k-1}, e_{2k}]_Q, Z) = -\frac{\sqrt{-1}}{2} d\theta_1(T_2 - \sqrt{-1}T_3, Z) \qquad (24)$$

hold for any $Z \in Q^{1,0} = \{Z \in Q\otimes\mathbb{C} \mid I_1 Z = \sqrt{-1}Z\}$. Here, $\{e_1,\ldots,e_{2n}\}$ is a local unitary frame field on $Q^{1,0}$ with respect to the Levi form $g = L_\theta$ satisfying $I_2 e_{2k-1} = \overline{e_{2k}}$, $I_2 e_{2k} = -\overline{e_{2k-1}}$, $k = 1,2,\ldots,n$.

The proof of this lemma is also postponed.

Now we show the vanishing of the right-hand side of (22) by using Lemma 5.2. Setting $Z = W - \sqrt{-1}IW \in Q^{1,0}$ for $W \in Q$ and plugging it into (23), we obtain

$$
\begin{aligned}
0 &= (d\theta_2 - \sqrt{-1}d\theta_3)(T_2 - \sqrt{-1}T_3, W - \sqrt{-1}IW) \\
&= (d\theta_2(T_2,W) - d\theta_3(T_3,W) - d\theta_2(T_3,IW) - d\theta_3(T_2,IW)) \\
&\quad -\sqrt{-1}(d\theta_2(T_2,IW) - d\theta_3(T_3,IW) + d\theta_2(T_3,W) + d\theta_3(T_2,W)),
\end{aligned}
$$

and this implies

$$
\begin{aligned}
&g((\nabla' - \nabla)_{T_3} I, \Phi) + g((\nabla' - \nabla)_{T_2} I, I\Phi) = \\
&\quad 4u\Big[\sum_{i=1}^{4n} g((D_{J\varepsilon_i} I)(JW), I\varepsilon_i) - (d\theta_1(T_2, KW) + d\theta_1(T_3, JW))\Big] \\
&+4v\Big[\sum_{i=1}^{4n} g((D_{K\varepsilon_i} I)(JW), I\varepsilon_i) - (-d\theta_1(T_2, JW) + d\theta_1(T_3, KW))\Big]. \qquad (25)
\end{aligned}
$$

Then, by replacing ε_i with $I\varepsilon_i$, one can rewrite the sum in the last line of (25) as

$$
\begin{aligned}
&\sum_{i=1}^{4n} g((D_{K\varepsilon_i} I)(JW), I\varepsilon_i) - (-d\theta_1(T_2, JW) + d\theta_1(T_3, KW)) \\
&= \sum_{i=1}^{2n} g((D_{J\varepsilon_i} I)(JW), I(I\varepsilon_i)) + (d\theta_1(T_2, KIW) + d\theta_1(T_3, JIW)) \\
&= -\Big[\sum_{i=1}^{2n} g((D_{J\varepsilon_i} I)(JIW), I\varepsilon_i) - (d\theta_1(T_2, KIW) + d\theta_1(T_3, JIW))\Big].
\end{aligned}
$$

For the last equality, we have used the fact that $(D_X I_a)(I_a Y) = -I_a(D_X I_a)(Y)$ holds for $X, Y \in Q$. To conclude the proof, it therefore suffices to show that

$$
\sum_{i=1}^{4n} g((D_{J\varepsilon_i} I)(JW), I\varepsilon_i) - (d\theta_1(T_2, KW) + d\theta_1(T_3, JW)) = 0 \qquad (26)
$$

for $W \in \Gamma(Q)$. Using a local unitary frame field $\{e_1, \ldots, e_{2n}\}$ for $Q^{1,0}$ with respect to g and (24) together with its conjugate, we compute the sum in the left-hand side of (26) as

$$
\begin{aligned}
&\sum_{i=1}^{4n} g((D_{J\varepsilon_i} I)(JW), I\varepsilon_i) = -\sum_{i=1}^{4n} g((D_{J\varepsilon_i} I)(\varepsilon_i), KW) \\
&= -\sum_{k=1}^{n} \{g((D_{\overline{e_{2k}}} I)(\overline{e_{2k-1}}), KW) - g((D_{\overline{e_{2k-1}}} I)(\overline{e_{2k}}), KW) \\
&\qquad + g((D_{e_{2k}} I)(e_{2k-1}), KW) - g((D_{e_{2k-1}} I)(e_{2k}), KW)\} \\
&= -\sum_{k=1}^{n} \{g([\overline{e_{2k-1}}, \overline{e_{2k}}]_Q, JZ) + g([e_{2k-1}, e_{2k}]_Q, J\overline{Z})\} \\
&= d\theta_1(T_2, KW) + d\theta_1(T_3, JW),
\end{aligned}
$$

where $Z = W - \sqrt{-1}IW \in Q^{1,0}$. This implies (26) and completes the proof of Theorem 5.1. □

Note.
- It remains to see whether $(\mathcal{P}, \mathcal{J})$ is integrable, and we address this problem in a forthcoming work.
- If the three almost CR structures constituting a hyper CR structure are integrable in the usual sense, then the product space $\mathcal{Z} = M \times S^2$ comes equipped with a natural CR structure.
- Marchiafava-Ornea-Pantilie [3] recently introduced the twistor theory for "CR quaternonic structure". It is desirable to clarify the relationship between the strongly integrable quaternionic CR structure and the CR quaternionic structure.

6. Comparison of quaternionic CR and quaternionic contact structures

Let Q be a rank $4n$ vector bundle on a manifold. A *$CSp(n)\cdot Sp(1)$-structure* on Q is a pair $([\gamma], \mathbb{I})$ of a $GL(n,\mathbb{H})\cdot Sp(1)$-structure $\mathbb{I}$ on Q and a conformal class $[\gamma]$ of $\mathbb{I}$-invariant metrics on Q.

Definition 6.1 (Quaternionic contact structure). Let M be a manifold of dimension $4n+3$. A *quaternionic contact structure* on M is a corank 3 subbundle $Q \subset TM$ equipped with a $CSp(n)\cdot Sp(1)$-structure $([\gamma], \mathbb{I})$ such that Q is locally the kernel of an $\mathbb{R}^3$-valued 1-form $\eta = (\eta_1, \eta_2, \eta_3)$ satisfying

$$d\eta_a(X, I_a Y) = \gamma(X, Y), \; X, Y \in Q, \; a = 1, 2, 3, \tag{27}$$

where I_a are local sections of $\mathbb{I}$ satisfying the quaternion relations.

Suppose that a quaternionic contact structure $(Q, [\gamma], \mathbb{I})$ be given. If one fixes a metric γ on Q, then each choice of a rank 3 bundle $Q^\perp$ transverse to Q determines a unique quaternionic pseudohermitian structure such that the associated Levi form is the metric γ, and therefore gives a quaternionic CR structure, having $(Q, \mathbb{I})$ as the underlying structure. Thus a quaternionic contact structure can always be "extended" to quaternionic CR structures in plenty of different manners. On the other hand, the following result asserts that the quaternionic contact structure is indeed more restrictive than the quaternionic CR structure.

Proposition 6.1 ([2], Proposition 5.3). *Let $(Q, \mathbb{I})$ be a $GL(n,\mathbb{H})\cdot Sp(1)$-structure underlying both a quaternionic contact structure $(Q, [\gamma], \mathbb{I})$ and a quaternionic CR structure. Let Θ be a pseudohermitian structure* *compatible with the quaternionic CR structure. Then* $[\gamma] = [L_\Theta]$ *and a local* $\mathbb{R}^3$*-valued* 1*-form* $\theta = (\theta_a)$ *in* Θ *satisfies*

$$L_\theta(X, Y) = d\theta_a(X, I_a Y), \ X, Y \in Q, \ a = 1, 2, 3.$$

Example 6.1. An ellipsoid in $\mathbb{H}^{n+1}$ has a canonical quaternionic CR structure as a real hypersurface, and therefore has a canonical $(Q, \mathbb{I})$. However, as follows from Proposition 6.1, the ellipsoid admits no quaternionic contact structure having $(Q, \mathbb{I})$ as the underlying structure, unless it is equivalent to the unit sphere by a quaternionic affine transformation.

For other such examples, we refer the reader to [2].

Given a quaternionic contact structure $(Q, [\gamma], \mathbb{I})$, as mentioned above, there are quaternionic CR structures having $(Q, \mathbb{I})$ as the underlying structure. While they are not strongly integrable in general, there is a canonical way of extending a quaternionic contact structure to a strongly integrable quaternionic CR structure. We here recall the following result from [1]. Let $\eta = (\eta_1, \eta_2, \eta_3)$ be a local $\mathbb{R}^3$-valued 1-form satisfying (27). If the dimension of the underlying manifold is greater than 7, then there exist vector fields R_a, $a = 1, 2, 3$, called the *Reeb vector field* for η, such that $\eta_a(R_a) = 1$, $\eta_a(R_b) = \eta_a(R_c) = 0$ and

$$d\eta_a(R_a, X) = 0, \quad d\eta_a(R_b, X) + d\eta_b(R_a, X) = 0, \quad X \in Q.$$

Note that though the vector fields R_a are defined only locally, the bundle $Q^\perp$ generated by them is well-defined globally.

Proposition 6.2. *Let* M *be a manifold of dimension* $4n + 3 > 7$*, and let* $(Q, [\gamma], \mathbb{I})$ *be a quaternionic contact structure on* M*. Then the quaternionic CR structure corresponding to the bundle* $Q^\perp$ *generated by the Reeb vector fields* R_a*,* $a = 1, 2, 3$*, is strongly integrable.*

Proof. Recall from [1] and [2] that there exists an affine connection D^B on M, called the *Biquard connection*, whose restriction to Q is an $Sp(n)\cdot Sp(1)$-connection. Our canonical connection D coincides with D^B as Q-partial connections. In particular, the connection D induces an $Sp(n)\cdot Sp(1)$-Q-partial connection on Q. The strong integarbility follows from this property of D together with the identity $D_X Y - D_Y X = [X, Y]_Q$, where $X, Y \in \Gamma(Q)$. □

7. Proofs of Lemmas 5.1 and 5.2

Proof of Lemma 5.1. Let $\{\varepsilon_1, \cdots, \varepsilon_{4n}\}$ be a local orthonormal frame field for Q with respect to g. By definition, we obtain

$$g(\nabla'_X I, \Phi) = \sum_{i=1}^{4n} \left(g(D'_X(I\varepsilon_i), \Phi\varepsilon_i) + g(D'_X\varepsilon_i, I\Phi\varepsilon_i) \right), \quad X \in Q.$$

We compute the right-hand side by using (6) and $\Phi I = -I\Phi$, and obtain

$$\begin{aligned} g(\nabla'_X I, \Phi) &= g(\nabla_X I, \Phi) - 4d_b f(I\Phi X) - 4\sum_{a=1}^{3} d\theta_a(X, I\Phi I_a W) \\ &\quad - \sum_{a=1}^{3}\sum_{i=1}^{4n} \left(d\theta_a(I\varepsilon_i, \Phi\varepsilon_i) + d\theta_a(\varepsilon_i, I\Phi\varepsilon_i) \right) g(I_a W, X). \end{aligned}$$

To simplify the last term of the right-hand side, we may suppose that $I_1 = I$. Write $J = I_2$, $K = I_3$ and $\Phi = uJ + vK$, and investigate

$$\sum_{i=1}^{4n} \left(d\theta_a(I\varepsilon_i, \Phi\varepsilon_i) + d\theta_a(\varepsilon_i, I\Phi\varepsilon_i) \right). \tag{28}$$

Since $d\theta_1$ is I-invariant, it is manifest that (28) is zero if $a = 1$. If $a = 2, 3$, then

$$\sum_{i=1}^{4n} \left(d\theta_a(I\varepsilon_i, \Phi\varepsilon_i) + d\theta_a(\varepsilon_i, I\Phi\varepsilon_i) \right) = -2\sum_{i=1}^{4n} g(I\varepsilon_i, I_a\Phi\varepsilon_i),$$

which can be verified by setting $X = I\varepsilon_i$ and $Y = \Phi\varepsilon_i$ in (18). Therefore,

$$\begin{aligned} &\sum_{a=1}^{3}\sum_{i=1}^{4n} \left(d\theta_a(I\varepsilon_i, \Phi\varepsilon_i) + d\theta_a(\varepsilon_i, I\Phi\varepsilon_i) \right) g(I_a W, X) \\ &= -8nv\, g(I_2 W, X) + 8nu\, g(I_3 W, X) = 8n\, g(I\Phi W, X). \end{aligned}$$

This completes the proof of Lemma 5.1. □

Proof of Lemma 5.2. Let X, Y, Z be sections of $Q^{1,0} = \{Z \in Q \otimes \mathbb{C} \mid I_1 Z = \sqrt{-1}Z\}$. Then

$$d\theta_a(X, Y) = -d\theta_a(X, I_a(I_a Y)) = \begin{cases} 0, & \text{if } a = 1, \\ -g(X, I_a Y), & \text{if } a = 2, 3. \end{cases} \tag{29}$$

Setting $a = 1$ in (13), we see that

$$I_1[X, Y] = \sqrt{-1}[X, Y] - ((\gamma_{12} - \sqrt{-1}\gamma_{13}) \wedge I_2)(X, Y). \tag{30}$$

Then we can obtain the following identities:

$$g([X,Y]_Q,Z) = (\sqrt{-1}/2)\big((\gamma_{12}-\sqrt{-1}\gamma_{13})\wedge(I_2Z)^{\flat_g}\big)(X,Y), \quad (31)$$
$$d\theta_1([X,Y]_Q,Z) = (\sqrt{-1}/2)\big((\gamma_{12}-\sqrt{-1}\gamma_{13})\wedge\iota_Z d\theta_1\circ I_2\big)(X,Y), (32)$$
$$d\theta_2([X,Y]_Q,I_2\overline{Z}) = g([X,Y]_Q,\overline{Z}) + (\sqrt{-1}/2)\big((\gamma_{12}-\sqrt{-1}\gamma_{13})\wedge\iota_{\overline{Z}}d\theta_2\big)(X,Y), \quad (33)$$
$$d\theta_3([X,Y]_Q,I_3\overline{Z}) = g([X,Y]_Q,\overline{Z}) - (1/2)\big((\gamma_{12}-\sqrt{-1}\gamma_{13})\wedge\iota_{\overline{Z}}d\theta_3\big)(X,Y), \quad (34)$$

where ι denotes the interior product and $\flat_g : Q\otimes\mathbb{C}\to Q^*\otimes\mathbb{C}$ denotes the metric dual with respect to g; e.g., $\iota_Z d\theta_1(X) = d\theta_1(Z,X)$ and $(I_2Z)^{\flat_g}(X) = g(I_2Z,X)$ for $X\in Q$. Indeed, taking account of the I_1-invariance of the Levi form g and $d\theta_1|_{Q\times Q}$ and using (30), one can verify (31) and (32). Noting (29) and using (30) and the I_a-invariance of $d\theta_a|_{Q\times Q}$, one can also verify (33) and (34). As an example, we display the process for (33):

$$\begin{aligned} g([X,Y]_Q,\overline{Z}) &= (1/2)\{d\theta_2([X,Y]_Q,I_2\overline{Z}) + d\theta_2(I_1[X,Y]_Q,I_2(I_1\overline{Z})\} \\ &= d\theta_2([X,Y]_Q,I_2\overline{Z}) \\ &\quad -(\sqrt{-1}/2)\big((\gamma_{12}-\sqrt{-1}\gamma_{13})\wedge\iota_{I_2\overline{Z}}d\theta_2\circ I_2\big)(X,Y) \\ &= d\theta_2([X,Y]_Q,I_2\overline{Z}) \\ &\quad -(\sqrt{-1}/2)\big((\gamma_{12}-\sqrt{-1}\gamma_{13})\wedge\iota_{\overline{Z}}d\theta_2\big)(X,Y). \end{aligned}$$

By making use of $dd\theta_a = 0$ for $a = 2,3$, we can verify

$$\begin{aligned} & d\theta_2(T_2-\sqrt{-1}T_3, Z - g(Z,\overline{e_{2k-1}})e_{2k-1} - g(Z,\overline{e_{2k}})e_{2k}) \\ &= e_{2k-1}g(Z,\overline{e_{2k-1}}) + e_{2k}g(Z,\overline{e_{2k}}) - g([e_{2k-1},e_{2k}]_Q,I_2Z) \\ &\quad -g([e_{2k},Z]_Q,\overline{e_{2k}}) - g([e_{2k-1},Z]_Q,\overline{e_{2k-1}}) \\ &= \sqrt{-1}d\theta_3(T_2-\sqrt{-1}T_3, Z - g(Z,\overline{e_{2k-1}})e_{2k-1} - g(Z,\overline{e_{2k}})e_{2k}). \quad (35) \end{aligned}$$

We assume (35) for the moment. Then using $g(Z,e_{2k-1}) = g(Z,e_{2k}) = 0$ and summing up (35) over $k = 1,2,\ldots,n$, we obtain

$$(n-1)d\theta_2(T_2-\sqrt{-1}T_3,Z) = (n-1)\sqrt{-1}d\theta_3(T_2-\sqrt{-1}T_3,Z).$$

In particular, (23) follows, as we are assuming $n>1$.

We can verify the first equality of (35) in the following way: For $Z\in\Gamma(Q^{1,0})$,

$$\begin{aligned} 0 &= dd\theta_2(e_{2k-1},e_{2k},Z) \\ &= e_{2k-1}d\theta_2(e_{2k},Z) + e_{2k}d\theta_2(Z,e_{2k-1}) + Zd\theta_2(e_{2k-1},e_{2k}) \\ &\quad - d\theta_2([e_{2k-1},e_{2k}],Z) - d\theta_2([e_{2k},Z],e_{2k-1}) - d\theta_2([Z,e_{2k-1}],e_{2k}) \end{aligned}$$

$$\begin{aligned}
&= -e_{2k-1}g(Z,\overline{e_{2k-1}}) - e_{2k}g(Z,\overline{e_{2k}}) + d\theta_2([e_{2k-1},e_{2k}]_Q, I_2(I_2Z))\\
&\quad +d\theta_2([e_{2k},Z]_Q, I_2\overline{e_{2k}}) - d\theta_2([Z,e_{2k-1}]_Q, I_2\overline{e_{2k-1}})\\
&\quad -d\theta_2([e_{2k-1},e_{2k}]_{Q^\perp}, Z) - d\theta_2([e_{2k},Z]_{Q^\perp}, e_{2k-1})\\
&\quad -d\theta_2([Z,e_{2k-1}]_{Q^\perp}, e_{2k}). \qquad (36)
\end{aligned}$$

Now recall $[X,Y]_{Q^\perp} = \sum_{a=1}^3 \theta_a([X,Y])T_a = -\sum_{a=1}^3 d\theta_a(X,Y)T_a$. Then, by (29), we can compute the sum of the last three terms in (36) as

$$\begin{aligned}
&\sum_{a=1}^{3}\{d\theta_a(e_{2k-1},e_{2k})d\theta_2(T_a,Z) + d\theta_a(e_{2k},Z)d\theta_2(T_a,e_{2k-1})\\
&\qquad +d\theta_a(Z,e_{2k-1})d\theta_2(T_a,e_{2k})\}\\
&= d\theta_2(T_2,Z) - g(Z,\overline{e_{2k-1}})d\theta_2(T_2,e_{2k-1}) - g(Z,\overline{e_{2k}})d\theta_2(T_2,e_{2k})\\
&\quad -\sqrt{-1}(d\theta_2(T_3,Z) - g(Z,\overline{e_{2k-1}})d\theta_2(T_3,e_{2k-1}) - g(Z,\overline{e_{2k}})d\theta_2(T_3,e_{2k})),
\end{aligned}$$

which is equal to the left-hand side of (35). By applying (33) to (36) and using the I_2-invariance of $d\theta_2|_{Q\times Q}$, we obtain

$$\begin{aligned}
&d\theta_2(T_2-\sqrt{-1}T_3, Z - g(Z,\overline{e_{2k-1}})e_{2k-1} - g(Z,\overline{e_{2k}})e_{2k})\\
&= e_{2k-1}g(Z,\overline{e_{2k-1}}) + e_{2k}g(Z,\overline{e_{2k}})\\
&\quad -g([e_{2k-1},e_{2k}]_Q, I_2Z) - (\sqrt{-1}/2)((\gamma_{12}-\sqrt{-1}\gamma_{13})\wedge \iota_{I_2Z}d\theta_2)(e_{2k-1},e_{2k})\\
&\quad -g([e_{2k},Z]_Q,\overline{e_{2k}}) - (\sqrt{-1}/2)((\gamma_{12}-\sqrt{-1}\gamma_{13})\wedge \iota_{\overline{e_{2k}}}d\theta_2)(e_{2k},Z)\\
&\quad -g([e_{2k-1},Z]_Q,\overline{e_{2k-1}}) - (\sqrt{-1}/2)((\gamma_{12}-\sqrt{-1}\gamma_{13})\wedge \iota_{\overline{e_{2k-1}}}d\theta_2)(e_{2k-1},Z)\\
&= e_{2k-1}g(Z,\overline{e_{2k-1}}) + e_{2k}g(Z,\overline{e_{2k}}) - g([e_{2k-1},e_{2k}]_Q, I_2Z)\\
&\quad -g([e_{2k},Z]_Q,\overline{e_{2k}}) - g([e_{2k-1},Z]_Q,\overline{e_{2k-1}}),
\end{aligned}$$

which shows the first equality of (35). Similarly, we can verify the second equality of (35).

We next prove (24). Plug $X = e_{2k-1}$ and $Y = e_{2k}$ into (31) and sum it up for $k = 1,2,\ldots,n$. Then we obtain

$$\sum_{k=1}^{n} g([e_{2k-1},e_{2k}]_Q, Z) = (\sqrt{-1}/2)(\gamma_{12}-\sqrt{-1}\gamma_{13})(Z). \qquad (37)$$

By taking the cyclic sums $\mathfrak{S}_{X,Y,Z}$ of (32) with respect to X,Y,Z and also noting the definition of g, we have

$$\begin{aligned}
&\mathfrak{S}_{X,Y,Z}\, d\theta_1([X,Y]_Q, Z)\\
&\quad = (\sqrt{-1}/2)\mathfrak{S}_{X,Y,Z}((\gamma_{12}-\sqrt{-1}\gamma_{13})\wedge \iota_Z d\theta_1\circ I_2)(X,Y)\\
&\quad = -(\sqrt{-1}/2)\mathfrak{S}_{X,Y,Z}(\gamma_{12}-\sqrt{-1}\gamma_{13})(X)(d\theta_1(I_2Y,Z) + d\theta_1(Y,I_2Z))\\
&\quad = \mathfrak{S}_{X,Y,Z}(\gamma_{12}-\sqrt{-1}\gamma_{13})(X)g(Y,I_2Z). \qquad (38)
\end{aligned}$$

Set $X = e_{2k-1}, Y = e_{2k}$ in (38) and sum up over k. We then obtain

$$\sum_{k=1}^{n} \{d\theta_1([e_{2k-1}, e_{2k}]_Q, Z) + d\theta_1([e_{2k}, Z]_Q, e_{2k-1}) + d\theta_1([Z, e_{2k-1}]_Q, e_{2k})\}$$
$$= -(n-1)(\gamma_{12} - \sqrt{-1}\gamma_{13})(Z). \tag{39}$$

On the other hand, by computations similar to those in and after (36), it follows from $dd\theta_1 = 0$ and the I_1-invariance of $d\theta_1|_{Q\times Q}$ that

$$\begin{aligned} 0 &= dd\theta_1(e_{2k-1}, e_{2k}, Z) \\ &= e_{2k-1}d\theta_1(e_{2k}, Z) + e_{2k}d\theta_1(Z, e_{2k-1}) + Zd\theta_1(e_{2k-1}, e_{2k}) \\ &\quad -d\theta_1([e_{2k-1}, e_{2k}], Z) - d\theta_1([e_{2k}, Z], e_{2k-1}) - d\theta_1([Z, e_{2k-1}], e_{2k}) \\ &= -d\theta_1([e_{2k-1}, e_{2k}]_Q, Z) - d\theta_1([e_{2k}, Z]_Q, e_{2k-1}) - d\theta_1([Z, e_{2k-1}]_Q, e_{2k}) \\ &\quad +d\theta_1(T_2 - \sqrt{-1}T_3, Z - g(Z, \overline{e_{2k-1}})e_{2k-1} - g(Z, \overline{e_{2k}})e_{2k}). \end{aligned}$$

Summing up these for $k = 1, 2, \ldots, n$ and using (39), we conclude

$$0 = (n-1)(\gamma_{12} - \sqrt{-1}\gamma_{13})(Z) + (n-1)d\theta_1(T_2 - \sqrt{-1}T_3, Z).$$

Since $n > 1$, we have $d\theta_1(T_2 - \sqrt{-1}T_3, Z) = -(\gamma_{12} - \sqrt{-1}\gamma_{13})(Z)$, and therefore (37) implies (24). □

Note added in proof. After the submission of this manuscript, we proved further that the almost CR structure $(\mathcal{P}, \mathcal{J})$ in Theorem 5.1 is partially integrable.

References

1. O. Biquard, *Métriques d'Einstein asymptotiquement symétriques.* Astérisque **265** (2000), vi+109 pp.
Asymptotically symmetric Einstein metrics, Translated by Stephen S. Wilson, SMF/AMS Texts and Monographs 13, 2006.
2. H. Kamada and S. Nayatani, Quaternionic CR geometry, *Hokkaido Math. J.* **42** (2013) 1–49.
3. S. Marchiafava, L. Ornea and R. Pantilie, Twistor theory for CR quaternionic manifolds and related structures, *Monatsh. Math.* **167** (2012) 531–545.

Received February 3, 2015
Revised April 21, 2015

Current Developments in Differential Geometry
and its Related Fields *pp. 115 – 128*

FIVE-DIMENSIONAL LIE GROUPS WHICH ARE ALMOST CONTACT B-METRIC MANIFOLDS WITH THREE NATURAL CONNECTIONS

Miroslava IVANOVA

Department of Informatics and Mathematics, Trakia University,
Stara Zagora, 6000, Bulgaria
E-mail: mivanova@uni-sz.bg

Hristo MANEV

Department of Pharmaceutical Sciences, Medical University of Plovdiv,
Plovdiv, 4002, Bulgaria;
Departament of Algebra and Geometry, University of Plovdiv,
Plovdiv, 4027, Bulgaria
E-mail: hmanev@uni-plovdiv.bg

We study three natural connections, linear connections preserving the structure tensors, on 5-dimensional Lie groups which turn to be almost contact B-metric manifolds. We give conditions that they are flat, Ricci-flat or have null scalar curvatures.

Keywords: Almost contact manifold, B-metric, natural connection, φKT-connection, φB-connection, φ-canonical connection, holomorphic sections, ξ-sections, Lie groups, Lie algebra.

1. Introduction

An almost contact B-metric manifold is an odd dimensional manifold admitting an almost contact structure and a pseudo-Riemannian metric of signature $(n+1, n)$ which are compatible to each other (see §2 for the definition). On such manifolds admitting additional geometric structures it is important to consider natural connections corresponding to their structures and to investigate their properties. Almost contact B-metric manifolds are classified into 11 basic classes with respect to a tensor of $(0, 3)$ type (see §2), and each class is denoted by $\mathcal{F}_i$ $(i = 1, \ldots, 11)$. Only on manifolds in the class $\mathcal{F}_3 \oplus \mathcal{F}_7$ which is characterized as manifolds having vanishing associated Nijenhuis tensor $\widehat{N}$, we have three natural connections which are called φKT-connection, φB-connection and φ-canonical connection (cf. [7]). Though we have φB-connections and φ-canonical connections on all almost contact B-metric manifolds, φKT-connections exist only on manifolds in the class $\mathcal{F}_3 \oplus \mathcal{F}_7$. Moreover, on each manifold which does not belong to $\mathcal{F}_3 \oplus \mathcal{F}_7$, its φB-connection and φ-canonical connection coincide with each other. Since natural connections on manifolds belonging to $\mathcal{F}_3$ are studied in [13] and some other articles in its references, we are interested in the class $\mathcal{F}_7$.

The aim of this paper is to study three natural connections on 5-dimensional Lie groups. With a global basis of left-invariant vector fields on a 5-dimensional Lie group we can define a natural almost contact B-metric structure of $\mathcal{F}_7$. Since 5 is the smallest dimension of manifolds in the class $\mathcal{F}_7$, our consideration will be a mile stone for future theoretical investigations on manifolds in the class $\mathcal{F}_7$. We calculate curvature and Ricci tensors and scalar curvatures for three natural connections on 5-dimensional Lie groups and show some properties. We also show some characterizations of 5-dimensional Lie groups to be isotropic-$\mathcal{F}_0$-manifolds in terms of torsion tensors and scalar curvatures of these natural connections.

2. Almost contact B-metric manifolds

We shall start by recalling some properties and notations concerned with almost contact manifolds. Let $(M, \varphi, \xi, \eta, g)$ be a $(2n+1)$-dimensional *almost contact B-metric manifold*. Here, (φ, ξ, η) is a triplet of a tensor field φ of type (1,1), a vector field ξ and its dual 1-form η satisfying

$$\varphi\xi = 0, \quad \varphi^2 = -\mathrm{Id} + \eta \otimes \xi, \quad \eta \circ \varphi = 0, \quad \eta(\xi) = 1,$$

where Id is the identity. Moreover, g is a pseudo-Riemannian metric, called a *B-metric*, satisfying

$$g(\varphi x, \varphi y) = -g(x, y) + \eta(x)\eta(y)$$

for arbitrary tangent vectors $x, y \in T_pM$ at an arbitrary point $p \in M$ ([4]). The triplet (φ, ξ, η) is called an almost contact structure on M. We note that the restriction of a B-metric on the contact distribution $H = \ker(\eta)$ coincides with the corresponding Norden metric with respect to the almost complex structure, the restriction of φ on H, acting as an anti-isometry on the metric on H, the restriction of g to H. Thus, we obtain a correlation between a $(2n+1)$-dimensional almost contact B-metric manifold and a $2n$-dimensional almost complex manifold with Norden metric (or an n-dimensional complex Riemannian manifold).

The associated metric $\tilde{g}$ of g on M is given by $\tilde{g}(x, y) = g(x, \varphi y) + \eta(x)\eta(y)$. It is also a B-metric, hence $(M, \varphi, \xi, \eta, \tilde{g})$ is also an almost contact B-metric manifold. Both metrics g and $\tilde{g}$ are indefinite of signature $(n+$

$1,n)$. The structure group of (M,φ,ξ,η,g) is $\mathcal{G}\times\mathcal{I}$, where $\mathcal{I}$ is the identity on span(ξ) and $\mathcal{G}=\mathcal{GL}(n;\mathbb{C})\cap\mathcal{O}(n,n)$.

By using the Levi-Civita connection ∇ of g, we define a tensor F of type (0,3) on M by $F(x,y,z)=g\big((\nabla_x\varphi)\,y,z\big)$. We then have

$$F(x,y,z)=F(x,z,y)=F(x,\varphi y,\varphi z)+\eta(y)F(x,\xi,z)+\eta(z)F(x,y,\xi)$$

(see [4]). Almost contact B-metric manifolds were classified into eleven basic classes $\mathcal{F}_i$ $(i=1,2,\ldots,11)$ with respect to F in [4]. The special class $\mathcal{F}_0$ is determined by the condition $F(x,y,z)=0$. Hence $\mathcal{F}_0$ is the class of almost contact B-metric manifolds having ∇-parallel structures, that is, $\nabla\varphi=\nabla\xi=\nabla\eta=\nabla g=\nabla\tilde{g}=0$. We should note that this special class $\mathcal{F}_0$ is contained in each class $\mathcal{F}_i$ $(i=1,\ldots,11)$. In this paper, we study almost contact manifolds with B-metric which belong to the class $\mathcal{F}_7$. This basic class is characterized by the following conditions:

$$F(x,y,z)=F(x,y,\xi)\eta(z)+F(x,z,\xi)\eta(y),$$
$$F(x,y,\xi)=-F(y,x,\xi)=-F(\varphi x,\varphi y,\xi).$$

A linear connection D is said to be a *natural connection* on (M,φ,ξ,η,g) if the almost contact structure (φ,ξ,η) and the B-metric g (consequently also $\tilde{g}$) are parallel with respect to D, that is, $D\varphi=D\xi=D\eta=Dg=D\tilde{g}=0$. It is known that a linear connection D is natural on (M,φ,ζ,η,g) if and only if $D\varphi=Dg=0$ ([12]). On each almost contact B-metric manifold we have a natural connection. It coincides with the Levi-Civita connection only on a $\mathcal{F}_0$-manifold. We define the torsion tensor T of D by $T(x,y)=D_xy-D_yx-[x,y]$. Here, D_xy-D_yx means D_XY-D_YX with some vector fields $X,Y\in\mathfrak{X}(M)$ on M which take x,y at the considered point, and $[x,y]=\nabla_XY-\nabla_YX$. We use such convention from now on. The corresponding tensor of type (0,3) is defined by $T(x,y,z)=g(T(x,y),z)$. We define the potential tensor Q of D with respect to ∇ by

$$Q(x,y)=D_xy-\nabla_xy, \tag{2.1}$$

and set its corresponding tensor of type (0,3) by $Q(x,y,z)=g(Q(x,y),z)$. According to [7], we know that a linear connection D is a natural connection on an almost contact B-metric manifold if and only if

$$Q(x,y,\varphi z)-Q(x,\varphi y,z)=F(x,y,z),$$
$$Q(x,y,z)=-Q(x,z,y).$$

We now restrict ourselves on almost contact B-metric manifolds belonging to the class $\mathcal{F}_7$, and give three natural connections on these manifolds which are induced by their almost contact structures. Let (M,φ,ξ,η,g) be an almost contact B-metric manifold in the class $\mathcal{F}_7$. We define a linear connection $D^{(1)}$ on M by $D^{(1)} = \nabla + T^{(1)}$ with the torsion tensor $T^{(1)}$ given by

$$T^{(1)}(x,y) = \tfrac{1}{2}\{(\nabla_x\varphi)\,\varphi y + (\nabla_x\eta)\,y\cdot\xi\} - \eta(y)\nabla_x\xi. \tag{2.2}$$

This connection is called a *φB-connection* ([10]). We can define this connection on an arbitrary almost contact B-metric manifold. It is the odd-dimensional counterpart of the B-connection on the corresponding almost complex manifold with Norden metric (see [2]). We define second linear connection $D^{(2)}$ on M by $D^{(2)} = \nabla + T^{(2)}$ with a totally skew symmetric torsion tensor $T^{(2)}$ given by

$$T^{(2)}(x,y) = 2\{\eta(x)\nabla_y\xi - \eta(y)\nabla_x\xi + (\nabla_x\eta)y\cdot\xi\} \tag{2.3}$$

(cf. [7]). It is called the *φKT-connection.* In Hermitian geometry, this connection is said to be the Bismut connection or the KT-connection. We define the third natural connection $D^{(3)}$ by $D^{(3)} = \nabla + T^{(3)}$ with the torsion tensor $T^{(3)}$ given by

$$T^{(3)}(x,y) = (\nabla_x\varphi)\,\varphi y - (\nabla_x\eta)\,y\cdot\xi - 2\eta(x)\nabla_y\xi. \tag{2.4}$$

This connection is called a *φ-canonical connection.* By (2.2), (2.3), (2.4) we find for an arbitrary manifold in the class $\mathcal{F}_7$ that $D^{(1)}$ is the average of $D^{(2)}$ and $D^{(3)}$, i.e. $2D^{(1)} = D^{(2)} + D^{(3)}$. We can define a φ-canonical connection on an arbitrary almost contact B-metric manifold with an appropriate torsion tensor (see [11]). But as we restrict ourselves on manifolds in the class $\mathcal{F}_7$, we can write down $T^{(3)}$ as above.

3. 5-dimensional Lie groups belonging to $\mathcal{F}_7$

In this paper we study 5-dimensional connected Lie groups with almost contact B-metric structure of class $\mathcal{F}_7$. On a 5-dimensional connected Lie group G we take a global basis of left-invariant vector fields $\{e_i\}_{i=1}^5$ on G. We define an almost contact structure on G by

$$\begin{aligned} &\varphi e_1 = e_3, \quad \varphi e_2 = e_4, \quad \varphi e_3 = -e_1, \quad \varphi e_4 = -e_2, \quad \varphi e_5 = 0;\\ &\xi = e_5; \qquad \eta(e_1) = \eta(e_2) = \eta(e_3) = \eta(e_4) = 0, \quad \eta(e_5) = 1; \end{aligned} \tag{3.1}$$

and define a B-metric on G by

$$\begin{gathered} g(e_1,e_1) = g(e_2,e_2) = -g(e_3,e_3) = -g(e_4,e_4) = g(e_5,e_5) = 1,\\ g(e_i,e_j) = 0, \qquad i\neq j, \quad i,j\in\{1,2,3,4,5\}; \end{gathered} \tag{3.2}$$

It was shown in [7] that (G,φ,ξ,η,g) belongs to the class $\mathcal{F}_7$. Given 6 real numbers λ_i $(i=1,2,3,4)$ and μ_1,μ_2 we define commutators as

$$\begin{aligned} [e_1,e_2] &= -[e_3,e_4] = -\lambda_1 e_1 - \lambda_2 e_2 + \lambda_3 e_3 + \lambda_4 e_4 + 2\mu_1 e_5, \\ [e_1,e_4] &= -[e_2,e_3] = -\lambda_3 e_1 - \lambda_4 e_2 - \lambda_1 e_3 - \lambda_2 e_4 + 2\mu_2 e_5. \end{aligned} \tag{3.3}$$

We then find that our Lie group does not belong to $\mathcal{F}_0$ if and only if $(\mu_1,\mu_2)\neq(0,0)$ holds (see [7]). Our results are the following.

Theorem 3.1. *Let (G,φ,ξ,η,g) be a 5-dimensional connected Lie group with almost contact B-metric structure which belongs to $\mathcal{F}_7$. Then the curvature tensors of the φB-connection $D^{(1)}$ and of the φ-canonical connection $D^{(3)}$ coincide with each other.*

A non-degenerate 2-plane α in the tangent space T_pG at a point $p\in G$ is said to be a φ-holomorphic section if it is invariant under the action of φ (i.e. $\varphi(\alpha)=\alpha$), and is said to be a ξ-section if it contains ξ_p. Sectional curvatures of φ-holomorphic sections (resp. ξ-sections) are said to be φ-holomorphic sectional curvatures (resp. ξ-sectional curvatures).

Theorem 3.2. *Let (G,φ,ξ,η,g) be a 5-dimensional connected Lie group with almost contact B-metric structure which belongs to $\mathcal{F}_7$.*

(1) *The φ-holomorphic sectional curvatures for ∇, $D^{(1)}, D^{(2)}$ and $D^{(3)}$ are zero.*
(2) *The ξ-sectional curvatures for $D^{(1)}, D^{(2)}$ and $D^{(3)}$ are zero.*
(3) *The associated Ricci tensor of ∇ is proportional to the metric g if and only if the following equalities hold:*

$$2(\lambda_1^2+\lambda_2^2-\lambda_3^2-\lambda_4^2)+3(\mu_1^2-\mu_2^2)=2\,(\lambda_1\lambda_3+\lambda_2\lambda_4)+5\mu_1\mu_2=0.$$

(4) *The manifold is scalar flat with respect to all of ∇, $D^{(1)}, D^{(2)}$ and $D^{(3)}$ if and only if the equalities $\lambda_1^2+\lambda_2^2-\lambda_3^2-\lambda_4^2=\mu_1^2-\mu_2^2=0$ hold.*
(5) *The manifold has vanishing associated scalar curvature with respect to ∇ (resp. $D^{(1)}$, $D^{(2)}$, $D^{(3)}$) if and only if the equality $\mu_1\mu_2=\nu(\lambda_1\lambda_3+\lambda_2\lambda_4)$ holds with $\nu=-\frac{4}{5}$ (resp. $\nu=-1$, $\nu=-\frac{1}{2}$, $\nu=-1$).*
(6) *Each of the natural connections $D^{(1)}, D^{(2)}, D^{(3)}$ coincide with the Levi-Civita connection ∇ if and only if our manifold belongs to $\mathcal{F}_0$;*
(7) *The manifold is flat with respect to ∇ if and only if it belongs to $\mathcal{F}_0$ and the equalities $\lambda_1^2+\lambda_2^2-\lambda_3^2-\lambda_4^2=\lambda_1\lambda_3+\lambda_2\lambda_4=0$ hold.*
(8) *For each $a=1,2,3$, the manifold is flat with respect to $D^{(a)}$ if and only if it is flat with respect to ∇.*

(9) *For each* $a = 1, 2, 3$, *the manifold is Ricci-flat with respect to* $D^{(a)}$ (*i.e.* $\rho^{(a)}$ *vanishes*) *if and only if it is associated Ricci flat with respect to* $D^{(a)}$ (*i.e.* ${}^*\rho^{(a)}$ *vanishes*).

(10) *When the manifold belongs to* $\mathcal{F}_0$, *the following conditions are mutually equivalent:*

1) *it is flat with respect to* ∇;
2) *it is Ricci flat with respect to* ∇ (*i.e.* ρ *vanishes*);
3) *it is associated Ricci flat with respect to* ∇ (*i.e.* ${}^*\rho$ *vanishes*);

(11) *The following conditions on the manifold are mutually equivalent:*

1) *it is flat with respect to* ∇;
2) *it is Ricci flat and associated Ricci flat with respect to* ∇;
3) *it is Ricci flat with respect to* ∇ *and* $D^{(2)}$;
4) *it is associated Ricci flat with respect to* ∇ *and* $D^{(2)}$.

4. Proofs of Theorems 3.1 and 3.2

Let $(G, \varphi, \xi, \eta, g)$ be a 5-dimensional connected Lie group with almost contact B-metric structure defined by make use of global basis $\{e_i\}_{i=1}^{5}$ of left-invariant vector fields on G.

First, we study the Levi-Civita connection ∇ of g on G. The following was obtained in [7]:

$$
\begin{aligned}
&\nabla_{e_1} e_1 = -\nabla_{e_3} e_3 = \lambda_1 e_2 - \lambda_3 e_4,\\
&\nabla_{e_1} e_2 = -\nabla_{e_3} e_4 = -\lambda_1 e_1 + \lambda_3 e_3 + \mu_1 e_5,\\
&\nabla_{e_1} e_3 = \nabla_{e_3} e_1 = \lambda_3 e_2 + \lambda_1 e_4,\\
&\nabla_{e_1} e_4 = \nabla_{e_3} e_2 = -\lambda_3 e_1 - \lambda_1 e_3 + \mu_2 e_5,\\
&\nabla_{e_2} e_1 = -\nabla_{e_4} e_3 = \lambda_2 e_2 - \lambda_4 e_4 - \mu_1 e_5,\\
&\nabla_{e_2} e_2 = -\nabla_{e_4} e_4 = -\lambda_2 e_1 + \lambda_4 e_3,\\
&\nabla_{e_2} e_3 = \nabla_{e_4} e_1 = \lambda_4 e_2 + \lambda_2 e_4 - \mu_2 e_5,\\
&\nabla_{e_2} e_4 = \nabla_{e_4} e_2 = -\lambda_4 e_1 - \lambda_2 e_3,\\
&\nabla_{e_1} e_5 = \nabla_{e_5} e_1 = -\mu_1 e_2 + \mu_2 e_4,\\
&\nabla_{e_2} e_5 = \nabla_{e_5} e_2 = \mu_1 e_1 - \mu_2 e_3,\\
&\nabla_{e_3} e_5 = \nabla_{e_5} e_3 = -\mu_2 e_2 - \mu_1 e_4,\\
&\nabla_{e_4} e_5 = \nabla_{e_5} e_4 = \mu_2 e_1 + \mu_1 e_3,\\
&\nabla_{e_5} e_5 = 0.
\end{aligned}
\tag{4.1}
$$

We set $R_{ijk\ell} = g\big(R(e_i, e_j)e_k, e_\ell\big)$ with the curvature tensor R associated with ∇ which is given by $R(X, Y)Z = \nabla_X \nabla_Y Z - \nabla_Y \nabla_X Z - \nabla_{[X,Y]} Z$ for arbitrary vector fields X, Y, Z on G. By (4.1) we can easily check the following:

$$\begin{aligned}
&R_{1212} = R_{3434} = \left(\lambda_1^2 + \lambda_2^2 - \lambda_3^2 - \lambda_4^2\right) + 3\mu_1^2,\\
&R_{1234} = -\left(\lambda_1^2 + \lambda_2^2 - \lambda_3^2 - \lambda_4^2\right) - 2\mu_1^2 + \mu_2^2,\\
&R_{1414} = R_{2323} = -\left(\lambda_1^2 + \lambda_2^2 - \lambda_3^2 - \lambda_4^2\right) + 3\mu_2^2,\\
&R_{1423} = \left(\lambda_1^2 + \lambda_2^2 - \lambda_3^2 - \lambda_4^2\right) + \mu_1^2 - 2\mu_2^2,\\
&R_{1214} = -R_{1223} = R_{2334} = -R_{1434} = 2\left(\lambda_1\lambda_3 + \lambda_2\lambda_4\right) + 3\mu_1\mu_2,\\
&R_{1324} = -\left(\mu_1^2 + \mu_2^2\right),\\
&R_{1535} = R_{2545} = -2\mu_1\mu_2,\\
&R_{1515} = R_{2525} = -R_{3535} = -R_{4545} = -\mu_1^2 + \mu_2^2.
\end{aligned} \tag{4.2}$$

These equalities show that our manifold is flat (i.e. $R_{ijk\ell} = 0$ for all i, j, k, ℓ) if and only if

$$\mu_1 = \mu_2 = 0, \quad \lambda_1^2 + \lambda_2^2 - \lambda_3^2 - \lambda_4^2 = 0, \quad \lambda_1\lambda_3 + \lambda_2\lambda_4 = 0$$

hold. This shows the seventh assertion of Theorem 3.2. By use of (4.2) we can compute sectional curvatures and some other curvatures. We denote by K_{ij} the sectional curvature of the section spanned by e_i, e_j $(i \neq j)$, that is $K_{ij} = \epsilon_i\epsilon_j g\big(R(e_i, e_j)e_j, e_i\big)$, where $\epsilon_k = g(e_k, e_k)$. We here note that φ-holomorphic sections are sections spanned by e_1, e_3 and by e_2, e_4, and that ξ-sections are sections containing e_5. By (4.2) we have

$$\begin{aligned}
&K_{12} = K_{34} = -\left(\lambda_1^2 + \lambda_2^2 - \lambda_3^2 - \lambda_4^2\right) - 3\mu_1^2,\\
&K_{14} = K_{23} = -\left(\lambda_1^2 + \lambda_2^2 - \lambda_3^2 - \lambda_4^2\right) + 3\mu_2^2;\\
&K_{13} = K_{24} = 0;\\
&K_{51} = K_{52} = K_{53} = K_{54} = \mu_1^2 - \mu_2^2.
\end{aligned} \tag{4.3}$$

In particular, we find that φ-holomorphic sectional curvatures for ∇ are zero.

We denote by $\rho_{ij} = \sum_{k=1}^{5} \epsilon_k g\big(R(e_k, e_i)e_j, e_k\big)$ the components of the Ricci tensor, and by $\tau = \sum_{k=1}^{5} \epsilon_k \rho(e_k, e_k)$ the scalar curvature. We then have by (4.2) that

$$\begin{aligned}
&\rho_{11} = \rho_{22} = -\rho_{33} = -\rho_{44} = -2\left(\lambda_1^2 + \lambda_2^2 - \lambda_3^2 - \lambda_4^2\right) - 2\left(\mu_1^2 - \mu_2^2\right),\\
&\rho_{13} = \rho_{24} = -4\left(\lambda_1\lambda_3 + \lambda_2\lambda_4\right) - 4\mu_1\mu_2,\\
&\rho_{55} = 4\left(\mu_1^2 - \mu_2^2\right);
\end{aligned} \tag{4.4}$$

$$\tau = -8\left(\lambda_1^2 + \lambda_2^2 - \lambda_3^2 - \lambda_4^2\right) - 4\left(\mu_1^2 - \mu_2^2\right), \tag{4.5}$$

and that the other components of ρ are zero. The associated Ricci tensor $^*\rho$ with respect to ∇ is defined by $^*\rho(x, y) = \sum_{i=1}^{5} \epsilon_k g\big(R(e_i, x)y, \varphi e_i\big)$ and the associated scalar curvature is given by ${}^*\tau = \sum_{i=1}^{5} \epsilon_k {}^*\rho(e_i, e_i)$. We set ${}^*\rho_{ij} = {}^*\rho(e_i, e_j)$. We note that the associated Ricci tensor and the associated scalar curvature are usually denoted by ρ^* and τ^*, respectively. But as we consider three more connections $D^{(a)}$ $(a = 1, 2, 3)$ and study their corresponding tensors and curvatures, we use ${}^*\rho$ and ${}^*\tau$ instead in this paper. By (4.2) we have

$$\begin{aligned} &{}^*\rho_{11} = {}^*\rho_{22} = -{}^*\rho_{33} = -{}^*\rho_{44} = -4\left(\lambda_1\lambda_3 + \lambda_2\lambda_4\right) - 6\mu_1\mu_2, \\ &{}^*\rho_{55} = 4\mu_1\mu_2, \qquad (4.6) \\ &{}^*\rho_{13} = {}^*\rho_{24} = {}^*\rho_{31} = {}^*\rho_{42} = 2\left(\lambda_1^2 + \lambda_2^2 - \lambda_3^2 - \lambda_4^2\right) + 3\left(\mu_1^2 - \mu_2^2\right); \\ &\quad {}^*\tau = -16\left(\lambda_1\lambda_3 + \lambda_2\lambda_4\right) - 20\mu_1\mu_2, \qquad (4.7) \end{aligned}$$

and the other components of ${}^*\rho$ are zero.

We here study the third and fifth assertions of Theorem 3.2. The fifth assertion on ∇ is a direct consequence of (4.7). If we set $g_{ij} = g(e_i, e_j)$, we see by (3.2) that $g_{11} = g_{22} = -g_{33} = -g_{44} = g_{55} = 1$ and others are zero. Thus, the associated Ricci tensor ${}^*\rho$ of ∇ is proportional to g if and only if

$$\begin{aligned} &2\mu_1\mu_2 = \kappa, \quad 2(\lambda_1\lambda_3 + \lambda_2\lambda_4) + 3\mu_1\mu_2 = -\kappa, \\ &2(\lambda_1^2 + \lambda_2^2 - \lambda_3^2 - \lambda_4^2) + 3(\mu_1^2 - \mu_2^2) = 0, \end{aligned}$$

hold with some constant κ. We hence get the third assertion of Theorem 3.2.

We next study the connection $D^{(2)}$. By the definition (2.3) of $T^{(2)}$ and by (3.1), (4.1), we can easily compute the component $T^{(2)}_{ijk} = g\big(T^{(2)}(e_i, e_j), e_k\big)$ of the torsion $T^{(2)}$ and get

$$\begin{aligned} T^{(2)}_{125} = -T^{(2)}_{215} = -T^{(2)}_{345} = T^{(2)}_{435} = -2\mu_1, \\ T^{(2)}_{145} = -T^{(2)}_{235} = T^{(2)}_{325} = -T^{(2)}_{415} = -2\mu_2. \end{aligned} \qquad (4.8)$$

The other components of $T^{(2)}$ are zero. Since $D^{(2)} = \nabla + T^{(2)}$, we find that $D^{(2)}$ coincides with ∇ if and only if $\mu_1 = \mu_2 = 0$, which is the case that our manifold belongs to $\mathcal{F}_0$.

We denote by $R^{(a)}_{ijk\ell}, K^{(a)}_{ij}, \rho^{(a)}_{ij}, \tau^{(a)}, {}^*\rho^{(a)}$ and ${}^*\tau^{(a)}$ the corresponding amounts for the connection $D^{(a)}$ $(a = 1, 2, 3)$. By (4.1), (4.8) and (3.3), we have

$$\begin{aligned} &R^{(2)}_{1212} = -R^{(2)}_{1234} = R^{(2)}_{3434} = \left(\lambda_1^2 + \lambda_2^2 - \lambda_3^2 - \lambda_4^2\right) + 4\mu_1^2, \\ &R^{(2)}_{1414} = -R^{(2)}_{1423} = R^{(2)}_{2323} = -\left(\lambda_1^2 + \lambda_2^2 - \lambda_3^2 - \lambda_4^2\right) + 4\mu_2^2, \qquad (4.9) \\ &R^{(2)}_{1214} = -R^{(2)}_{1223} = R^{(2)}_{2334} = -R^{(2)}_{1434} = 2\left(\lambda_1\lambda_3 + \lambda_2\lambda_4\right) + 4\mu_1\mu_2, \end{aligned}$$

and the other components are zero. These equalities in (4.9) show that our manifold is flat with respect to $D^{(2)}$ if and only if

$$\mu_1 = \mu_2 = 0, \quad \lambda_1^2 + \lambda_2^2 - \lambda_3^2 - \lambda_4^2 = 0, \quad \lambda_1\lambda_3 + \lambda_2\lambda_4 = 0$$

hold. Hence it is flat with respect to $D^{(2)}$ if and only if it is flat with respect to ∇. By (4.9) we obtain

$$\begin{aligned} K_{12}^{(2)} &= K_{34}^{(2)} = -\left(\lambda_1^2 + \lambda_2^2 - \lambda_3^2 - \lambda_4^2\right) - 4\mu_1^2, \\ K_{14}^{(2)} &= K_{23}^{(2)} = -\left(\lambda_1^2 + \lambda_2^2 - \lambda_3^2 - \lambda_4^2\right) + 4\mu_2^2; \\ K_{13}^{(2)} &= K_{24}^{(2)} = K_{51}^{(2)} = K_{52}^{(2)} = K_{53}^{(2)} = K_{54}^{(2)} = 0. \end{aligned} \tag{4.10}$$

We therefore find that φ-holomorphic sectional curvatures and ζ-curvatures for $D^{(2)}$ are zero. Also we have

$$\begin{aligned} \rho_{11}^{(2)} &= \rho_{22}^{(2)} = -\rho_{33}^{(2)} = -\rho_{44}^{(2)} = -2\left(\lambda_1^2+\lambda_2^2-\lambda_3^2-\lambda_4^2\right) - 4\left(\mu_1^2-\mu_2^2\right), \\ \rho_{13}^{(2)} &= \rho_{24}^{(2)} = -4\left(\lambda_1\lambda_3 + \lambda_2\lambda_4\right) - 8\mu_1\mu_2; \end{aligned} \tag{4.11}$$

$$\tau^{(2)} = -8\left(\lambda_1^2 + \lambda_2^2 - \lambda_3^2 - \lambda_4^2\right) - 16\left(\mu_1^2 - \mu_2^2\right). \tag{4.12}$$

Similarly we have

$$\begin{aligned} {}^*\rho_{11}^{(2)} &= {}^*\rho_{22}^{(2)} = -{}^*\rho_{33}^{(2)} = -{}^*\rho_{44}^{(2)} = -4\left(\lambda_1\lambda_3 + \lambda_2\lambda_4\right) - 8\mu_1\mu_2, \\ {}^*\rho_{13}^{(2)} &= {}^*\rho_{24}^{(2)} = {}^*\rho_{31}^{(2)} = {}^*\rho_{42}^{(2)} = 2\left(\lambda_1^2 + \lambda_2^2 - \lambda_3^2 - \lambda_4^2\right) + 4\left(\mu_1^2 - \mu_2^2\right); \end{aligned} \tag{4.13}$$

$${}^*\tau^{(2)} = -16\left(\lambda_1\lambda_3 + \lambda_2\lambda_4\right) - 32\mu_1\mu_2, \tag{4.14}$$

and the other components of $\rho^{(2)}$ and ${}^*\rho^{(2)}$. Thus we find that $\rho^{(2)}$ vanishes if and only if ${}^*\rho^{(2)}$ vanishes.

In the third place, we study the connection $D^{(1)}$. By the definition (2.2) of $T^{(1)}$ we can easily get the component $T_{ijk}^{(1)} = g\big(T^{(1)}(e_i, e_j), e_k\big)$ of the torsion $T^{(1)}$ as follows:

$$\begin{aligned} T_{125}^{(1)} &= -T_{215}^{(1)} = 2T_{251}^{(1)} = -T_{345}^{(1)} \\ &= 2T_{354}^{(1)} = T_{435}^{(1)} = -2T_{521}^{(1)} = -2T_{534}^{(1)} = -2\mu_1, \\ T_{145}^{(1)} &= -T_{235}^{(1)} = 2T_{253}^{(1)} = T_{325}^{(1)} \\ &= -T_{415}^{(1)} = 2T_{451}^{(1)} = -2T_{523}^{(1)} = -2T_{541}^{(1)} = -2\mu_2, \end{aligned} \tag{4.15}$$

and the other components are zero. Since we have $D^{(1)} = \nabla + T^{(1)}$, we find the following by make use of (3.3):

$$\begin{aligned} R_{1212}^{(1)} &= -R_{1234}^{(1)} = R_{3434}^{(1)} = \left(\lambda_1^2 + \lambda_2^2 - \lambda_3^2 - \lambda_4^2\right) + 2\mu_1^2, \\ R_{1414}^{(1)} &= -R_{1423}^{(1)} = R_{2323}^{(1)} = -\left(\lambda_1^2 + \lambda_2^2 - \lambda_3^2 - \lambda_4^2\right) + 2\mu_2^2, \\ R_{1214}^{(1)} &= -R_{1223}^{(1)} = -R_{1434}^{(1)} = R_{2334}^{(1)} = 2\left(\lambda_1\lambda_3 + \lambda_2\lambda_4\right) + 2\mu_1\mu_2, \end{aligned} \tag{4.16}$$

and the other components are zero. These equalities show that our manifold is flat with respect to $D^{(1)}$ if and only if

$$\mu_1 = \mu_2 = 0, \quad \lambda_1^2 + \lambda_2^2 - \lambda_3^2 - \lambda_4^2 = 0, \quad \lambda_1\lambda_3 + \lambda_2\lambda_4 = 0$$

hold. Hence it is flat with respect to $D^{(1)}$ if and only if it is flat with respect to ∇. As a direct consequence of (4.16) we have

$$\begin{aligned} K_{12}^{(1)} &= K_{34}^{(1)} = -\left(\lambda_1^2 + \lambda_2^2 - \lambda_3^2 - \lambda_4^2\right) - 2\mu_1^2, \\ K_{14}^{(1)} &= K_{23}^{(1)} = -\left(\lambda_1^2 + \lambda_2^2 - \lambda_3^2 - \lambda_4^2\right) + 2\mu_2^2; \\ K_{13}^{(1)} &= K_{24}^{(1)} = K_{51}^{(1)} = K_{52}^{(1)} = K_{53}^{(1)} = K_{54}^{(1)} = 0. \end{aligned} \tag{4.17}$$

We hence find that φ-holomorphic sectional curvatures and ξ-curvatures for $D^{(1)}$ are zero. Also we have

$$\begin{aligned} \rho_{11}^{(1)} &= \rho_{22}^{(1)} = -\rho_{33}^{(1)} = -\rho_{44}^{(1)} = -{}^*\rho_{13}^{(1)} = -{}^*\rho_{24}^{(1)} = -{}^*\rho_{31}^{(1)} = -{}^*\rho_{42}^{(1)} \\ &= -2\left(\lambda_1^2 + \lambda_2^2 - \lambda_3^2 - \lambda_4^2\right) - 2\left(\mu_1^2 - \mu_2^2\right), \\ \rho_{13}^{(1)} &= \rho_{24}^{(1)} = \rho_{31}^{(1)} = \rho_{42}^{(1)} = {}^*\rho_{11}^{(1)} = {}^*\rho_{22}^{(1)} = -{}^*\rho_{33}^{(1)} = -{}^*\rho_{44}^{(1)} \\ &= -4\left(\lambda_1\lambda_3 + \lambda_2\lambda_4\right) - 4\mu_1\mu_2; \end{aligned} \tag{4.18}$$

$$\tau^{(1)} = -8\left(\lambda_1^2 + \lambda_2^2 - \lambda_3^2 - \lambda_4^2\right) - 8\left(\mu_1^2 - \mu_2^2\right), \tag{4.19}$$

$${}^*\tau^{(1)} = -16\left(\lambda_1\lambda_3 + \lambda_2\lambda_4\right) - 16\mu_1\mu_2, \tag{4.20}$$

and the other components of $\rho^{(1)}$ and ${}^*\rho^{(1)}$ are zero. In particular, $\rho^{(1)}$ vanishes if and only if ${}^*\rho^{(1)}$ vanishes.

We finally study the connection $D^{(3)}$. By the definition (2.4) of $T^{(3)}$ we can easily compute its component $T_{ijk}^{(3)}$ and get

$$\begin{aligned} T_{125}^{(3)} &= -T_{215}^{(3)} = T_{251}^{(3)} = -T_{345}^{(3)} \\ &= T_{354}^{(3)} = T_{435}^{(3)} = -T_{521}^{(3)} = -T_{534}^{(3)} = -2\mu_1, \\ T_{145}^{(3)} &= -T_{235}^{(3)} = T_{253}^{(3)} = T_{325}^{(3)} \\ &= -T_{415}^{(3)} = T_{451}^{(3)} = -T_{523}^{(3)} = -T_{541}^{(3)} = -2\mu_2, \end{aligned} \tag{4.21}$$

and the other components are zero. Thus we find $D_{e_i}^{(3)} e_5 = D_{e_5}^{(3)} e_i = 0$ for $i = 1, \ldots, 5$. Moreover, by comparing (4.8) and (4.15) we find that

$$D_{e_i}^{(3)} e_j = D_{e_i}^{(1)} e_j, \quad i, j \in \{1, 2, 3, 4\}.$$

These lead us to

$$R_{ijk\ell}^{(1)} = R_{ijk\ell}^{(3)}, \quad i, j, k, \ell \in \{1, 2, 3, 4, 5\},$$

which shows Theorem 3.1. Also, we get the assertions on $D^{(3)}$ in Theorem 3.2 with the aid of the assertions on $D^{(1)}$.

We here show the tenth and eleventh assertions of Theorem 3.2. We recall that our manifold is flat if and only if

$$\mu_1 = \mu_2 = 0, \quad \lambda_1^2 + \lambda_2^2 - \lambda_3^2 - \lambda_4^2 = 0, \quad \lambda_1\lambda_3 + \lambda_2\lambda_4 = 0$$

hold. In view of our computations (4.4), (4.6), (4.11), (4.13) and (4.18) on its Ricci tensor and its associated Ricci tensor, we find that the key is to show $\mu_1 = \mu_2 = 0$, in other words, to show that our manifold belongs to $\mathcal{F}_0$. Thus we get the tenth assertion by first two of these relations, and get the eleventh assertion by using (4.4) and (4.6), (4,4) and (4.11), or (4.6) and (4.13).

At last we study the scalar curvatures of ∇ and $D^{(a)}$. By (4.5), (4.12) and (4.19) we find that if $\lambda_1^2 + \lambda_2^2 - \lambda_3^2 - \lambda_4^2 = \mu_1^2 - \mu_2^2 = 0$ hold then our manifold is scalar flat with respect to all of $\nabla, D^{(1)}, D^{(2)}$ and $D^{(3)}$. On the other hand if our manifold is scalar flat with respect to two of $\nabla, D^{(1)}$ and $D^{(2)}$, then these two equalities hold. We hence get the fourth assertion in Theorem 3.2. This complete the proof of Theorem 3.2.

5. Isotropic-$\mathcal{F}_0$-manifolds

Let $(M, \varphi, \xi, \eta, g)$ be an almost contact B-metric manifold. We take a basis $\{e_i\}_{i=1}^{2n+1}$ of the tangent space T_pM of M at an arbitrary point $p \in M$ which satisfies $g(e_i, e_i) = -g(e_{i+1}, e_{i+1}) = 1$ for $i = 1, \ldots, n$, $g(e_{2n+1}, e_{2n+1}) = 1$ and $g(e_j, e_k) = 0$ for $j \neq k$. We define the *square norm* of $\nabla\varphi$ by

$$\|\nabla\varphi\|^2 = \sum_{i,j-1}^{2n+1} \epsilon_i\epsilon_j g\big((\nabla_{e_i}\varphi)\, e_j, (\nabla_{e_i}\varphi)\, e_j\big),$$

where $\epsilon_k = g(e_k, e_k)$. It is clear that $\|\nabla\varphi\|^2 = 0$ holds if $(M, \varphi, \xi, \eta, g)$ is in the class $\mathcal{F}_0$ but the converse does not hold in general. We say an almost contact B-metric manifold satisfying $\|\nabla\varphi\|^2 = 0$ to be an *isotropic-$\mathcal{F}_0$-manifold* ([7]). We shall call a tensor S isotropic if its square norm defined as above vanishes. The Nijenhuis tensor N of the almost contact structure is defined by $N = [\varphi, \varphi] + \mathrm{d}\eta \otimes \xi$, where $[\varphi, \varphi](x, y) = [\varphi x, \varphi y] + \varphi^2\,[x, y] - \varphi\,[\varphi x, y] - \varphi\,[x, \varphi y]$ and $\mathrm{d}\eta$ is the exterior derivative of η. Thus, N is expressed as

$$\begin{aligned} N(x, y) = {} & (\nabla_{\varphi x}\varphi)\, y - \varphi\,(\nabla_x\varphi)\, y + (\nabla_x\eta)\, y \cdot \xi \\ & - (\nabla_{\varphi y}\varphi)\, x + \varphi\,(\nabla_y\varphi)\, x - (\nabla_y\eta)\, x \cdot \xi. \end{aligned}$$

In this section we study an isotropic properties for 5-dimensional connected Lie groups.

Theorem 5.1. *Let $(G, \varphi, \xi, \eta, g)$ be a 5-dimensional connected Lie group with almost contact B-metric structure which belongs to $\mathcal{F}_7$. The following conditions are mutually equivalent:*

1) *Our manifold is an isotropic-$\mathcal{F}_0$-manifold;*
2) *The scalar curvatures for ∇, $D^{(1)}, D^{(2)}, D^{(3)}$ coincide with each other;*
3) *One of the vectors $\nabla_{e_i}\xi$ $(i = 1, 2, 3, 4)$ is isotropic;*
4) *All the vectors $\nabla_{e_i}\xi$ $(i = 1, 2, 3, 4)$ are isotropic;*
5) *The Nijenhuis tensor N is isotropic;*
6) *One of the torsion tensors $T^{(1)}, T^{(2)}, T^{(3)}$ for $D^{(1)}, D^{(2)}, D^{(3)}$ is isotropic;*
7) *All the torsion tensors $T^{(1)}, T^{(2)}, T^{(3)}$ for $D^{(1)}, D^{(2)}, D^{(3)}$ are isotropic;*
8) *The sectional curvatures of the ξ-sections with respect to ∇ vanish;*
9) *The equality $\mu_1 = \pm\mu_2$ holds.*

Proof. We take the basis $\{e_i\}_{i=1}^5$ as in §3. We first compute the square norm of $\nabla\varphi$. By (3.1) we have

$$\|\nabla\varphi\|^2 = -8(\mu_1^2 - \mu_2^2)$$

(see [7]). Therefore we find that our manifold is an isotropic-$\mathcal{F}_0$-manifold if and only if $\mu_1^2 = \mu_2^2$ holds.

If we compute $\|\nabla_{e_i}e_5\|^2 = g\big(\nabla_{e_i}e_5, \nabla_{e_i}e_5\big)$ by using (4.1) we have

$$\|\nabla_{e_1}e_5\|^2 = \|\nabla_{e_2}e_5\|^2 = \mu_1^2 - \mu_2^2 = -\|\nabla_{e_3}e_5\|^2 = -\|\nabla_{e_4}e_5\|^2.$$

Thus we see that $\nabla_{e_i}e_5$ is isotropic for some $i = 1, 2, 3, 4$ hence all of them are isotropic if and only if G is an isotropic-$\mathcal{F}_0$-manifold.

Next we compute the Nijenhuis tensor. Since $(G, \varphi, \xi, \eta, g)$ belongs to $\mathcal{F}_7$, we find that the Nijenhuis tensor is given as

$$N(x, y) = 4\,(\nabla_x\eta)\,y\cdot\xi.$$

We therefore get by (4.1) and (3.1) that the components $N_{ij} = N(e_i, e_j)$ of N is given as

$$N_{12} = -N_{21} = -N_{34} = N_{43} = -4\mu_1\xi,$$
$$N_{14} = -N_{23} = N_{32} = -N_{41} = -4\mu_2\xi,$$

and obtain

$$\|N\|^2 := \sum_{i,j=1}^{5} \epsilon_i\epsilon_j g(N_{ij}, N_{ij}) = 64(\mu_1^2 - \mu_2^2).$$

We hence find that the Nijenhuis tensor is isotropic if and only if G is an isotropic-$\mathcal{F}_0$-manifold.

In the third place, we compute the square norm $\|T^{(a)}\|^2 := \sum_{i,j=1}^{5} \epsilon_i\epsilon_j g\big(T_{ij}^{(a)}, T_{ij}^{(a)}\big)$ of the torsion tensor $T^{(a)}$ ($a = 1, 2, 3$). By (4.8), (4.15) and (4.21) we get

$$\|T^{(1)}\| = 20(\mu_1^2 - \mu_2^2), \quad \|T^{(2)}\| = 16(\mu_1^2 - \mu_2^2), \quad \|T^{(3)}\| = 32(\mu_1^2 - \mu_2^2).$$

Thus we find that one of $T^{(a)}$ is isotropic hence all of $T^{(a)}$ are isotropic if and only if G is an isotropic-$\mathcal{F}_0$-manifold.

The equivalency of the first and the eighth assertions is the direct consequence of (4.2). We hence get the conclusion. □

At last we make mention of the classes $\mathcal{F}_3$ and $\mathcal{F}_7$ a bit more. The class $\mathcal{F}_3 \oplus \mathcal{F}_7$ is characterized by the property of admitting vanishing associated Nijenhuis tensor $\widehat{N}$ which is given by $\widehat{N} = \{\varphi, \varphi\} + (\mathcal{L}_\xi g) \otimes \xi$ or by

$$\begin{aligned}\widehat{N}(x, y) = (\nabla_{\varphi x}\varphi)\, y - \varphi\, (\nabla_x \varphi)\, y + (\nabla_x \eta)\, y \cdot \xi \\ + (\nabla_{\varphi y}\varphi)\, x - \varphi\, (\nabla_y \varphi)\, x + (\nabla_y \eta)\, x \cdot \xi\end{aligned}$$

(see [12]). The basic classes $\mathcal{F}_3$ and $\mathcal{F}_7$ are the horizontal and the vertical components of $\mathcal{F}_3 \oplus \mathcal{F}_7$, respectively. The contact distribution $\ker(\eta)$ of an arbitrary $\mathcal{F}_3$-manifold is an almost complex manifold with Norden metric belonging to the class $\mathcal{W}_3$. Here, the class $\mathcal{W}_3$ is a class of quasi-Kähler manifolds with Norden metric. Relations between curvature tensors and natural connections on manifolds in the class $\mathcal{W}_3$ are well-known (see [13], for example).

References

1. V. Alexiev and G. Ganchev, Canonical connection on a conformal almost contact metric manifolds, *Ann. Univ. Sofia Fac. Math. Inform.* **81**, 1 (1987) 29–38.
2. G. Ganchev, K. Gribachev and V. Mihova, B-connections and their conformal invariants on conformally Kaehler manifolds with B-metric, *Publ. Inst. Math., Beograd (N.S.)* **42**, 56 (1987) 107–121.
3. G. Ganchev and V. Mihova, Canonical connection and the canonical conformal group on an almost complex manifold with B-metric, *Ann. Univ. Sofia Fac. Math. Inform.* **81**, (1987) 195–206.
4. G. Ganchev, V. Mihova and K. Gribachev, Almost contact manifolds with B-metric, *Math. Balkanica (N.S.)* **7**, (1993) 261–276.

5. S. Kobayshi and K. Nomizu, Foundations of differential geometry, *Intersc. Publ., New York* **1, 2**, 1963 (1969).
6. M. Manev, Almost contact B-metric hypersurfaces of Kaehlerian manifolds with B-metric, In: Perspectives of Complex analysis, Differential Geometry and Mathematical Physics, Eds. St. Dimiev and K. Sekigawa, World Sci. Publ., Singapore, 2001, pp. 159–170.
7. M. Manev, A connection with totally skew-symmetric torsion on almost contact manifolds with B-metric, *Int. J. Geom. Methods Mod. Phys.* **9**, 5 (2012).
8. M. Manev, Curvature properties on some classes of almost contact manifolds with B-metric, *C. R. Acad. Bulg. Sci.* **65**, 3 (2012) 283–290.
9. M. Manev and K. Gribachev, Conformally invariant tensors on almost contact manifolds with B-metric, *Serdica Math. J.* **20**, (1994) 133–147.
10. M. Manev and M. Ivanova, A natural connection on some classes of almost contact manifolds with B-metric, *C. R. Acad. Bulg. Sci.* **65**, 4 (2012) 429–436.
11. M. Manev and M. Ivanova, Canonical-type connection on almost contact manifolds with B-metric, *Ann. Global Anal. Geom.* **43**, 4 (2013) 397–408.
12. M. Manev and M. Ivanova, A classification of the torsions on almost contact manifolds with B-metric, *Cent. Eur. J. Math.* **12**, 10 (2014) 1416–1432.
13. D. Mekerov and M. Manev, On the geometry of quasi-Kähler manifolds with Norden metric, *Nihonkai Math. J.* **16**, 2 (2005) 89–93.

Received October 18, 2014
Revised March 30, 2015

Current Developments in Differential Geometry and its Related Fields pp. 129 – 138

ON HYPERELLIPTIC MINIMAL SURFACES WITH EVEN GENUS

Norio EJIRI *

Department of Mathematics, Meijo University, Tempaku, Nagoya, 468-8502 Japan
E-mail: ejiri@meijo-u.ac.jp

Toshihiro SHODA †

Faculty of Culture and Education, Saga University, 1 Honzyo, Saga 840-8502 Japan
E-mail: tshoda@cc.saga-u.ac.jp

In this paper, we announce our results for compact oriented minimal surfaces in flat 4-tori which are given in our preprint [7]. We consider (i) the Morse index of a one-parameter family, (ii) the existence of non-holomorphic hyperelliptic minimal surfaces with even genus. Key tools are established by the first author [3,4], and we continue the work related to our previous papers [5,6] for the higher genus and codimensional case. Our key tools can be applied for the higher genus and codimensional case, but the situation is different from that of the previous paper. It is crucial to solve such difficulties, and we give a partial answer for it. We will give a clear explanation about the long story.

Keywords: Hyperelliptic minimal surfaces, flat tori, Morse index.

1. Introduction

Let $f : M \to \mathbb{R}^n/\Lambda$ be a minimal immersion of a compact oriented surface of genus γ into a flat n-torus. The isothermal coordinates of M give rise to a complex structure on M, and we call f a *conformal minimal immersion*. For a conformal minimal immersion, the following is well-known:

Theorem 1.1 (Weierstrass representation formula). *Let $f : M \to \mathbb{R}^n/\Lambda$ be a conformal minimal immersion of a compact Riemann surface of genus γ into a flat n-torus. Then, up to translations, f can be represented by the following path-integrals*:

$$f(p) \equiv \Re \int_{p_0}^{p} {}^t(\omega_1, \cdots, \omega_n) \mod \Lambda,$$

*Partially supported by JSPS Grant-in-Aid for Scientific Research (C) 22540103.
†Partially supported by JSPS Grant-in-Aid for Young Scientists (B) 24740047.

where p_0 is a fixed point on M and the ω_i's are holomorphic differentials on M satisfying the following three conditions.

$$\omega_1^2 + \cdots + \omega_n^2 = 0, \tag{1}$$

$$\omega_1, \cdots, \omega_n \text{ have no common zeros,} \tag{2}$$

$$\left\{ \Re \int_C {}^t(\omega_1, \cdots, \omega_n) \,\middle|\, C \in H_1(M, \mathbb{Z}) \right\} \text{ is a sublattice of } \Lambda. \tag{3}$$

Conversely, the real part of path-integrals of holomorphic differentials satisfying the above three conditions defines a conformal minimal immersion.

We now introduce the complex hyperquadric

$$Q_{n-2} = \{[z_1, \cdots, z_n] \in \mathbb{C}P^{n-1} \mid z_1^2 + \cdots + z_n^2 = 0\}.$$

For $\{\omega_i\}_{i=1}^n$ which satisfies (1) and (2), we can define a holomorphic map

$$M \ni p \to [\omega_1, \cdots, \omega_n]_p \in Q_{n-2}. \tag{4}$$

Throughout this paper, we call the map (4) the *Gauss map* of M.

We recall the second variation of the area function given by J. Simons [15]. For a minimal immersion $f : M \to \mathbb{R}^n/\Lambda$, let ∇ be the normal connection, $\triangle$ the Laplacian given by ∇. Let A be the shape operator of f, and we can construct ${}^tA$, the transpose of A by the L^2-norm on the normal bundle. Then the second variation of the area function is obtained as follows.

$$\int_M \{|\nabla x|^2 - |Ax|^2\} dv = -\int_M \langle\{\triangle + ({}^tA \circ A)\}x, x\rangle dv, \tag{5}$$

where x is a normal vector field of f. $\triangle + ({}^tA \circ A)$ is a self-adjoint strongly elliptic differential operator and we call it the *Jacobi operator*. Also, a number λ is called an *eigenvalue* of $\triangle + ({}^tA \circ A)$ if there exists a non-zero normal vector field x satisfying $(\triangle + ({}^tA \circ A) + \lambda)x = 0$.

The Morse index (resp. the nullity) of a compact oriented minimal surface in a flat torus is defined as the number of negative eigenvalues (resp. zero eigenvalue) of the Jacobi operator of the area function, counted with multiplicity. It is well-known that translations in $\mathbb{R}^n/\Lambda$ (the Killing vector fields) give rise to the nullity, and so the nullity is greater than or equal to n. We call translations in $\mathbb{T}^n$ *trivial Jacobi fields*, and the nullity is equal to n if and only if a minimal surface has only trivial Jacobi fields. Note that the Morse index and the nullity can be defined for the Gauss map of M, and the Morse index (resp. the nullity) for the Gauss map coincides with the Morse index (resp. the nullity) for a minimal surface. In fact, if we write $\omega_i = f_i(z)dz$ for a local complex coordinate z on M and substitute $\Re^t(f_1, \cdots, f_n)$, $\Im^t(f_1, \cdots, f_n)$ to (5) as tangent vectors, then we can apply the same arguments for the Gauss map.

In the previous work [5], we computed the Morse indices and the nullities for some one-parameter families of compact oriented minimal surfaces of genus 3 in flat 3-tori. We continue this work in the higher genus and codimensional case, namely the case $\gamma = n = 4$. As related topics, Simons [15] gave the lower bound of the Morse index for a submanifold in S^n, in particular, he computed the Morse index for a totally geodesic sphere in S^n. But there are few other works for this problem. One of our main results is as follows.

Theorem 1.2. *For $0 < a < 1$, let M be a hyperelliptic Riemann surface of genus 4 defined by $w^2 = z(z^4 + a^4)(z^4 + \frac{1}{a^4})$. Set*

$$f(p) = \Re \int_{p_0}^{p} {}^t(1 - z^3,\, i\,(1 + z^3),\, z^2 + z,\, i\,(z^2 - z))\frac{dz}{w}$$

on M. Then, there exist $0 < a_1 < a_2 < a_3 < a_4 < 1$ and a discrete set $S \subset (0, a_2) \cup (a_3, 1)$ satisfying the following properties:

(i) *f has Morse index 5 and nullity 4 for $a \in (0, a_1) \cup (a_4, 1) \setminus S$;*
(ii) *f has Morse index 3 and nullity 4 for $a \in (a_1, a_2) \cup (a_3, a_4) \setminus S$;*
(iii) *f has Morse index 2 and nullity 4 for $a \in (a_2, a_3)$.*

Note that f in Theorem 1.2 might not be well-defined, but we can consider the Morse index of f as seen above. Main tools to prove Theorem 1.2 are (a) an algorithm to compute the Morse index and the nullity which is given by the first author [3,4] (see § 2 and the previous paper [6]), (b) periods of the Abelian differentials of the second kind, that is, meromorphic differentials with zero residues on a minimal surface. The situation is quite different from the genus 3 case and we solve the difficulties by Micallef and Wolfson's technique [11].

Next we refer to the existence of hyperelliptic minimal surfaces. It is well-known that a hyperelliptic Riemann surface of even genus cannnot be minimally immersed into any flat 3-torus (see Theorem 3.3 in Meeks' paper [10] or Nagano-Smyth's paper [12]). On the other hand, there are many examples of hyperelliptic minimal surfaces of odd genus in flat 3-tori [8,9]. For the higher codimensional case, C. Arezzo and G. P. Pirola [2] proved the existence of hyperelliptic minimal surfaces with odd genus in flat n-tori ($n \geq 4$). Their procedure is as follows. First we consider hyperelliptic minimal surfaces in flat 3-tori with only trivial Jacobi fields, and deform them as minimal surfaces in flat n-tori. From the above topological obstruction for hyperelliptic minimal surfaces in flat 3-tori, we have to restrict the odd genus case. So we focus on the existence of hyperelliptic minimal surfaces for the even genus case. In particular, we treat the simplest case, that is, the case $n = 4$. Recall that every conformal minimal immersion of a hyperelliptic Riemann surface of genus 2 is holomorphic with respect to some complex structure compatible with the metric on the torus [1,3]. Thus we assume that genus is greater than or equal to 4, and moreover, we only treat non-holomorphic minimal immersions. Our another main result is as follows.

Theorem 1.3. *There exist countably many non-holomorphic hyperelliptic minimal surfaces of genus* 4 *in flat* 4*-tori.*

Remark 1.1. If one is dealing with a compact minimal surface in a flat n-torus, then the set of compact minimal surfaces, parameterized by homotheties and $O(n)$, constitute the collection of comact minimal surfaces in flat n-tori. Our construction in Theorem 1.3 is distinct from such a construction.

2. Outline of an algorithm

In this section, we shall introduce a part of theory given by the first author [3,4]. It includes two subjects: (i) an algorithm to compute the Morse index of a minimal surface, (ii) the existence theorem of compact minimal surfaces in flat n-tori. We fix an integer $\gamma \geq 3$ and assume that a minimal surface has genus γ.

Applying R. Schoen and S. T. Yau's theory [14] to a fixed flat n-torus $\mathbb{T}^n$, we can define an energy $E_{\mathbb{T}^n}$ on the Teichmüller space $\mathcal{T}$ (A choice of the energy $E_{\mathbb{T}^n}$ depends on an action on the fundamental group but we will skip the details). First we can translate a compact minimal surface in $\mathbb{T}^n$ into a critical point of $E_{\mathbb{T}^n}$ as follows.

Theorem 2.1 ([13,14]). *Every critical point of $E_{\mathbb{T}^n}$ is one to one corresponding to a compact oriented minimal surface in $\mathbb{T}^n$, up to translations in $\mathbb{T}^n$.*

Recall that the Teichmüller space $\mathcal{T}$ is diffeomorphic to $\mathbb{R}^{6\gamma-6}$. Thus we can define the Morse index and the nullity for $E_{\mathbb{T}^n}$ by the Hessian. Let $index_a$ (resp. $nullity_a$) denote the Morse index (resp. the nullity) of a minimal surface, $index_E$ (resp. $nullity_E$) the Morse index (resp. the nullity) of $E_{\mathbb{T}^n}$. The following result is a first step to compute the Morse index and the nullity of a minimal surface [3].

Theorem 2.2. *At a critical point of* $E_{\mathbb{T}^n}$,

$$index_E = index_a, \quad nullity_E = nullity_a - n.$$

Straightforward applications of Theorem 2.2 yields

Corollary 2.1. *The following are equivalent.*

(1) $E_{\mathbb{T}^n}$ *is non-degenerate at a critical point;*
(2) *The corresponding minimal surface satisfies* $nullity_a = n$;
(3) *The corresponding minimal surface has only trivial Jacobi fields.*

Next we introduce a new energy derived from $E_{\mathbb{T}^n}$. To do this, we review some fundamental arguments on compact Riemann surfaces. Let $\mathcal{H}_\gamma$ denote the Siegel upper half space of genus γ, that is, the space of complex symmetric $\gamma \times \gamma$ matrices with positive definite imaginary part. Let M be a compact Riemann surface of genus γ. Let $\{A_i, B_i\}_{i=1}^{\gamma}$ be a canonical homology basis on M, and let $\Phi = {}^t(\varphi_1, \cdots, \varphi_\gamma)$ be a basis for the space of holomorphic differentials on M which is dual to $\{A_i\}_{i=1}^{\gamma}$. We hence obtain

$$\left(\int_{A_1} \Phi \cdots \int_{A_\gamma} \Phi \int_{B_1} \Phi \cdots \int_{B_\gamma} \Phi\right) = \left(I_\gamma \; \tau\right),$$

where I_γ is the identity matrix of degree γ and $\tau \in \mathcal{H}_\gamma$. τ is called the *Riemann matrix* of M. Recall that $\mathcal{T}$ can be represented by an equivalent class of a pair (M, h) where M is a compact Riemann surface and $h : M_0 \to M$ is an orientation preserving diffeomorphism from a base surface M_0 to M. We difine a map $\kappa : \mathcal{T} \to \mathcal{H}_\gamma$ of the form $\kappa([M, h]) = \tau$, and it is well-known that there exists a unique complex structure on $\mathcal{T}$ such that κ is holomorphic.

Theorem 2.3. *Let* L *be an* $n \times 2\gamma$ *real matrix which generates the lattice of* $\mathbb{T}^n$. *Then* $E_{\mathbb{T}^n}$ *factors through* $\mathcal{H}_\gamma$, *i.e. there exists a map* $E_L : \mathcal{H}_\gamma \to \mathbb{R}$ *such that* $E_{\mathbb{T}^n} = E_L \circ \kappa$.

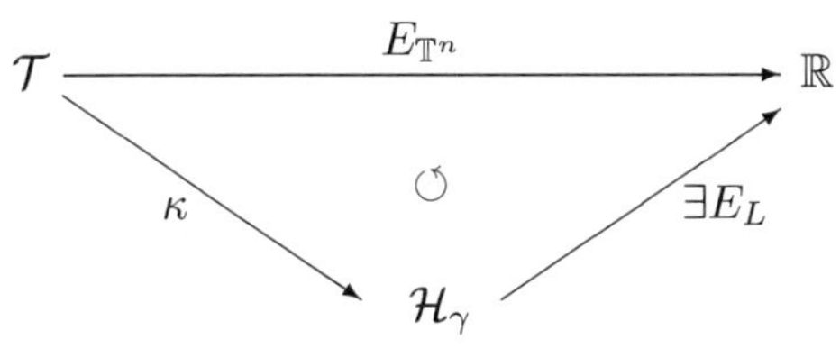

We can define a new energy E_L satisfying the condition in Theorem 2.3 for a fixed $n \times 2\gamma$ real matrix L by

$$E_L(\tau) = \frac{1}{2}\mathrm{trace}(P(\tau)\,{}^tLL),$$

where

$$P(\tau) = \begin{pmatrix} (\Im\tau) + (\Re\tau)(\Im\tau)^{-1}(\Re\tau) & -(\Re\tau)(\Im\tau)^{-1} \\ -(\Im\tau)^{-1}(\Re\tau) & (\Im\tau)^{-1} \end{pmatrix}.$$

The next theorems suggest that E_L is an important object instead of $E_{\mathbb{T}^n}$.

Theorem 2.4 ([3]). *Suppose that L generates a lattice Λ in $\mathbb{R}^n$. Every critical point of E_L in $\kappa(\mathcal{T})$ is corresponding to a minimal surface in $\mathbb{R}^n/\Lambda$.*

Theorem 2.5 ([4]). *If E_L is non-degenerate at a critical point, then there exist countably many minimal surfaces in flat n-tori.*

We now focus on hyperelliptic minimal surfaces. It is well-known that the hyperelliptic locus is a $2\gamma - 1$ complex dimensional submanifold in $\mathcal{T}$, and we consider the Hessian of E_L on the hyperelliptic locus in $\mathcal{T}$. Let $\mathrm{Hess}(E_{\mathbb{T}^n})$ denote the Hessian of $E_{\mathbb{T}^n}$, and $\mathrm{Hess}(E_L)$ the Hessian of E_L. Note that $\mathrm{Hess}(E_{\mathbb{T}^n})$ (resp. $\mathrm{Hess}(E_L)$) is a real symmetric matrix of degree $6\gamma - 6$ (resp. $4\gamma - 2$). $\mathrm{Hess}(E_{\mathbb{T}^n})$ can be decomposed into the tangent part and the normal part of the hyperelliptic locus as follows [3].

Theorem 2.6. *For a hyperelliptic minimal surface in $\mathbb{T}^n$, the Hessian of $E_{\mathbb{T}^n}$ is described by the Hessian of E_L and real values $\lambda_1, \cdots, \lambda_{\gamma-2}$ as follows:*

$$\mathrm{Hess}(E_{\mathbb{T}^n}) = \begin{pmatrix} \mathrm{Hess}(E_L) & & & & & O \\ & \lambda_1 & & & & \\ & & -\lambda_1 & & & \\ & O & & \ddots & & \\ & & & & \lambda_{\gamma-2} & \\ & & & & & -\lambda_{\gamma-2} \end{pmatrix}.$$

The locus $\lambda_1 \cdots \lambda_{\gamma-2} = 0$ is analytic [4], and it remains to be done to study the locus $\lambda_1 \cdots \lambda_{\gamma-2} = 0$. The following result is one of fascinating aspects for it.

Theorem 2.7 ([3]). *If every λ_i in Theorem 2.6 vanishes, then the corresponding hyperelliptic minimal surface lies in an even dimensional subtorus* *of $\mathbb{T}^n$ and is holomorphic with respect to some orthogonal complex structure on this even dimensional subtorus.*

Remark 2.1. For a compact minimal surface in $\mathbb{T}^3$, the minimal surface lies in a two dimensional subtorus of $\mathbb{T}^3$ and is holomorphic with respect to some orthogonal complex structure on the two dimensional subtorus if and only if $\gamma = 1$. Moreover, if $\gamma = 3$, then a minimal surface must be hyperelliptic (see Corollary 3.2 in Meeks' paper [10] or Nagano-Smyth's paper [12]). Hence, for $\gamma = 3$, we can apply Theorem 2.6, and a minimal surface does not lie in any totally geodesic subtorus. So we find $\lambda_1 \neq 0$ by Theorem 2.7. But, for a hyperelliptic minimal surface with $\gamma \geq 4$, we have to determine whether $\lambda_i = 0$ or $\lambda_i \neq 0$.

Remark 2.2. Theorem 2.5, 2.6 hold even if L does not generate a lattice in $\mathbb{R}^n$.

3. A construction of the Gauss map for the genus 4 case

In this section, we shall construct the Gauss map of a hyperelliptic Riemann surface of genus 4.

For distinct complex numbers $a_1, \cdots, a_9$, let M be a hyperelliptic Riemann surface of genus 4 defined by

$$w^2 = (z - a_1)(z - a_2)\cdots(z - a_9).$$

Then the four differentials

$$\frac{dz}{w}, \ \frac{z}{w}dz, \ \frac{z^2}{w}dz, \ \frac{z^3}{w}dz$$

form a basis for the holomorphic differentials on M.

We now consider the Gauss map $\omega : M \to Q_2$. Setting $\Phi = {}^t(1, z, z^2, z^3)\frac{dz}{w}$, we have $\omega = A\Phi$ where $A \in GL(4, \mathbb{C})$. (1) can be rewritten as

$$(1, z, z^2, z^3)\,{}^tAA\,{}^t(1, z, z^2, z^3) = 0. \tag{6}$$

Let $(\alpha_{ij}) = {}^tAA$. Then (6) takes the form

$$\alpha_{ij} = \alpha_{ji}, \quad \sum_{i,j=0}^{3} \alpha_{ij} z^i z^j = \sum_{k=0}^{6} \left(\sum_{i+j=k,\, 0\leq i,\, j\leq k} \alpha_{ij} \right) z^k = 0,$$

and hence, we find

$$\alpha_{ij} = \alpha_{ji}, \qquad \sum_{i+j=k,\, 0\leq i,\, j\leq k} \alpha_{ij} = 0$$

for $k = 0, 1, \cdots, 6$. Thus, we have

$$(\alpha_{ij}) = \begin{pmatrix} 0 & 0 & \alpha & \beta \\ 0 & -2\alpha & -\beta & \gamma \\ \alpha & -\beta & -2\gamma & 0 \\ \beta & \gamma & 0 & 0 \end{pmatrix} \tag{7}$$

for $\alpha, \beta, \gamma \in \mathbb{C}$.

Next, let S_g denote the set of $g \times g$ complex symmetric matrices and RS_g the subset of S_g which consists of regular matrices. Define $\pi : GL(g, \mathbb{C}) \ni A \mapsto {}^tAA \in RS_g$, and the following are immediate consequences.

Proposition 3.1. *Let Ξ' be the set of (α_{ij}) given by (7). Then Ξ' is a complex 3-dimensional subspace in S_g.*

Proposition 3.2. *Set $\Xi = \Xi' \cap RS_g$. Then $\pi^{-1}(\Xi)$ is the space of Gauss maps $M \to Q_2$.*

For f in Theorem 1.2, we choose $\alpha = \gamma = 0$, $\beta = -2$ as α_{ij} in (7). Moreover, $\pi : GL(4, \mathbb{C}) \to RS_4$ is given by

$$\begin{pmatrix} 1 & 0 & 0 & -1 \\ i & 0 & 0 & i \\ 0 & 1 & 1 & 0 \\ 0 & -i & i & 0 \end{pmatrix} \mapsto \begin{pmatrix} 0 & 0 & 0 & -2 \\ 0 & 0 & 2 & 0 \\ 0 & 2 & 0 & 0 \\ -2 & 0 & 0 & 0 \end{pmatrix}.$$

4. Stories of the proofs of the Main Theorems

To compute the Morse index of a hyperelliptic minimal surface, we have to investigate the number of negative eigenvalues of $\mathrm{Hess}(E_L)$ and that of non-zero λ_i by Theorem 2.6. Let *index* denote the Morse index of E_L.

We now consider M and f as in Theorem 1.2. The next theorem is one of key results.

Theorem 4.1. *There exist $0 < a_1 < a_2 < a_3 < a_4 < 1$ satisfying the following properties:*

$$\begin{cases} \textit{index} = 3 & (a \in (0, a_1) \cup (a_4, 1)) \\ \textit{index} = 1 & (a \in (a_1, a_2) \cup (a_3, a_4)) \\ \textit{index} = 0 & (a \in (a_2, a_3)) \end{cases} \tag{8}$$

By Theorem 4.1, E_L is non-degenerate for $a \notin \{0, a_1, a_2, a_3, a_4, 1\}$. Therefore, Theorem 1.3 follows from Theorem 2.5.

Theorem 4.1 implies that we have to determine $a \in (0, 1)$ at which $\lambda_1\lambda_2 \neq 0$. The next theorem gives a partial answer for it.

Theorem 4.2. *$\lambda_1\lambda_2 \neq 0$ on $(0, 1)$ except a discrete set. In particular, $\lambda_1\lambda_2 \neq 0$ on $[a_2, a_3]$.*

The first statement of Theorem 4.2 follows immediately from the analyticity of the locus $\lambda_1\lambda_2 = 0$. To prove the remaining statement of Theorem 4.2, we apply Micallef and Wolfson's technique to f. We shall now explain about the arguments.

We first consider a complex version of (5). Since M is a compact Riemann surface, we have the decomposition

$$T^*_{\mathbb{C}}M = \Lambda^{1,0}M \oplus \Lambda^{0,1}M, \tag{9}$$

where $T^*_{\mathbb{C}}M = T^*M \otimes_{\mathbb{R}} \mathbb{C}$. By extending ∇ and A complex linearly to $N^*_{\mathbb{C}}M = N^*M \otimes_{\mathbb{R}} \mathbb{C}$, and using the splitting (9), we obtain

$$\nabla^{0,1} : \Gamma(N_{\mathbb{C}}M) \to \Gamma(\Lambda^{0,1}M \otimes N_{\mathbb{C}}M),\ A^{1,0} : \Gamma(N_{\mathbb{C}}M) \to \Gamma(\Lambda^{1,0}M \otimes T_{\mathbb{C}}M),$$

where $T_{\mathbb{C}}M = TM \otimes_{\mathbb{R}} \mathbb{C}$. Then (5) can be rewritten as

$$2\int_M \{|\nabla^{0,1}x|^2 - |A^{1,0}x|^2\}dv \tag{10}$$

for $x \in \Gamma(N_{\mathbb{C}}M)$.

By Newlander-Nirenberg or Koszul-Malgrange Theorem, $N_{\mathbb{C}}M$ can be made into a holomorphic bundle whose $\bar{\partial}$-operator is $\nabla^{0,1}$. It follows from (10) that the dimension of holomorphic sections of $N_{\mathbb{C}}M$ gives the lower bound of the Morse index of a minimal surface. The following theorem is a next key result for us.

Theorem 4.3. *$N_{\mathbb{C}}M$ has two linearly independent holomorphic sections, and therefore, we have $index_a \geq 2$.*

We now prove the remaining statement of Theorem 4.2. By (8), Theorem 2.6, and Theorem 4.3, we obtain $\lambda_1\lambda_2 \neq 0$ on (a_2, a_3). Hence $index_a = 2$ on (a_2, a_3). Note that 2 is the minimum value of $index_a$ for f by Theorem 4.3. We now prove $index_a = 2$ at $a = a_2, a_3$ by contradiction. Let $\mu_1(a), \cdots, \mu_k(a)$ be negative eigenvalues of the Jacobi operator at $a \in (0, 1)$. If $index_a \neq 2$ at $a = a_2$, then $k \geq 3$. Since every $\mu(a)$ is continuous for a, $k \geq 3$ on a sufficiently small interval which contains a_2. It contradicts $k = 2$ on (a_2, a_3). The same argument yields $k = 2$ for $a = a_3$, and our claim follows. Similarly, we have $index = 0$ at $a = a_2, a_3$. We have Theorem 4.2 as an immediate consequence.

References

1. C. Arezzo and M. Micallef, Minimal surfaces in flat tori, *Geom. Funct. Anal.* **10**, No.4 (2000), 679–701.
2. C. Arezzo and G. P. Pirola, On the existence of periodic minimal surfaces, *J. Algebraic. Geom.* **8** (1999), 765-785.
3. N. Ejiri, Differential-Geometric Schottky Problem and Minimal Surfaces in Tori, *Contemporary. Math.* **308** (2002), 101-144.
4. N. Ejiri, A generating function of a complex Lagrangian cone in $\mathbb{H}^n$, preprint.
5. N. Ejiri and T. Shoda, The Morse index of a triply periodic minimal surface, preprint.
6. N. Ejiri and T. Shoda, On a moduli theory of minimal surfaces, *Prospects of Differential Geometry and its related fields*, T. Adachi, H. Hashimoto and M. Hristov eds., World Scientific (2014), 155–172.
7. N. Ejiri and T. Shoda, The existence of hyperelliptic minimal surfaces with even genus and their geometric invariants, preprint.
8. A. Fogden and S. T. Hyde, Parametrization of triply periodic minimal surfaces. II. Regular class solutions, *Acta. Cryst. A* **48** (1992), 575-591.
9. A. Fogden, and M. Haeberlein, and S. Lidin, Generalizations of the gyroid surface, *J. Phys. I. France.* **3** (1993), 2371-2385.
10. M. H. Meeks III, *The theory of triply periodic minimal surfaces*, Indiana. Univ. Math. J. **39** (1990), 877–936.
11. W. Micallef and J. Wolfson, The second variation of area of minimal surfaces in four-manifolds, *Math. Ann.* **295** (1993), 245–267.
12. T. Nagano and B. Smyth, *Sur les surfaces minimales hyperelliptiques dans un tore*, C. R. Acad. Sci. Paris, Serie A **280** (1975), 1527–1529.
13. J. Sacks and K. Uhlenbeck, The Existence of Minimal Immersions of 2-spheres, *Ann. of Math.* **113** (1981), 1-24.
14. R. Schoen and S. T. Yau, Existence of Incompressible Minimal Surfaces and the Topology of Three Dimensional Manifolds with Non-Negative Scalar Curvature, *Ann. of Math.* **110** (1979), 127-142.
15. J. Simons, Minimal Varieties in Riemannian Manifolds, *Ann. of Math.* **88** (1968), 62–105.

Received March 3, 2015
Revised April 9, 2015

Current Developments in Differential Geometry
and its Related Fields pp. 139 – 154

LAPLACIANS OF KÄHLER GRAPHS

YAERMAIMAITI Tuerxunmaimaiti *
(YARMAMAT Tursun)

Division of Mathematics and Mathematical Science,
Nagoya Institute of Technology,
Nagoya 466-8555, Japan
E-mail: yarimamat@gmail.com

A Kähler graph is a compound of two graphs having the same set of vertices. We consider Kähler graphs as discrete models of Kähler manifolds admitting magnetic fields, and regard paths consisting of two kinds of edges as trajectories for magnetic fields. In this paper we give a report on eigenvalues of $(1,1)$-Laplacians, which are generating operators of random walks through trajectories, for some finite Kähler graphs.

Keywords: Kähler graphs, bicolored paths, $(1,1)$-Laplacians, complement-filled Kähler graphs, Kähler graphs of product type, Petersen graph, Heawood graph, isospectral.

1. Introduction

The aim of this paper is to give a report on eigenvalues of Laplacians for finite Kähler graphs based on the recent joint works [5,6,8] of Adachi and the author.

A graph consists of a set of vertices and a set of edges and is represented as a 1-dimensional CW-complex. We call a graph Kähler if its edges are classified into two classes, principal edges and auxiliary edges. Kähler graphs were introduced by Adachi [2] as a discrete model of a Kähler manifold admitting magnetic fields. A magnetic field on a Riemannian manifold is a closed 2-form. Since geodesics on a Riemannian manifold can be regarded as motions of electric charged particles without influence of magnetic fields, it is natural to consider that trajectories for magnetic fields are extension of geodesics and that they show some properties of base manifolds (see [1]). In Riemannian geometry, graphs are considered as discrete models of Riemannian manifolds. Paths on a graph, which are chain of edges, are treated as corresponding objects of geodesics.

On a Kähler graph, we consider that paths having principal edges and auxiliary edges alternatively correspond to trajectories for a magnetic field,

*The author is partially supported by Hirose International Scholarship Foundation

though we can not yet define magnetic fields on graphs. Making use of such paths we define the Laplacian, or more precisely $(1, 1)$-Laplacian, for a Kähler graph by just the same way as we define the discrete Laplacian for a graph. We study eigenvalues of Laplacians for finite Kähler graphs; complement-filled Kähler graphs, Kähler graphs of product types and a Cayley Kähler graph of a dihedral group. For these Kähler graphs their Laplacians are selfadjoint operators. Also, we study eigenvalues of Kähler graphs made by a Petersen graph, a cube and a Heawood graph.

The author is grateful to his supervisor Professor Adachi for his encouragement and his advice.

2. Kähler graphs

A graph $G = (V, E)$ is a pair of a set V of vertices and a set E of edges. We suppose that each edge does not have an orientation. Hence a graph is represented as a 1-dimensional CW-complex. We say two vertices $v, v' \in V$ to be adjacent to each other if there is an edge joining them. We denote as $v \sim v'$ in this case. When an edge $e \in E$ joins two vertices v, v' we call them ends or termini of e. An edge is said to be a loop if it joins one vertex and itself, that is its two ends coincide. We say two and more edges to be multiple if they have the same pair of ends. Through out of this paper we suppose that every graph is simple, that is, it does not have loops and multiple edges. Given a vertex $v \in V$ we denote by $d_G(v)$ the cardinality of the set $\{v' \in V \mid v' \sim v\}$, and call it the degree at v. We call a graph $G = (V, E)$ to be locally finite if $d_G(v)$ is finite for all v, and call it finite if both of V and E are finite sets. In this case we denote by n_G the cardinality of V. A graph G is said to be regular if the degree $d_G(v)$ does not depend on the choice of v, and is said to be complete if all pairs of distinct vertices are adjacent to each other. Clearly a complete graph is regular.

A graph $G = (V, E)$ is said to be *Kähler* if E is divided into two subsets $E^{(p)}$, $E^{(a)}$ and satisfies that at each v there are two edges in $E^{(p)}$ and two edges in $E^{(a)}$ which have v as their termini. In order to distinguish Kähler graphs from other graphs, we sometimes call a graph as an ordinary graph. For a Kähler graph $G = (V, E^{(p)} + E^{(a)})$ we call the graph $G^{(p)} = (V, E^{(p)})$ its *principal graph* and $G^{(a)} = (V, E^{(a)})$ its *auxiliary graph*. By putting $d_G^{(p)}(v) := d_{G^{(p)}}(v)$ and $d_G^{(a)}(v) := d_{G^{(a)}}(v)$, we call them the principal and the auxiliary degrees at v. By our definition we have $d_G^{(p)}(v) \geq 2$ and $d_G^{(a)}(v) \geq 2$ at each $v \in V$. For distinct vertices $v, v' \in V$ we denote by $v \sim_p v'$ if they are adjacent to each other in $G^{(p)}$, and denote by $v \sim_a v'$

if they are adjacent to each other in $G^{(a)}$. We here show some ways for constructing Kähler graphs.

Example 2.1. Let $G = (V, E)$ be an ordinary finite graph. We take its complement graph $G^c = (V, E^c)$ which is defined so that two distinct vertices are adjacent to each other in G^c if and only if they are not adjacent to each other in G. When G satisfies $2 \le \min_{v \in V} d_G(v) \le \max_{v \in V} d_G(v) \le n_G - 3$, we get a Kähler graph $G^K = (V, E + E^c)$ which is complete as an ordinary graph. We call this a *complement-filled* Kähler graph.

We call a Kähler graph G *regular* if both of its principal and its auxiliary graphs are regular, and call it *complete* if it is a regular Kähler graph and is complete as an ordinary graph. When G is a finite regular graph with $2 \le d_G \le n_G - 3$, then its complement-filled Kähler graph is a complete Kähler graph.

Example 2.2. Let $\mathcal{G}$ be a group and $\mathcal{S}_p, \mathcal{S}_a$ are finite nonempty subsets satisfying that they do not contain the identity, are symmetric with respect to inverse operation (i.e. $g \in \mathcal{S}_p$ if and only if $g^{-1} \in \mathcal{S}_p$ and so does $\mathcal{S}_a$), and $\mathcal{S}_p \cap \mathcal{S}_a = \emptyset$. For two distinct $g, g' \in \mathcal{G}$, we define $g \sim_p g'$ if and only if $g^{-1}g' \in \mathcal{S}_p$ and define $g \sim_a g'$ if and only if $g^{-1}g' \in \mathcal{S}_a$. Then we get a Kähler graph $(\mathcal{G}, E^{(p)} + E^{(a)})$. We call this a Cayley Kähler graph.

In graph theory we have four kinds of major product operations; Cartesian product, tensor product, strong product and lexicographical product. Corresponding to these we can define four kinds of Kähler graphs of product type. Given two ordinary graphs $G = (V, E)$, $H = (W, F)$ whose degrees are not smaller than 2, we define Kähler graphs of Cartesian product type $G \widehat{\square} H$, of strong product type $G \widehat{\boxtimes} H$, of tensor product type $G \widehat{\otimes} H$, and of lexicographic product type $G \rhd H$ as follows:

i) Their sets of vertices are the product $V \times W$;
ii) Two vertices (v, w), $(v', w') \in V \times W$ are adjacent in their principal graph if and only if v, v' are adjacent in G and $w = w'$;
iii) Two vertices (v, w), $(v', w') \in V \times W$ are adjacent in their principal graph if and only if they satisfy the following condition;
   a) $v = v'$ and w, w' are adjacent in H, for $G \widehat{\square} H$,
   b) either $v = v'$ and w, w' are adjacent in H, or v, v' are adjacent in G and w, w' are adjacent in H, for $G \widehat{\boxtimes} H$,
   c) v, v' are adjacent in G and w, w' are adjacent in H, for $G \widehat{\otimes} H$,
   d) w, w' are adjacent in H, for $G \rhd H$.

It is clear that these four graphs are Kähler. When G and H are finite, we see their Kähler graphs of product type are finite and find that

$$\begin{aligned}
&d^{(p)}_{G\widehat{\Box}H}(v,w) = d^{(p)}_{G\widehat{\boxtimes}H}(v,w) = d^{(p)}_{G\widehat{\otimes}H}(v,w) = d^{(p)}_{G\triangleright H}(v,w) = d_G(v),\\
&d^{(a)}_{G\widehat{\Box}H}(v,w) = d_H(w), \qquad\qquad d^{(a)}_{G\widehat{\boxtimes}H}(v,w) = d_H(w)\bigl(d_G(v)+1\bigr),\\
&d^{(a)}_{G\widehat{\otimes}H}(v,w) = d_G(v)d_H(w), \quad d^{(a)}_{G\triangleright H}(v,w) = n_G d_H(w).
\end{aligned}$$

In order to make clear these operations we here give examples.

Example 2.3. Let $\mathbb{Z}$ be the set of all integers. For $n, n' \in \mathbb{Z}$ we define $n \sim n'$ if and only if $n' = n \pm 1$. We call this $(\mathbb{Z}, E)$ a graph of real lattice.

We take G and H as graphs of real lattice. We express G by a horizontal lattice and H by a vertical lattice. Hence the set of vertices of each Kähler graph of their product type is expressed as a lattice in a plane $\mathbb{R}^2$. The adjacency at an arbitrary vertex of their Kähler graphs of product type are like the following figures.

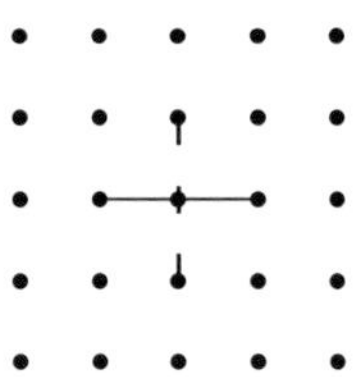

Fig. 1. $G\widehat{\Box}H$

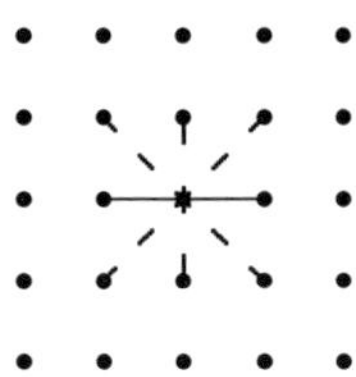

Fig. 2. $G\widehat{\boxtimes}H$

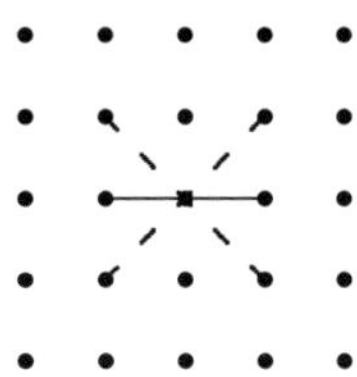

Fig. 3. $G\widehat{\otimes}H$

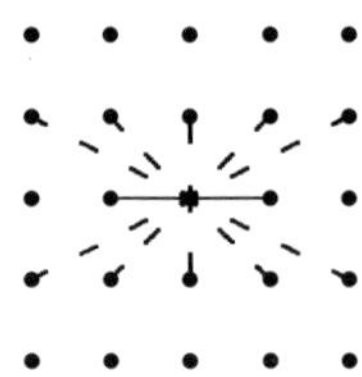

Fig. 4. $G \triangleright H$

A graph is said to be connected if it is connected as a 1-dimensional CW-complex. In view of figures in Example 2.3, we see that principal graphs of four Kähler graphs of product type are not connected. Moreover, the above operations for getting Kähler graphs are not commutative in general. That is, $G\widehat{\Box}H \neq H\widehat{\Box}G$ in general, for example. This point is quite different from the product operations of ordinary graphs called Cartesian product, tensor product and strong product. Therefore we here give other three ways of product operations. Given two graphs $G = (V, E)$, $H = (W, F)$ we define their Kähler graphs $G \boxplus H$, $G \boxdot H$, $G \Diamond H$ of Cartesian-tensor product type, of Cartesian-complement product type and of Cartesian-lexicographical product type as follows:

i) the set of vertices is the product $V \times W$;
ii) two vertices $(v, w), (v', w') \in V \times W$ are $(v, w) \sim_p (v', w')$ if and only if either $v = v'$ and $w \sim w'$ in H or $w = w'$ and $v \sim v'$ in G;

iii) two vertices (v, w), $(v', w') \in V \times W$ are $(v, w) \sim_a (v', w')$ if and only if the following conditions hold;

a) $v \sim v'$ in G and $w \sim w'$ in H, for $G \boxplus H$,

b) either $v \neq v'$, $v \not\sim v'$ in G and $w \sim w'$ in H, or $w \neq w'$, $w \not\sim w'$ in H and $v \sim v'$ in G, for $G \boxdot H$,

c) either $v \neq v'$ and $w \sim w'$ in H, or $w \neq w'$ and $v \sim v'$ in G, for $G \Diamond H$.

If we suppose either $\max_{v\in V} d_G(v) \leq n_G - 2$ or $\max_{w\in W} d_H(w) \leq n_H - 2$, in the case of getting $G \boxdot H$, these three graphs are Kähler. When G and H connected, we find that principal and auxiliary graphs of these product type are also connected. Moreover, we find that these operations are commutative. When G and H are finite, at $(v, w) \in V \times W$ we have

$$d^{(p)}_{G\boxplus H}(v, w) = d^{(p)}_{G\boxdot H}(v, w) = d^{(p)}_{G\Diamond H}(v, w) = d_G(v) + d_H(w),$$
$$d^{(a)}_{G\boxplus H}(v, w) = d_G(v) d_H(w),$$
$$d^{(a)}_{G\boxdot H}(v, w) = (n_G - d_G(v) - 1) d_H(w) + (n_H - d_H(w) - 1) d_G(v),$$
$$d^{(a)}_{G\Diamond H}(v, w) = (n_G - 1) d_H(w) + (n_H - 1) d_G(v).$$

Example 2.4. When G and H are graphs of real lattice, then the adjacency at an arbitrary vertex of their Kähler graphs of connected product type are like the following figures by the same expression of the sets of vertices as in Example 2.3.

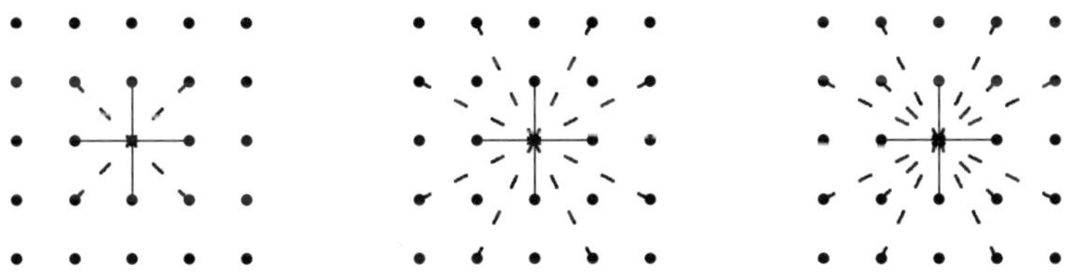

Fig. 5. $G \boxplus H$ Fig. 6. $G \boxdot H$ Fig. 7. $G \Diamond H$

3. Bicolored paths and Laplacians

In this section we first define trajectories for magnetic fields on Kähler graphs. But as graphs are of 1-dimensional, we do not define objects which correspond to magnetic fields but only give candidates corresponding to trajectories on Riemannian manifolds.

On a graph $G = (V, E)$, a sequence of vertices $\gamma = (v_0, \ldots, v_n)$ is said to be a path or an n-step path, more precisely, if $v_{i-1} \sim v_i$ for $i = 1, \ldots, n$.

We set $o(\gamma) = v_0$, $t(\gamma) = v_n$ and call them the origin and the terminus of γ, respectively. We say γ contains a backtracking if there is i_0 with $v_{i_0-1} = v_{i_0+1}$. Regarding a graph as a discrete model of a Riemannian manifold, we consider paths without backtracking as geodesics.

On a Kähler graph G we say an even-step path $\gamma = (v_0, \ldots, v_{2m})$ without backtracking to be a bicolored path, or more precisely a $(1,1)$-bicolored path, if (v_{2j-2}, v_{2j-1}) is a principal edge and (v_{2j-1}, v_{2j}) is an auxiliary path for $j = 1, \ldots, m$. We regard paths without backtracking on the principal graph as geodesics, which are motions of electric charged particles without getting influence of magnetic fields, and consider that they are bended under a magnetic field and turn to bicolored paths. At this point we need to take care of one thing. As a graph is of 1-dimensional, we can not show the direction of the action of magnetic fields. Therefore, if we consider bicolored paths, paths on the principal graph seem to be bifurcated. In order to get rid of this we treat bicolored paths probabilistically. For a bicolored path $\gamma = (v_0, \ldots, v_{2m})$, we define its probabilistic weight $\omega(\gamma)$ by $\omega(\gamma) = 1/\prod_{j=1}^{m} d_G^{(a)}(v_{2j-1})$. In this paper we only treat paths where principal and auxiliary edges appear alternatively, but we can treat p-step principal paths followed by q-step auxiliary paths. We consider such paths correspond to trajectories for a magnetic field whose Lorentz force is of strength q/p.

Next we define a (magnetic) Laplacian for a Kähler graph which is the generating operators for random walk made by bicolored paths. When $G = (V, E)$ is an ordinary graph, its adjacency operator $\mathcal{A}_G$ and its transition operator $\mathcal{P}_G$ acting on the set $C(V)$ of all functions on V by

$$\mathcal{A}_G f(v) = \sum_{v':v'\sim v} f(v'), \qquad \mathcal{P}_G f(v) = \frac{1}{d_G(v)} \sum_{v':v'\sim v} f(v')$$

for each $f \in C(V)$, respectively. By using the degree operator $\mathcal{D}_G$ given by $\mathcal{D}_G f(v) = d_G(v) f(v)$, we define the combinatorial Laplacian $\Delta_{\mathcal{A}_G}$ and the transitional Laplacian $\Delta_{\mathcal{P}_G}$ by $\Delta_{\mathcal{A}_G} = \mathcal{D}_G - \mathcal{A}_G$ and $\Delta_{\mathcal{P}_G} = I - \mathcal{P}_G$, where I denotes the identity. When G is regular, we see that $\Delta_{\mathcal{A}_G} = d_G \Delta_{\mathcal{P}_G}$ and $\Delta_{\mathcal{A}_G} = d_G \Delta_{\mathcal{P}_G}$.

We now take a Kähler graph $G = (V, E^{(p)} + E^{(a)})$. We define its $(1,1)$-adjacency operator $\mathcal{A}_G^{(1,1)}$ and its $(1,1)$-transition operator $\mathcal{P}_G^{(1,1)}$ acting on $C(V)$ by

$$\mathcal{A}_G^{(1,1)} f(v) = \sum_{\gamma} \omega(\gamma) f\bigl(t(\gamma)\bigr), \quad \mathcal{P}_G^{(1,1)} f(v) = \frac{1}{d_G^{(p)}(v)} \sum_{\gamma} \omega(\gamma) f\bigl(t(\gamma)\bigr),$$

where γ runs over the set of all 2-step bicolored paths of origin v. As we have

$$\mathcal{A}_G^{(1,1)} f(v) = \sum_{\substack{(v,v',v'') \\ v \sim_p v' \sim_a v''}} \frac{1}{d_G^{(a)}(v')} f(v'')$$

$$= \sum_{v':v\sim_p v} \sum_{v'':v''\sim_a v'} \frac{1}{d_G^{(a)}(v')} f(v'') = \mathcal{A}_{G^{(p)}} \mathcal{P}_{G^{(a)}} f(v),$$

we obtain the following

Lemma 3.1. *For a Kähler graph G we have*

$$\mathcal{A}_G^{(1,1)} = \mathcal{A}_{G^{(p)}} \mathcal{P}_{G^{(a)}}, \quad \mathcal{P}_G^{(1,1)} = \mathcal{P}_{G^{(p)}} \mathcal{P}_{G^{(a)}}.$$

This lemma shows that the adjacency operator and the transition operator for a Kähler graph are not selfadjoint in general. The $(1,1)$-combinatorial Laplacian $\Delta_{\mathcal{A}_G^{(1,1)}}$ and the $(1,1)$-transitional Laplacian $\Delta_{\mathcal{P}_G^{(1,1)}}$ are defined by $\Delta_{\mathcal{A}_G^{(1,1)}} = \mathcal{D}_{G^{(p)}} - \mathcal{A}_G^{(1,1)}$ and $\Delta_{\mathcal{P}_G^{(1,1)}} = I - \mathcal{P}_G^{(1,1)}$. Trivially they are not selfadjoint in general. In the following sections we study eigenvalues of Laplacians for finite Kähler graphs we mentioned in §2.

4. Eigenvalues of complement-filled Kähler graphs

In this section we treat complement-filled Kähler graphs made by finite ordinary graphs.

Theorem 4.1 ([6]). *Let $G = (V, E)$ be a connected finite ordinary graph whose degree function satisfies $2 \leq d_G \leq n_G - 2$. We make its complement-filled Kähler graph G^K.*

(1) $\mathcal{A}_{G^K}^{(1,1)}$ *and* $\mathcal{P}_{G^K}^{(1,1)}$ *are selfadjoint with respect to the ordinary inner product on $C(V)$ given by $\langle f, g\rangle = \sum_{v\in V} f(v)g(v)$.*
(2) *If we denote the eigenvalues of $\Delta_{\mathcal{A}_G}$ of G by λ_j $(j = 1, \dots, n_G)$ with $0 = \lambda_1 \leq \cdots \leq \lambda_{n_G}$, then the eigenvalues of $\Delta_{\mathcal{A}_{G^K}^{(1,1)}}$ are*

$$0, \ \{\lambda_j^2 - \lambda_j(2d_G + 1) + n_G d_G\}/(n_G - d_G - 1) \quad (j = 2, \dots, n_G).$$

Proof. Let $\mathcal{M}_G$ denotes the operator defined by $\mathcal{M}_G f = \sum_{v\in V} f(v)$. We then have $\mathcal{A}_{G^c} = \mathcal{M}_G - I - \mathcal{A}_G$ and $d_{G^c} = n_G - d_G - 1$. Since G is regular, we have $\mathcal{M}_G\mathcal{A}_G = d_G\mathcal{M}_G = \mathcal{A}_G\mathcal{M}_G$, hence obtain that $\mathcal{A}_G\mathcal{A}_{G^c} = \mathcal{A}_G\mathcal{A}_{G^c}$.

As the adjacency operator of an ordinary graph is selfadjoint we get the first assertion.

We take an eigenfunction f_j for λ_j, that is, a function satisfying $\Delta_{\mathcal{A}_G} f_j = \lambda_j f_j$. As G is connected we see f_1 is a constant function, hence $\mathcal{M}_G f_1 = n_G f_1$. On contrary, as f_2 is orthogonal to f_1 and G is regular, we have $\mathcal{M}_G f_j = 0$ for $j \geq 2$. Thus we get the second assertion by direct computation. □

Given two ordinary graphs $G_1 = (V_1, E_1)$, $G_2 = (V_2, E_2)$ we call a map $\varphi : V_1 \to V_2$ a homomorphism if it satisfies that $\varphi(v) \sim \varphi(v')$ when $v \sim v'$. A bijection $\varphi : V_1 \to V_2$ is called an isomorphism if both φ and φ^{-1} are homomorphism. When there is an isomorphism between G_1 and G_2 we say that they are isomorphic. We call two ordinary finite graphs G_1, G_2 combinatorially *isospectral* (resp. transitionally isospectral) if they are not isomorphic to each other and if their combinatorial Laplacians (resp. transitional Laplacians) have the same eigenvalues by taking account of their multiplicities. It is well known that there are many pairs of isospectral regular graphs (see [4]).

When $G_1 = (V_1, E_1^{(p)} + E_1^{(a)})$, $G_2 = (V_1, E_2^{(p)} + E_2^{(a)})$ are Kähler graphs, a map $\varphi : V_1 \to V_2$ is said to be a *homomorphism* if it satisfied that $\varphi(v) \sim_p \varphi(v')$ when $v \sim_p v'$ and that $\varphi(v) \sim_a \varphi(v')$ when $v \sim_a v'$. One can easily guess the definition of isomorphic Kähler graphs. We say that two finite Kähler graphs G_1, G_2 are $(1,1)$-combinatorially isospectral (resp. $(1,1)$-transitionally isospectral) if their $(1,1)$-combinatorial Laplacians (resp. $(1,1)$-transitional Laplacians) have the same eigenvalues by taking account of their multiplicities and if their principal graphs $G_1^{(p)}, G_2^{(p)}$ are isospectral. If we take Kähler graphs having regular principal graphs, it is clear that they are $(1,1)$-combinatorially isospectral if and only if they are $(1,1)$-transitionally isospectral. As a consequence of Theorem 4.1 we have

Corollary 4.1 ([6]). *If we take two finite connected regular ordinary graphs G_1, G_2 which are isospectral, then their compliment-filled Kähler graphs G_1^K, G_2^K are isospectral as Kähler graphs.*

Here we give an example. It is known that there exist two pairs of isospectral regular graphs having ten vertices. Also, it is known that there are no pairs of regular graphs whose cardinalities of the set of vertices are less than 10.

Example 4.1. We take the following Kähler graphs in Figures 8 and 9 where we denote their principal and auxiliary graphs separately. Then they are isospectral. Their eigenvalues of combinatorial Laplacians for principal graphs and of $(1,1)$-laplacians are

$$\mathrm{Spec}(\Delta_{\mathcal{A}_{G^{(p)}}}) = \left[\, 0, 3, 5, 5, 5, 5, 4-\sqrt{5}, 4+\sqrt{5}, (9-\sqrt{17})/2, (9+\sqrt{17})/2 \,\right],$$
$$\mathrm{Spec}(\Delta_{\mathcal{A}_G^{(1,1)}}) = \left[\, 0, 4, 4, 4, 4, 22/5, 24/5, 24/5, (25-\sqrt{5})/5, (25+\sqrt{5})/5 \,\right].$$

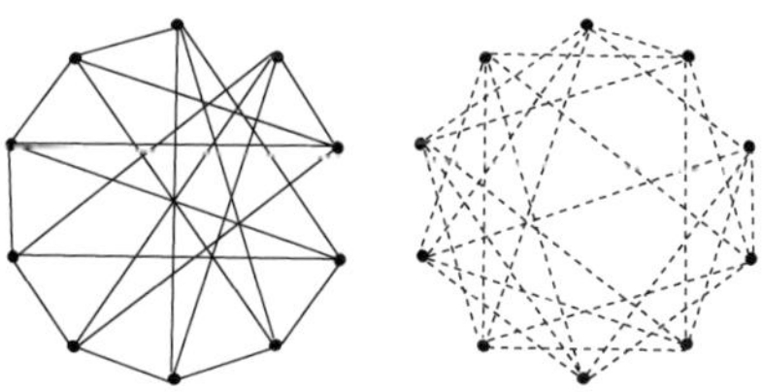

Fig. 8.

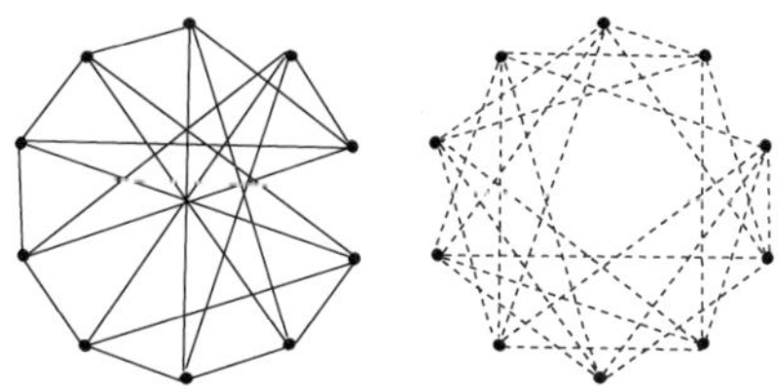

Fig. 9.

Given a Kähler graph $G = (V, E^{(p)} + E^{(a)})$ we set $E^{(1,1)}$ as the set of all $(1,1)$-bicolored paths and define its $(1,1)$-derived (oriented) graph by $G_{1,1} = (V, E^{(1,1)})$. It is clear that if two Kähler graphs have isomorphic $(1,1)$-derived graphs then their $(1,1)$-combinatorial Laplacians (resp. $(1,1)$-transitional Laplacians) have the same eigenvalues. We have pairs of Kähler graphs satisfying that their principal graph are not isomorphic to each other but their $(1,1)$-derived graphs are isomorphic (see Fig. 10, 11). The principal graphs in Fig. 10, 11 are not isospectral. Therefore, when we study isospectral Kähler graphs, the condition that principal graphs are isospectral is important.

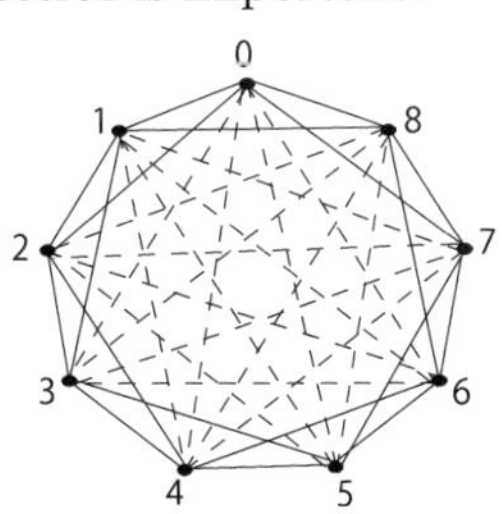

Fig. 10. $K^1_{(9;4,4)}$

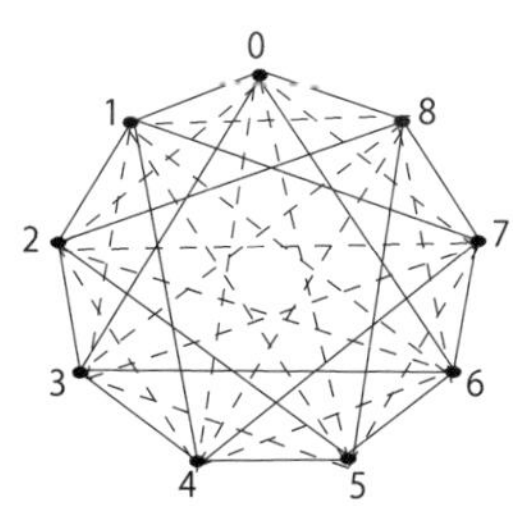

Fig. 11. $K^2_{(9;4,4)}$

The eigenvalues of these Kähler graphs are

$$\mathrm{Spec}(\Delta_{\mathcal{A}^{(1,1)}}) = \begin{bmatrix} 0, 9/2, 9/2, (9+\sqrt{3}\cos\frac{\pi}{18})/2, (9+\sqrt{3}\cos\frac{\pi}{18})/2, \\ (9-\sqrt{3}\cos\frac{5}{18}\pi)/2, (9-\sqrt{3}\cos\frac{5}{18}\pi)/2, \\ (9-\sqrt{3}\cos\frac{7}{18}\pi)/2, (9-\sqrt{3}\cos\frac{7}{18}\pi)/2 \end{bmatrix}.$$

In Theorem 4.1 we used regular graphs. We here make mention of complement-filled Kähler graphs of non-regular graphs. For such a Kähler graph, the adjacency operator of its principal graph and the transition operator of its auxiliary graph are not simultaneously diagonalizable. As a matter of fact, for each $f \in C(V)$ the function $\mathcal{M}_G\mathcal{A}_G f$ is constant;

$$\mathcal{M}_G\mathcal{A}_G f(v) = \sum_{w\in V}\sum_{w':w'\sim w} f(w') = \sum_{w'\in V}\sum_{w:w\sim w'} f(w') = \sum_{w'\in V} d_G(w')f(w').$$

On the other hand, we have $\mathcal{A}_G\mathcal{M}f(v) = d_G(v)\sum_{w\in V} f(w)$, which is not constant.

Example 4.2. We take a complement-filled Kähler graph G whose principal graph is not regular as in Fig. 12. Its eigenvalues of the principal graph and of Kähler graph are as follows:

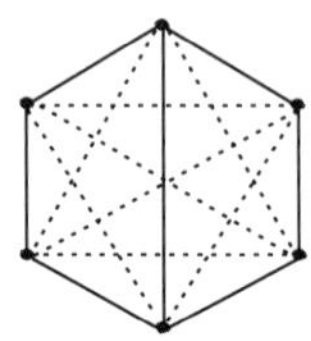

Fig. 12.

$$\mathrm{Spec}(\Delta_{\mathcal{A}_{G^{(p)}}}) = \big[\, 0, 1, 2, 3, 3, 5 \,\big],$$
$$\mathrm{Spec}(\Delta_{\mathcal{A}_G^{(1,1)}}) = \big[\, 0, 2, 4, 8/3, 8/3, 8/3 \,\big],$$
$$\mathrm{Spec}(\Delta_{\mathcal{P}_G^{(1,1)}}) = \big[\, 0, 1, 4/3, 17/18, (49+\sqrt{97})/36, (49-\sqrt{97})/36 \,\big].$$

Clearly, these eigenvalues do not satisfy the relation in Theorem 4.1.

5. Eigenvalues of typical Kähler graphs

In this section we give some examples of Kähler graphs and study their eigenvalues.

Example 5.1. A Petersen Kähler graph is a regular graph of $d^{(p)} = d^{(a)} = 3$ which is given as Fig. 13. Its $(1,1)$-adjacency operator and its $(1,1)$-transition operator are not selfasdjoint. The eigenvalues

$$\mathrm{Spec}(\Delta_{\mathcal{A}_{G^{(p)}}}) = \big[\, 0, 2, 2, 2, 2, 2, 5, 5, 5, 5 \,\big],$$
$$\mathrm{Spec}(\Delta_{\mathcal{A}_G^{(1,1)}}) = \big[\, 0, \frac{8}{3}, \epsilon, \epsilon, \rho, \rho, \varrho, \varrho, \varsigma, \varsigma \,\big],$$

where

$$\begin{cases} \epsilon = \dfrac{-41-\sqrt{5}}{12} - \dfrac{\sqrt{-1}}{12}\sqrt{34-10\sqrt{5}}, \\ \rho = \dfrac{-41-\sqrt{5}}{12} + \dfrac{\sqrt{-1}}{12}\sqrt{34-10\sqrt{5}} \end{cases} \qquad \begin{cases} \varrho = \dfrac{-41+\sqrt{5}}{12} - \dfrac{\sqrt{-1}}{12}\sqrt{34+10\sqrt{5}}, \\ \varsigma = \dfrac{-41+\sqrt{5}}{12} + \dfrac{\sqrt{-1}}{12}\sqrt{34+10\sqrt{5}}. \end{cases}$$

Example 5.2. A Kähler Petersen graph is a regular graph of $d^{(p)}=3$ and $d^{(a)}=4$ which is given as Fig. 14. Its $(1,1)$-adjacency operator and its $(1,1)$-transition operator are not selfadjoint. Its eigenvalues are

$$\mathrm{Spec}(\Delta_{\mathcal{A}_{G^{(p)}}}) = \big[\,0,2,2,2,2,2,5,5,5,5\,\big],$$
$$\mathrm{Spec}(\Delta_{\mathcal{A}_G^{(1,1)}}) = \big[\,0,3,\epsilon,\epsilon,\rho,\rho,\varrho,\varrho,\varsigma,\varsigma\,\big],$$

where

$$\begin{cases} \epsilon = \dfrac{27+\sqrt{5}}{8} + \dfrac{\sqrt{-1}}{8}(\sqrt{5}-1), \\ \rho = \dfrac{27+\sqrt{5}}{8} - \dfrac{\sqrt{-1}}{8}(\sqrt{5}-1) \end{cases} \qquad \begin{cases} \varrho = \dfrac{27-\sqrt{5}}{8} + \dfrac{\sqrt{-1}}{8}(\sqrt{5}+1), \\ \varsigma = \dfrac{27-\sqrt{5}}{8} - \dfrac{\sqrt{-1}}{8}(\sqrt{5}+1). \end{cases}$$

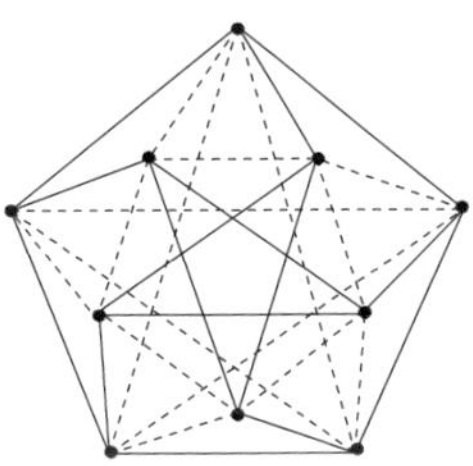

Fig. 13. Petersen Kähler graph

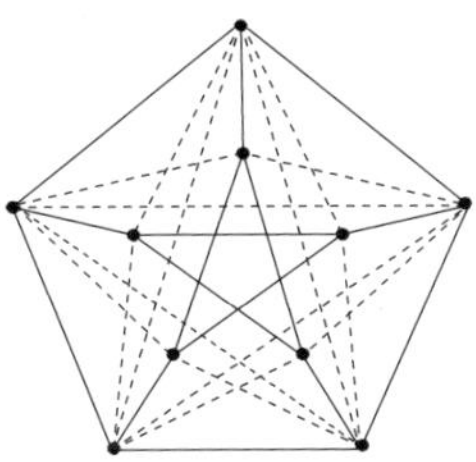

Fig. 14. Kähler Petersen graph

Example 5.3. 3-Kähler cube is a Kähler graph which is given as follows. The set of vertices is $Q_3 = \{(a_1,a_2,a_3) \mid a_i \in \{0,1\}\}$. For $v=(a_1,a_2,a_3)$, $w=(b_1,b_2,b_3)\in Q_3$, we define $v\sim_p w$ if and only if there is i_0 with $a_{i_0}\neq b_{i_0}$ and $a_i=b_i$ for $i\neq i_0$, and define $v\sim_a w$ if and only if there is i_1,i_2 $(i_1<i_2)$ with $a_{i_1}\neq b_{i_1}$, $a_{i_2}\neq b_{i_2}$ and $a_i=b_i$ for $i\neq i_1,i_2$. Its eigenvalues are

$$\mathrm{Spec}(\Delta_{\mathcal{A}_{Q_3^{(p)}}}) = \big[\,0,2,2,2,4,4,4,6\,\big],$$
$$\mathrm{Spec}(\Delta_{\mathcal{A}_{Q_3}^{(1,1)}}) = \big[\,0,8/3,3/8,3/8,10/3,10/3,10/3,6\,\big]$$

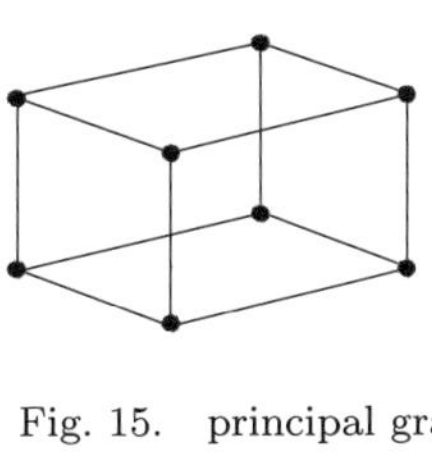

Fig. 15. principal graph

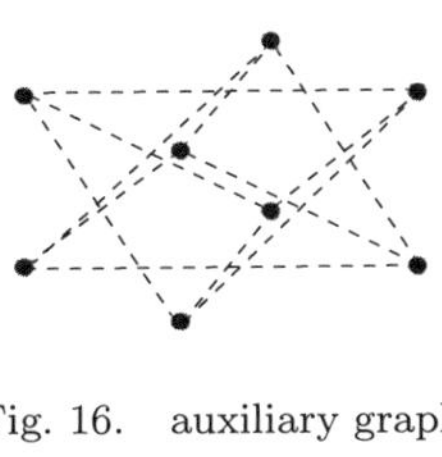

Fig. 16. auxiliary graph

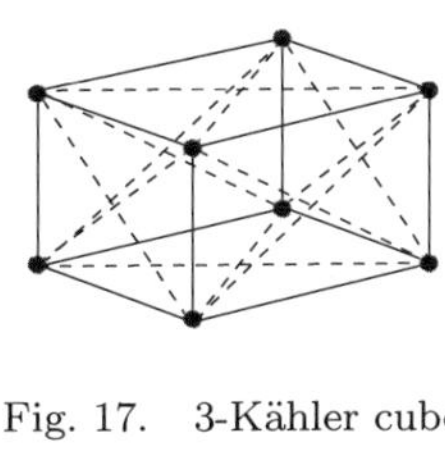

Fig. 17. 3-Kähler cube

Example 5.4. We take a dihadral group

$$D_4 = \langle a, b \mid a^4 = b^2 = 1,\ ab = ba^3 \rangle$$
$$= \langle b, c \mid b^2 = c^2 = 1,\ bcbc = cbcb \rangle,$$

where $c = ab$. If we take $\mathcal{S}^{(p)} = \{b, c\}$ and $\mathcal{S}^{(a)} = \{a, a^3\}$, we get a regular Kähler graph as in Fig. 18. Clearly, we find that its principal and auxiliary degrees are 2 and that the adjacent operators of the principal and the auxiliary graphs are commutative. Its eigenvalues are

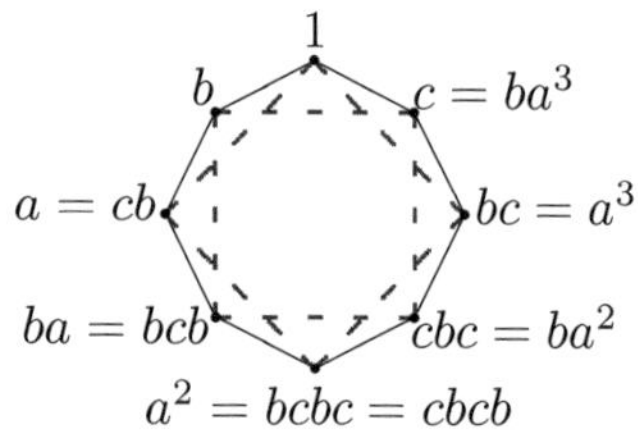

Fig. 18.

$$\mathrm{Spec}(\Delta_{\mathcal{A}_{G^{(p)}}}) = \left[\, 0, 2, 2, 2+\sqrt{2}, 2+\sqrt{2}, 2+\sqrt{2}, 2+\sqrt{2}, 4 \,\right],$$
$$\mathrm{Spec}(\Delta_{\mathcal{A}_G^{(1,1)}}) = \left[\, 0, 2, 2, 2, 2, 2, 2, 4 \,\right].$$

Example 5.5. An original Heawood graph is like Fig. 19 and has 14 vertices. We can construct many regular Kähler graphs; five kinds of Kähler graphs of auxiliary degree 2, eight kinds of Kähler graphs of auxiliary degree 3, and so on. We here study eigenvalues of the Kähler Heawood graph given as follows (see Fig. 20): We take $V = \{v_0, v_1, \cdots, v_{13}\}$ and define principal and auxiliary edges as

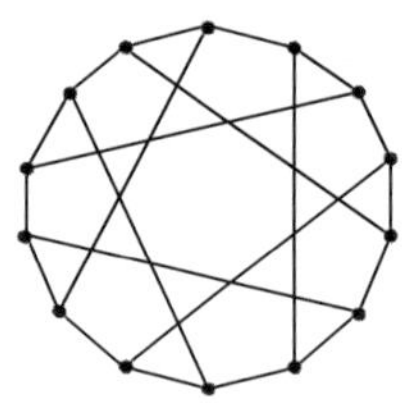
Fig. 19. original

$$E^{(p)} = \left\{ \begin{array}{l} (v_i, v_{i+1}) \quad (0 \le i \le 13), \\ (v_0, v_5), (v_2, v_7), (v_4, v_9), (v_6, v_{11}), \\ (v_8, v_{13}), (v_{10}, v_1), (v_{12}, v_3) \end{array} \right\},$$
$$E^{(a)} = \{(v_i, v_{i+4}) \mid 0 \le i \le 13\}.$$

Its eigenvalues are

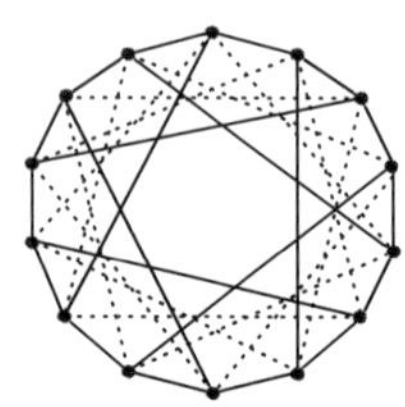
Fig. 20.

$$\mathrm{Spec}(\Delta_{\mathcal{A}_{G^{(p)}}}) = \left[\, 0, 6, a, a, a, a, a, a, b, b, b, b, b, b \,\right],$$
$$\mathrm{Spec}(\Delta_{\mathcal{A}_G^{(1,1)}}) = \left[\, 0, 6, 3-\epsilon, 3-\epsilon, 3+\epsilon, 3+\epsilon, 3-\rho, 3-\rho, 3+\rho, 3+\rho, 3-\varrho, 3-\varrho, 3+\varrho, 3+\varrho \,\right],$$

where

$$\begin{cases} a = 3 + \sqrt{2}, \\ b = 3 - \sqrt{2}, \end{cases} \qquad \begin{cases} \epsilon = \dfrac{1}{6}\Big(30 + 12\sqrt{7}\cos\dfrac{\theta}{3}\Big)^{1/2}, \\ \rho = \dfrac{1}{6}\Big\{30 - 12\sqrt{7}\sin\big(\dfrac{\theta}{3} + \dfrac{\pi}{6}\big)\Big\}^{1/2}, \\ \varrho = \dfrac{1}{6}\Big\{30 + 12\sqrt{7}\sin\big(\dfrac{\theta}{3} - \dfrac{\pi}{6}\big)\Big\}^{1/2}, \end{cases}$$

with θ determined by $\sin\theta = 3\sqrt{3}/28$, $\cos\theta = 1/28$.

6. Eigenvalues of Kähler graphs of product type

In this last section, we study eigenvalues of Kähler graphs obtained by product operations given in §2.

Theorem 6.1 ([6]). *Let $G = (V, E)$ and $H = (W, F)$ be finite ordinary graphs.*

(1) $\mathcal{P}^{(1,1)}_{G\widehat{\square}H}$ *is selfadjoint with respect to the weighted inner product on $C(V \times W)$ given by $\langle\langle\phi,\psi\rangle\rangle = \sum_{(v,w)\in V\times W} d_G(v)d_H(w)\phi(v,w)\psi(v,w)$. Here, $d_G(v)$, $d_H(w)$ are the principal and the auxiliary degrees at (v,w).*
(2) *If we denote the eigenvalues of $\Delta_{\mathcal{P}_G}$ of G by μ_j $(j = 1,\dots,n_G)$ and the eigenvalues of $\Delta_{\mathcal{P}_H}$ of H by ν_α $(\alpha = 1,\dots,n_G)$, then the eigenvalues of $\Delta_{\mathcal{P}^{(1,1)}_{G\widehat{\square}H}}$ are $\mu_j + \nu_\alpha - \mu_j\nu_\alpha$ $(j = 1,\dots,n_G,\ \alpha = 1,\dots,n_H)$.*

Proof. Since we have

$$\mathcal{P}^{(p)}_{G\widehat{\square}H}\mathcal{P}^{(a)}_{G\widehat{\square}H}\phi(v,w) = \frac{1}{d_G(v)d_H(w)}\sum_{v':v'\sim v(G)}\ \sum_{w':w'\sim w(H)}\phi(v',w')$$
$$= \mathcal{P}^{(a)}_{G\widehat{\square}H}\mathcal{P}^{(p)}_{G\widehat{\square}H}\phi(v,w),$$

we get the first assertion.

If we take functions $f \in C(V)$, $g \in C(W)$ with $\Delta_{\mathcal{P}_G}f = \mu f$ and $\Delta_{\mathcal{P}_H}g = \nu g$, we see $\mathcal{P}_G f = (1-\mu)f$ and $\mathcal{P}_H g = (1-\nu)g$. By taking a function $(f,g) \in C(V\times W)$ defined by $(f,g)(v,w) = f(v)g(w)$, we have $\mathcal{P}^{(1,1)}_{G\widehat{\square}H}(f,g) = (\mathcal{P}_G f, \mathcal{P}_H g)$, hence get the conclusion. □

Along the same lines we have the following.

Theorem 6.2 ([6]). *Let $G = (V, E)$ and $H = (W, F)$ be finite ordinary graphs. Suppose G be regular.*

(1) $\mathcal{P}^{(1,1)}_{G\widehat{\boxtimes}H}$ *is selfadjoint with respect to the weighted inner product on $C(V \times W)$.*
(2) *By use of the same notations as in Theorem 6.1, the eigenvalues of $\Delta_{\mathcal{P}^{(1,1)}_{G\widehat{\boxtimes}H}}$ are $(1+d_G-d_G\mu_j)(\mu_j+\nu_\alpha-\mu_j\nu_\alpha)/(d_G+1)$ $(j = 1,\dots,n_G,\ \alpha = 1,\dots,n_H)$.*

Proof. We have

$$\begin{aligned}
&\mathcal{P}^{(p)}_{G\widehat{\boxtimes}H}\mathcal{P}^{(a)}_{G\widehat{\boxtimes}H}\phi(v,w)\\
&\quad=\frac{1}{d_H(w)}\sum_{w':w'\sim w(H)}\left\{\frac{1}{d_G(d_G+1)}\Big(\sum_{v':v'\sim v(G)}\phi(v',w')+\sum_{\gamma}\phi\big(t(\gamma),w'\big)\Big)\right\}\\
&\quad=\mathcal{P}^{(a)}_{G\widehat{\boxtimes}H}\mathcal{P}^{(p)}_{G\widehat{\boxtimes}H}\phi(v,w),
\end{aligned}$$

where γ runs over the set of all 2-step paths in G of origin v. We hence get the first assertion.

As we have $\mathcal{P}^{(1,1)}_{G\widehat{\boxtimes}H}(f,g)=(\frac{\mathcal{P}_G(\mathcal{A}_G+1)}{d_G+1}f,\mathcal{P}_Hg)$, we get the conclusion. □

Theorem 6.3 ([6]). *Let $G=(V,E)$ and $H=(W,F)$ be finite ordinary graphs.*

(1) $\mathcal{P}^{(1,1)}_{G\widehat{\otimes}H}$ *is selfadjoint with respect to the weighted inner product on $C(V\times W)$.*
(2) *By use of the same notations as in Theorem 6.1, the eigenvalues of $\Delta_{\mathcal{P}^{(1,1)}_{G\widehat{\otimes}H}}$ are $(1-\mu_j)(\mu_j+\nu_\alpha-\mu_j\nu_\alpha)+\mu_j$ ($j=1,\ldots,n_G$, $\alpha=1,\ldots,n_H$).*

Proof. Since we have

$$\begin{aligned}
\mathcal{P}^{(p)}_{G\widehat{\otimes}H}\mathcal{P}^{(a)}_{G\widehat{\otimes}H}\phi(v,w)&=\frac{1}{d_H(w)}\sum_{w':w'\sim w(H)}\sum_{\gamma=(v,v',v'')}\frac{1}{d_G(v)d_G(v')}\phi(v'',w')\\
&=\mathcal{P}^{(a)}_{G\widehat{\otimes}H}\mathcal{P}^{(p)}_{G\widehat{\otimes}H}\phi(v,w)
\end{aligned}$$

and $\mathcal{P}^{(1,1)}_{G\widehat{\otimes}H}(f,g)=(\mathcal{P}_G^2f,\mathcal{P}_Hg)$, we get the conclusion. □

Theorem 6.4 ([6]). *Let $G=(V,E)$ and $H=(W,F)$ be finite ordinary graphs.*

(1) $\mathcal{P}^{(1,1)}_{G\triangleright H}$ *is selfadjoint with respect to the weighted inner product on $C(V\times W)$.*
(2) *If we denote the eigenvalues of $\Delta_{\mathcal{P}_H}$ of H by ν_α ($\alpha=1,\ldots,n_G$), then the eigenvalues of $\Delta_{\mathcal{P}^{(1,1)}_{G\triangleright H}}$ are $n_G(n_G-1)$ times of 1 and ν_α ($\alpha=1,\ldots,n_H$).*

Proof. We have

$$\begin{aligned}
&\mathcal{P}^{(p)}_{G\triangleright H}\mathcal{P}^{(a)}_{G\triangleright H}\phi(v,w)\\
&=\mathcal{P}^{(p)}_{G\triangleright H}\left(\frac{1}{n_G d_H(w)}\sum_{u\in V}\sum_{w':w'\sim w(H)}\phi(u,w')\right)\\
&=\frac{1}{n_G d_H(w)}\sum_{u\in V}\sum_{w':w'\sim w(H)}\phi(u,w')=\mathcal{P}^{(a)}_{G\triangleright H}\mathcal{P}^{(p)}_{G\triangleright H}\phi(v,w),
\end{aligned}$$

hence get the first assertion. The above also shows $\mathcal{P}^{(1,1)}_{G\triangleright H}(f,g) = \left(\frac{1}{n_G}\mathcal{M}_G f, \mathcal{P}_H g\right)$. We set $V = \{v_1,\dots,v_{n_G}\}$ and take functions ε_k $(k = 1,\dots,n_G)$ defined by $\varepsilon_1 = \delta_{v_i}$ $\varepsilon_k = \delta_{v_1} - \delta_{v_k}$ $(k \geq 2)$ with characteristic functions δ_{v_k}. Since we have $\mathcal{M}_G\varepsilon_1 = n_G\varepsilon_1$ and $\mathcal{M}_G\varepsilon_k = 0$ $(k \geq 2)$, we have $\Delta_{\mathcal{P}^{(1,1)}_{G\triangleright H}}(\varepsilon_1, g) = (\varepsilon_1, \Delta_{\mathcal{P}_H} g)$ and $\Delta_{\mathcal{P}^{(1,1)}_{G\triangleright H}}(\varepsilon_k, g) = (\varepsilon_k, g)$ for $k \geq 2$. We hence get the conclusion. □

Theorem 6.5 ([8]). *Let $G = (V,E)$ and $H = (W,F)$ be finite regular ordinary graphs.*

(1) $\mathcal{P}^{(1,1)}_{G\boxplus H}$ *is selfadjoint with respect to the canonical inner product on* $C(V \times W)$.
(2) *By use of the same notations as in Theorem 6.1, the eigenvalues of* $\Delta_{\mathcal{P}^{(1,1)}_{G\boxplus H}}$ *are* $1 - (1-\mu_j)(1-\nu_\alpha)\{d_G(1-\mu_j) + d_H(1-\nu_\alpha)\}$ $(j = 1,\dots,n_G,\ \alpha = 1,\dots,n_H)$.

In order to see this the following lemma is useful.

Lemma 6.1. *Let G be a finite regular Kähler graph. Suppose the adjacency operators $\mathcal{A}_{G^{(p)}}$ and $\mathcal{A}_{G^{(a)}}$ are simultaneously diagonalizable. If we denote the corresponding eigenvalues of $\mathcal{A}_{G^{(p)}}$ and $\mathcal{A}_{G^{(a)}}$ by λ_k, ρ_k $(k = 1,\dots,n_G)$, then the eigenvalues of $\Delta_{\mathcal{P}_{G(1,1)}}$ are of the form $1 - \lambda_k\rho_k/\{d_G^{(p)}d_G^{(a)}\}$ $(k = 1,\dots,n_G)$.*

By the definition of $G \boxplus H$ it is easy to find that its adjacency operators of principal and auxiliary graphs are simultaneously diagonalizable. As we have $\mathcal{A}_{(G\boxplus H)^{(p)}}(f,g) = (\mathcal{A}_G f, g) + (f, \mathcal{A}_H g)$ and $\mathcal{A}_{(G\boxplus H)^{(a)}}(f,g) = (\mathcal{A}_G f, \mathcal{A}_H g)$, we get the conclusion of Theorem 6.5.

Similarly as we have $\mathcal{A}_{(G\boxdot H)^{(a)}}(f,g) = (\mathcal{A}_G f, \mathcal{A}_{H^c} g) + (\mathcal{A}_{G^c} f, \mathcal{A}_H g)$ and $\mathcal{A}_{(G\lozenge H)^{(a)}}(f,g) = (\mathcal{A}_G f, \mathcal{M}_H g - g) + (\mathcal{M}_G f - f, \mathcal{A}_H g)$, we can get the eigenvalues of $(1,1)$-laplacians for $G \boxdot H$ and $G\lozenge H$.

Since we can write down the eigenvalues of $(1,1)$-Laplacians for Kähler graphs of product types by eigenvalues of their factors, we get the following.

Corollary 6.1. *Let G_1, G_2 and H_1, H_2 are two pairs of isospectral ordinary graphs.*

(1) $G_1 \widehat{\Box} H_1$, $G_2 \widehat{\Box} H_2$ *and* $G_1 \widehat{\otimes} H_1$, $G_2 \widehat{\otimes} H_2$ *are two pairs of isospectral Kähler graphs.*
(2) *If* G_1, G_2 *are regular, then* $G_1 \widehat{\boxtimes} H_1$ *and* $G_2 \widehat{\boxtimes} H_2$ *are isospectral Kähler graphs.*
(3) *If* G_1, G_2 *are connected and regular, then* $G_1 \triangleright H_1$ *and* $G_2 \triangleright H_2$ *are isospectral Kähler graphs.*
(4) *If* G_1, G_2, H_1, H_2 *are regular, then* $G_1 \boxplus H_1$ *and* $G_2 \boxplus H_2$ *are isospectral Kähler graphs.*
(5) *If* G_1, G_2, H_1, H_2 *are connected and regular, then* $G_1 \boxdot H_1$, $G_2 \boxdot H_2$ *and* $G_1 \Diamond H_1$, $G_2 \Diamond H_2$ *are two pairs of isospectral Kähler graphs.*

References

1. T. Adachi, Kähler magnetic flows on a manifold of constant holomorphic sectional curvature, *Tokyo J. Math.* **18**, 473–483 (1995).
2. ______, A discrete model for Kähler magnetic fields on a complex hyperbolic space, in *Trends in Differential Geometry, Complex Analysis and Mathematical Physics*, K. Sekigawa, V.S. Gerdjikov & S. Dimiev eds, 1–9 (World Scientific, Singapore, 2009).
3. ______, Laplacians for regular Kähler graphs and their duals, this volume.
4. A.E. Brouwer & W.H. Haemers, *Spectra of Graphs*, Springer 2012.
5. Yaermaimaiti T. & T. Adachi, A note on vertex-transitive Kähler graphs, to appear in *Hokkaido Math. J.*.
6. ______, Isospectral Kähler graphs, *Kodai Math. J.* **38**, 560–580 (2015).
7. ______, Laplacians for derived graphs of a regular Kähler graph, *C. R. Math. Acad. Sci. Soc. R. Canada* **37**, 141–156 (2015).
8. ______, Kähler graphs of connected product type, *Applied Mathematics and Information Sciences* **9**, 2767–2773 (2015).

Received January 16, 2015
Revised April 21, 2015

Current Developments in Differential Geometry
and its Related Fields *pp. 155 – 163*

HOPF FIBRATION AND CARTAN IMBEDDING OF TYPE *AI*

Hideya HASHIMOTO *

Department of Mathematics, Meijo University,
Nagoya 468-8502, Japan
E-mail: hhashi@meijo-u.ac.jp

Kazuhiro SUZUKI

Division of Mathematics and Mathematical Science,
Nagoya Institute of Technology,
Nagoya 466-8555, Japan
E-mail: cjv17505@stn.nitech.ac.jp

The Hopf fibration $\pi : S^3 \to S^2$ over two dimensional sphere S^2 with S^1 fibre is related to the Cartan imbedding of type AI. Taking the composition of the Cartan imbedding and non-flat totally geodesic imbedding from S^2 to $SU(n)/SO(n)$ of symmetric space of type AI, we give the explicit representation of the totally geodesic imbedding from S^2 to $SU(n)$.

Keywords: Hopf fibration, Cartan imbedding of type AI, non-flat totally geodesic surfaces.

1. Introduction

Recently, in [1], K. Mashimo classified non-flat totally geodesic surfaces in symmetric spaces of classical type. The composition of his totally geodesic imbedding and the Cartan imbedding is a totally geodesic imbedding from two dimensional sphere S^2 to $SU(n)$. In this paper, we give the concrete expression of these imbeddings by using the polynomials and calculate the sectional curvature with respect to the induced metric.

2. On symmetric space of type *AI*

First we recall the fundamental facts about Cartan imbeddings of type AI. Let $\langle\ ,\ \rangle$ be the Hermitian metric on $\mathbb{C}^n$. Let $SU(n)$, and $SO(n)$ be the special unitary group and special orthogonal group, respectively. These groups are defined by

$$SU(n) = \{g \in M_{n\times n}(\mathbb{C}) \mid \langle gu,\ gv\rangle = \langle u,\ v\rangle \text{ for any } u, v \in \mathbb{C}^n\}$$

*The first author is partially supported by JSPS KAKENHI Grant Number 15K04860.

and

$$SO(n) = \{g \in SU(n) \mid \bar{g} = g\},$$

where $\bar{g}$ denote the conjugation of $g \in M_{n\times n}(\mathbb{C})$. We define the involution σ on $SU(n)$ as

$$\sigma(g) = \bar{g}.$$

The map $\widetilde{\Phi_\sigma} : SU(n) \to SU(n)$ defined by

$$\widetilde{\Phi_\sigma}(g) = g\sigma(g^{-1}) = g\,{}^t g,$$

induces the imbedding

$$\Phi_\sigma : SU(n)/SO(n) \to SU(n),$$

which is called Cartan imbedding. This imbedding is a totally geodesic one. Here the quotient space $SU(n)/SO(n)$ is a Riemannian symmetric space of type AI with respect to the induced metric by Φ_σ. We note that the metric has the ambiguity of the constant factor. To see the geometric properties of the Riemannian symmetric space $SU(n)/SO(n)$ of type AI, we give the decomposition of the tangent space $\mathfrak{su}(n) = T_eSU(n)$ at the identity $e = I_n \in M_{n\times n}$ of $SU(n)$ by σ. Since σ is an involution on $SU(n)$, the linear map

$$\sigma_*|_e : \mathfrak{su}(n) \to \mathfrak{su}(n)$$

has two eigenvalues $\{\pm 1\}$. Let $\mathfrak{p}$ and $\mathfrak{k}$ be the eigen-vector space corresponding to the eigenvalue -1 and $+1$, respectively. Then

$$\begin{aligned}\mathfrak{p} &= \{X \in \mathfrak{su}(n) \mid \sigma_* X = -X\}\\ &= \{iU \in M_{n\times n}(\mathbb{C}) \mid U \in M_{n\times n}(\mathbb{R}), {}^t U = U\}\end{aligned}$$

and

$$\begin{aligned}\mathfrak{k} &= \{X \in \mathfrak{su}(n) \mid \sigma_* X = X\}\\ &= \{X \in M_{n\times n}(\mathbb{R}) \mid X + {}^t X = O_{n\times n}\} = \mathfrak{so}(n).\end{aligned}$$

The subspace $\mathfrak{p}$ can be identified with the tangent space

$$T_{eSO(n)}SU(n)/SO(n).$$

We note that the dimension of $\mathfrak{p}$ is $n(n+1)/2 - 1$.

3. On irreducible representations of $SU(2)$

We recall construction of complex irreducible representations of $SU(2)$. Let $V(d)$ be the complex vector space of homogeneous polynomials $P(z, w)$ of degree d in two variables, (z, w) of $\mathbb{C}^2$. Namely

$$V(d) = span_{\mathbb{C}}\{z^{d-j}\, w^j \mid j \in \mathbb{N}_0, 0 \leq j \leq d\}.$$

Let $\rho_d : SU(2) \to End\{V(d)\}$ be the representation, which is defined by

$$(\rho_d(g)P)(z,w) = P\left({}^t(g^{-1} \begin{pmatrix} z \\ w \end{pmatrix}) \right) = P\left((z,w) \begin{pmatrix} \bar{a} & -b \\ \bar{b} & a \end{pmatrix} \right),$$

for any function $P \in V(d)$ and $g = \begin{pmatrix} a & -\bar{b} \\ b & \bar{a} \end{pmatrix} \in SU(2)$. Here $a, b \in \mathbb{C}$ satisfy $|a|^2 + |b|^2 = 1$. Then each ρ_d is a complex irreducible representation of $SU(2)$.

Example 3.1. For $V(2)$, we give the representation matrix with respect to the basis $\{P_0(z,w) = w^2, P_1(z,w) = \sqrt{2}wz, P_2(z,w) = z^2\}$. To compute this, we have

$$\begin{aligned}
(\rho(g)P_0)(z,w) &= (-bz + aw)^2 = w^2a^2 + wz(-2ab) + z^2b^2, \\
(\rho(g)P_1)(z,w) &= \sqrt{2}\,(-bz + aw)(\bar{a}z + \bar{b}w) \\
&= \sqrt{2}\{w^2a\bar{b} + wz(|a|^2 - |b|^2) + z^2(-\bar{a}b)\}, \\
(\rho(g)P_2)(z,w) &= (\bar{a}z + \bar{b}w)^2 = w^2\bar{b}^2 + wz(2\bar{a}\bar{b}) + z^2\bar{a}^2.
\end{aligned}$$

Let A_g be the representation matrix defined by

$$(\rho(g)P_0\ \rho(g)P_1\ \rho(g)P_2) = (P_0\ P_1\ P_2)A_g.$$

Then this matrix A_g is given by

$$A_g = \begin{pmatrix} a^2 & \sqrt{2}a\bar{b} & \bar{b}^2 \\ -\sqrt{2}ab & |a|^2 - |b|^2 & \sqrt{2}\bar{a}\bar{b} \\ b^2 & -\sqrt{2}\bar{a}b & \bar{a}^2 \end{pmatrix} \in SU(3). \tag{3.1}$$

Example 3.2. For $V(3)$, we give the representation matrix with respect to the basis $\{P_0(z,w) = w^3, P_1(z,w) = \sqrt{3}w^2z, P_2(z,w) = \sqrt{3}wz^2, P_0(z,w) = z^3\}$. By the computation similar to the above one, we obtain the representation matrix

$$A_g = \begin{pmatrix} a^3 & \sqrt{3}a^2\bar{b} & \sqrt{3}a\bar{b}^2 & \bar{b}^3 \\ -\sqrt{3}a^2b & a(|a|^2 - 2|b|^2) & -\bar{b}(|b|^2 - 2|a|^2) & \sqrt{3}\bar{a}\bar{b}^2 \\ \sqrt{3}ab^2 & b(|b|^2 - 2|a|^2) & \bar{a}(|a|^2 - 2|b|^2) & \sqrt{3}\bar{a}^2\bar{b} \\ -b^3 & \sqrt{3}\bar{a}b^2 & -\sqrt{3}\bar{a}^2b & \bar{a}^3 \end{pmatrix} \in SU(4). \tag{3.2}$$

4. Composition Cartan imbeddings and irreducible representations of $SU(2)$

Since the image of the Cartan imbedding Φ_σ is a totally geodesic submanifold and the irreducible representations ρ_{n-1} (considered as a mapping from $SU(2)$ to $SU(n)$) is the totally geodesic immersion from $SU(2)$ into $SU(n)/SO(n)$, the composition $\Phi_\sigma \circ \rho_n$ is also totally geodesic imbedding. By using this composition, we can calculate directly the Gauss curvature of S^2 by taking account of submanifold theory. The image of the irreducible representation is a non-flat surface (which is isometric to S^2 or $\mathbb{R}P^2$, this fact comes from Gauss-Bonnet theorem or see Wolf [3]). In his paper [1], K. Mashimo classified non-flat totally geodesic surfaces in symmetric spaces of classical type. In this article, we give the expression of these maps by polynomials, explicitly.

Example 4.1. When $n = 2$, the image of the Cartan imbedding

$$\Phi_\sigma : SU(2)/SO(2) \to SU(2)$$

is a totally geodesic surface of $S^3 \simeq SU(2)$. Its tangent space at the identity can be identified with $\mathrm{span}_{\mathbb{R}}\left\{\begin{pmatrix} i & 0 \\ 0 & -i \end{pmatrix}, \begin{pmatrix} 0 & i \\ i & 0 \end{pmatrix}\right\}$, and the base of tangent space (at the identity) of the fibre $SO(2)$ of the projection $SU(2) \to SU(2)/SO(2)$ is $\begin{pmatrix} 0 & -1 \\ 1 & 0 \end{pmatrix}$. In fact, for any $g = \begin{pmatrix} a & -\bar{b} \\ b & \bar{a} \end{pmatrix} \in SU(2)$, we have

$$\Phi_\sigma(g) = g\,{}^t g = \begin{pmatrix} a & -\bar{b} \\ b & \bar{a} \end{pmatrix} \begin{pmatrix} a & b \\ -\bar{b} & \bar{a} \end{pmatrix} = \begin{pmatrix} a^2 + \bar{b}^2 & ab - \bar{a}\bar{b} \\ ab - \bar{a}\bar{b} & b^2 + \bar{a}^2 \end{pmatrix}.$$

If we identify $\begin{pmatrix} z & -\bar{w} \\ w & \bar{z} \end{pmatrix} \in SU(2)$ with $\begin{pmatrix} z \\ w \end{pmatrix} \in S^3$, then the image

$$\left\{ \begin{pmatrix} a^2 + \bar{b}^2 \\ ab - \bar{a}\bar{b} \end{pmatrix} \,\middle|\, \begin{pmatrix} a \\ b \end{pmatrix} \in S^3 \right\}$$

of Φ_σ is a totally geodesic 2-sphere S^2 in S^3. In fact, the image is a intersection of (Real part of w) $= 0$ and S^3. We note that the inverse image $\Phi_\sigma^{-1}(I_{2\times 2}) = SO(2)$ where

$$SO(2) = \left\{ \begin{pmatrix} \cos t & -\sin t \\ \sin t & \cos t \end{pmatrix} \,\middle|\, t \in \mathbb{R} \right\}.$$

More precisely, let π be the map from S^3 to S^2 defined by

$$\pi(\begin{pmatrix} a \\ b \end{pmatrix}) = \begin{pmatrix} a^2 + \bar{b}^2 \\ ab - \bar{a}\bar{b} \end{pmatrix} = \begin{pmatrix} \alpha \\ iu \end{pmatrix}$$

where $\begin{pmatrix} a \\ b \end{pmatrix} \in S^3 \subset \mathbb{C}^2$. The map π coincides with the Hopf fibration $\pi : S^3 \to S^2$ with fibre S^1. In fact we put $\alpha = a^2 + \bar{b}^2$ and $iu = ab - \bar{a}\bar{b}$, then $\alpha \in \mathbb{C}$ and $u \in \mathbb{R}$ which satisfy $|\alpha|^2 + u^2 = 1$.

Example 4.2. By (3.1), we can obtain the map from $SU(2)/SO(2)$ to $SU(3)$ as

$$\tilde{\varphi_3}([g]) = A_g\ {}^tA_g = \begin{pmatrix} \alpha^2 & -\sqrt{2}iu\alpha & -u^2 \\ -\sqrt{2}iu\alpha & |\alpha|^2 - u^2 & -\sqrt{2}iu\bar{\alpha} \\ -u^2 & -\sqrt{2}iu\bar{\alpha} & \bar{\alpha}^2 \end{pmatrix} \in SU(3), \qquad (4.1)$$

where $\alpha = a^2 + \bar{b}^2$ and $iu = ab - \bar{a}\bar{b}$ as above and $[g] \in SU(2)/SO(2)$. We put $\varphi_3(\alpha, u)$ the right hand side of (4.1). Then φ_3 is an immersion from S^2 to $SU(3)$. Since $\varphi_3(\alpha, u) = \varphi_3(-\alpha, -u)$, we can show that the image $\varphi_3(S^2)$ is diffeomorphic to a 2-dimensional real projective space $\mathbb{R}P^2$. In particular,

$$\varphi_3(e^{it}, 0) = \begin{pmatrix} e^{2it} & 0 & 0 \\ 0 & 1 & 0 \\ 0 & 0 & e^{-2it} \end{pmatrix}.$$

Note that a curve $a = e^{it/2}$, $b = 0$ of S^3 corresponds to a geodesic $(e^{it}, 0)$ in S^2.

$$\varphi_3(\cos s, \sin s) = \begin{pmatrix} \cos^2 s & -\sqrt{2}i \sin s \cos s & -\sin^2 s \\ -\sqrt{2}i \sin s \cos s & \cos 2s & -\sqrt{2}i \sin s \cos s \\ -\sin^2 s & -\sqrt{2}i \sin s \cos s & \cos^2 s \end{pmatrix},$$

and that a curve $a = \cos s/2$, $b = i \sin s/2$ of S^3 corresponds to a geodesic $(\cos s, \sin s)$ in S^2.

Example 4.3. By (3.2), we can obtain the map from $SU(2)/SO(2)$ to $SU(4)$ as

$$\tilde{\varphi_4}([g]) = A_g\ {}^tA_g$$
$$= \begin{pmatrix} \alpha^3 & -\sqrt{3}iu\alpha^2 & -\sqrt{3}u^2\alpha & iu^3 \\ -\sqrt{3}iu\alpha^2 & \alpha(1-3u^2) & iu(3u^2-2) & -\sqrt{3}u^2\bar{\alpha} \\ -\sqrt{3}u^2\alpha & iu(3u^2-2) & \bar{\alpha}(1-3u^2) & -\sqrt{3}iu\bar{\alpha}^2 \\ iu^3 & -\sqrt{3}u^2\bar{\alpha} & -\sqrt{3}iu\bar{\alpha}^2 & \bar{\alpha}^3 \end{pmatrix} \in SU(4), \qquad (4.2)$$

where $\alpha = a^2 + \bar{b}^2$ and $iu = ab - \bar{a}\bar{b}$ as above $[g] \in SU(2)/SO(2)$. We put $\varphi_4(\alpha, u)$ the right hand side of (4.2). Then φ_4 is an immersion from S^2 to $SU(4)$. Since $\varphi_4(-\alpha, -u) = -\varphi_4(\alpha, u)$, we can show that the image $\varphi_4(S^2)$ is diffeomorphic to a 2-dimensional sphere and the immersion is an imbedding. In particular, if we put $\alpha = e^{it}, u = 0$, then

$$\varphi_4(e^{it}, 0) = \begin{pmatrix} e^{3it} & 0 & 0 & 0 \\ 0 & e^{it} & 0 & 0 \\ 0 & 0 & e^{-it} & 0 \\ 0 & 0 & 0 & e^{-3it} \end{pmatrix}.$$

Note that a curve $a = e^{it/2}$, $b = 0$ of S^3 corresponds to a geodesic $(e^{it}, 0)$ in S^2. Also we have

$$\varphi_4(\cos s, \sin s) =$$
$$\begin{pmatrix} \cos^3 s & -\sqrt{3}i \sin s \cos^2 s & -\sqrt{3} \sin^2 s \cos s & i \sin^3 s \\ -\sqrt{3}i \sin s \cos^2 s & \cos s(1 - 3\sin^2 s) & i \sin s(3\sin^2 s - 2) & -\sqrt{3} \sin^2 s \cos s \\ -\sqrt{3} \sin^2 s \cos s & i \sin s(3\sin^2 s - 2) & \cos s(1 - 3\sin^2 s) & -\sqrt{3}i \sin s \cos^2 s \\ i \sin^3 s & -\sqrt{3} \sin^2 s \cos s & -\sqrt{3}i \sin s \cos^2 s & \cos^3 s \end{pmatrix}$$

and that a curve $a = \cos s/2$, $b = i \sin s/2$ of S^3 corresponds to a geodesic $(\cos s, \sin s)$ in S^2.

5. Induced metrics by φ_3 and φ_4

Let $\langle\ ,\ \rangle_0$ be a bi-invariant Riemannian metric on $SU(n)$ by

$$\langle X, Y \rangle_0 = \mathrm{Re}(\mathrm{tr}({}^t\bar{X}Y)),$$

for any $X, Y \in T_g SU(n)$. We give the local parametrization of the immersion φ_3 by

$$\tau_3(t, \tilde{s})$$
$$= A(t) \begin{pmatrix} \cos^2 2\tilde{s} & -\sqrt{2}i \sin 2\tilde{s} \cos 2\tilde{s} & -\sin^2 2\tilde{s} \\ -\sqrt{2}i \sin 2\tilde{s} \cos 2\tilde{s} & \cos 4\tilde{s} & -\sqrt{2}i \sin 2\tilde{s} \cos 2\tilde{s} \\ -\sin^2 2\tilde{s} & -\sqrt{2}i \sin 2\tilde{s} \cos 2\tilde{s} & \cos^2 2\tilde{s} \end{pmatrix} A(t)$$
$$= \begin{pmatrix} e^{4it} \cos^2 2\tilde{s} & -e^{2it}\sqrt{2}i \sin 2\tilde{s} \cos 2\tilde{s} & -\sin^2 2\tilde{s} \\ -e^{2it}\sqrt{2}i \sin 2\tilde{s} \cos 2\tilde{s} & \cos 4\tilde{s} & -e^{-2it}\sqrt{2}i \sin 2\tilde{s} \cos 2\tilde{s} \\ -\sin^2 2\tilde{s} & -e^{-2it}\sqrt{2}i \sin 2\tilde{s} \cos 2\tilde{s} & e^{-4it} \cos^2 2\tilde{s} \end{pmatrix}$$

where $A(t) = \begin{pmatrix} e^{2it} & 0 & 0 \\ 0 & 1 & 0 \\ 0 & 0 & e^{-2it} \end{pmatrix}$. Put $A = 2\begin{pmatrix} i & 0 & 0 \\ 0 & 0 & 0 \\ 0 & 0 & -i \end{pmatrix}$, $B = -\sqrt{2}i\begin{pmatrix} 0 & 1 & 0 \\ 1 & 0 & 1 \\ 0 & 1 & 0 \end{pmatrix}$,
and we define the map $\varphi_{AB} : \mathbb{R}^2 \to SU(3)$ by

$$\varphi_{AB}(t, \tilde{s}) = \exp(tA)\exp(2\tilde{s}B)\exp(tA)$$

We can easily check that

$$\varphi_{AB}(t, \tilde{s}) = \tau_3(t, \tilde{s})$$

for any $(t, \tilde{s}) \in \mathbb{R}^2$. In order to compute the induced metric, we use the tangent vector fields along the immersion φ_{AB} given by

$$\varphi_{AB*}\left(\frac{\partial}{\partial t}\right) = A \cdot \varphi_{AB}(t, \tilde{s}) + \varphi_{AB}(t, \tilde{s}) \cdot A, \tag{5.1}$$

and

$$\begin{aligned} \varphi_{AB*}\left(\frac{\partial}{\partial \tilde{s}}\right) &= 2Ad(\exp tA)B \cdot \varphi_{AB}(t, \tilde{s}) \\ &= 2\varphi_{AB}(t, \tilde{s}) \cdot Ad(\exp(-tA))B \end{aligned} \tag{5.2}$$

Proposition 5.1. *Let $\varphi_{AB} : \mathbb{R}^2 \to SU(3)$ be the immersion defined as above. Then the induced metric $\varphi_{AB}^*\langle\ ,\ \rangle_0$ is given by*

$$\varphi_{AB}^*\langle\ ,\ \rangle_0 = 2\{||A||_0^2 + \langle A, Ad(\exp(-2\tilde{s}B)A)\rangle_0\}\, dt^2 + 4||B||_0^2\, d\tilde{s}^2. \tag{5.3}$$

The Gaussian curvature K of image of φ_{AB} with respect to the induced metric is given by

$$K = \frac{||[A, B]||_0^2}{4\ ||A||_0^2\ ||B||_0^2} \tag{5.4}$$

where $[A, B] = AB - BA$.

Usually the computation of the curvature of symmetric spaces is by using the notion of transvection. In this article, we give a more direct method to compute the curvatures. Namely the method by using the submanifold theory. Now we give a proof of Proposition 5.1.

Proof. By (5.1), (5.2), we define the dual 1-forms ω_1, ω_2 on $\mathbb{R}^2$ by

$$\omega_1 = \sqrt{f(\tilde{s})}\, dt, \tag{5.5}$$

$$\omega_2 = 2||B||_0\, d\tilde{s}, \tag{5.6}$$

where $f(\tilde{s}) = 2\{||A||_0^2 + \langle A, Ad(\exp(-2\tilde{s}B)A)\rangle_0\}$. Let D be a differential of $M_{3\times 3}(\mathbb{C}) \simeq \mathbb{C}^9$. By surface theory, we can define the connection 1-form ω_{12} as

$$\omega_{12}(X) = \langle D_X \varphi_{AB*} e_2, \varphi_{AB*} e_1 \rangle_0, \tag{5.7}$$

where $e_1 = \dfrac{1}{\sqrt{f(\tilde{s})}}\dfrac{\partial}{\partial t}, e_2 = \dfrac{1}{||B||_0}\dfrac{\partial}{\partial \tilde{s}}$ and X is a tangent vector at any point of $\mathbb{R}^2$. Then we have

$$d\omega_1 + \omega_{12} \wedge \omega_2 = 0.$$

By (5.7), we see that

$$\omega_{12} = \frac{1}{4||B||_0\sqrt{f(\tilde{s})}}\frac{df}{d\tilde{s}}(\tilde{s})dt \tag{5.8}$$

Then the Gaussian curvature $K(t, \tilde{s})$ of this immersion is defined by the following equation

$$d\omega_{12} = K(t, \tilde{s})\omega_1 \wedge \omega_2 \tag{5.9}$$

By (5.8), (5.9), we get

$$\begin{aligned} d\omega_{12} &= \frac{1}{4||B||_0} \cdot \frac{d}{d\tilde{s}}\left(\sqrt{f(\tilde{s})}\frac{df}{d\tilde{s}}(\tilde{s})\right) d\tilde{s} \wedge dt \\ &= \frac{-1}{8||B||_0^2\sqrt{f(\tilde{s})}} \cdot \frac{d}{d\tilde{s}}\left(\sqrt{f(\tilde{s})}\frac{df}{d\tilde{s}}(\tilde{s})\right) \omega_1 \wedge \omega_2 \end{aligned} \tag{5.10}$$

By (5.9) and (5.10), we have

$$K(t, \tilde{s}) = \frac{1}{16||B||_0^2 f(\tilde{s})^2}\left\{\left(\frac{df}{d\tilde{s}}(\tilde{s})\right)^2 - 2f(\tilde{s})\frac{d^2 f}{d\tilde{s}^2}(\tilde{s})\right\} \tag{5.11}$$

We note that $K(t, \tilde{s})$ depends only on $\tilde{s}$. Also we see that

$$f(0) = 4||A||_0^2, \quad \frac{df}{d\tilde{s}}(0) = 0, \quad \frac{d^2 f}{d\tilde{s}^2}(0) = -8||\,[A, B]\,||_0^2. \tag{5.12}$$

Therefore we obtain

$$K(0, 0) = \frac{||\,[A, B]\,||_0^2}{4\,||A||_0^2\,||B||_0^2}.$$

Since the immersion φ_{AB} is defined as an orbit of the $SU(2)$−action, the Gaussian curvature is constant. Hence we get the desired result. □

Acknowledgements

The author is grateful to the referee for his careful reading.

References

1. K. Mashimo. Non-flat totally geodesic surfaces of symmetric space of classical type, preprint.
2. M. Takeuchi. *Lie groups II*, (in Japanese) Iwanami Shoten, Tokyo, 1984
3. J.A. Wolf. *Spaces of constant curvature.* Sixth edition. A.M.S. Chelsa Publishing, Providence, RI, 2011.

Received January 22, 2015
Revised April 25, 2015

Current Developments in Differential Geometry and its Related Fields pp. 165 – 182

ON TOTALLY UMBILICAL AND SCREEN TOTALLY UMBILICAL RADICAL TRANSVERSAL LIGHTLIKE HYPERSURFACES OF KÄHLER-NORDEN MANIFOLDS

Galia NAKOVA *

Department of Algebra and Geometry, Faculty of Mathematics and Informatics, University of Veliko Turnovo "St. Cyril and St. Methodius", T. Tarnovski 2 str., 5003 Veliko Tarnovo, Bulgaria
E-mail: gnakova@gmail.com

In this paper we study totally umbilical and screen totally umbilical radical transversal lightlike hypersurfaces of Kähler-Norden manifolds of constant totally real sectional curvatures. We investigate two classes of radical transversal lightlike hypersurfaces in such a Kähler-Norden manifold, and show conditions for non-existence of totally umbilical or screen totally umbilical hypersurfaces in these classes. We construct examples of four classes of totally umbilical and screen totally umbilical radical transversal lightlike hypersurfaces.

Keywords: Kähler-Norden manifold, radical transversal lightlike hypersurfaces.

1. Introduction

Recently, many authors have studied lightlike submanifolds of semi-Riemannian manifolds. The general theory of these submanifolds has been developed in [3,4]. The geometries of non-degenerate and lightlike submanifolds are different from each other because in lightlike case the tangent bundle and the normal bundle have a non empty intersection.

A $2n$-dimensional almost complex manifold with Norden metric $(\overline{M},\overline{J},\overline{g},\overline{\overline{g}})$ is a semi-Riemannian manifold equipped with an almost complex structure $\overline{J}$ which is an anti-isometry with respect to the metric $\overline{g}$. Due to this, there exist two Norden metrics (or B-metrics) $\overline{g}$ and $\overline{\overline{g}}$ on $\overline{M}$ which are necessarily indefinite of signature (n,n). We studied submanifolds of an almost complex manifold with Norden metric which are non-degenerate with respect to one Norden metric and lightlike with respect to the other one in [9]. We introduced in [10] radical transversal lightlike hypersurfaces

*The author is partially supported by Scientific researches fund of "St. Cyril and St. Methodius" University of Veliko Tarnovo under contract RD-09-422-13 / 09.04.2014.

of $(\overline{M}, \overline{J}, \overline{g}, \overline{\overline{g}})$. The paper [11] was devoted to study curvature conditions of semi-symmetric type of totally umbilical radical transversal lightlike hypersurfaces of Kähler-Norden manifolds of constant totally real sectional curvatures $\overline{\nu}$ and $\overline{\overline{\nu}}$.

In this paper we study totally umbilical and screen totally umbilical radical transversal lightlike hypersurfaces of Kähler-Norden manifolds of constant $\overline{\nu}$ and $\overline{\overline{\nu}}$. Since there exist two Norden metrics, we consider two classes of such hypersurfaces - with respect to the induced metrics g and $\widetilde{g}$ by $\overline{g}$ and $\overline{\overline{g}}$, respectively. According to the values of $\overline{\nu}$ and $\overline{\overline{\nu}}$ we obtain classes of Kähler-Norden manifolds which do not contain such hypersurfaces. On the other hand, if we consider a 4-parametric family of Lie groups as 4-dimensional Kähler-Norden manifolds which was given in [8], we can construct examples of totally umbilical and screen totally umbilical radical transversal lightlike hypersurfaces of Lie groups belonging to this family.

2. Preliminaries

2.1. *Almost complex manifolds with Norden metric*

A $2n$-dimensional semi-Riemannian manifold $\overline{M}$ is said to be *an almost complex manifold with Norden metric* (or *an almost complex manifold with B-metric*) if it is equipped with an almost complex structure $\overline{J}$ and a semi-Riemannian metric $\overline{g}$ satisfying $\overline{J}^2 X = -X$ and $\overline{g}(\overline{J}X, \overline{J}Y) = -\overline{g}(X, Y)$ for X, Y are arbitrary vector fields on $\overline{M}$ (see [5]). The tensor field $\overline{\overline{g}}$ on $\overline{M}$ defined by $\overline{\overline{g}}(X, Y) = \overline{g}(\overline{J}X, Y)$ is a Norden metric on $\overline{M}$, which is said to be *an associated metric* of $\overline{M}$. Both metrics $\overline{g}$ and $\overline{\overline{g}}$ are necessarily indefinite of signature (n, n). The Levi-Civita connection of $\overline{g}$ and $\overline{\overline{g}}$ are denoted by $\overline{\nabla}$ and $\overline{\overline{\nabla}}$, respectively. We define a tensor field F of type $(0, 3)$ on $\overline{M}$ by $F(X, Y, Z) = \overline{g}((\overline{\nabla}_X \overline{J})Y, Z)$. With respect to this tensor, almost complex manifolds with Norden metric are classified into eight classes in [5]. Further, we consider the class W_0 of the Kähler-Norden manifolds which is determined by the condition $F(X, Y, Z) = 0$. This class is commonly contained in all the eight classes.

2.2. *Kähler-Norden manifolds of constant totally real sectional curvatures*

The eight classes of almost complex manifolds with Norden metric are characterized by conditions on the tensor $\Phi(X, Y) = \overline{\overline{\nabla}}_X Y - \overline{\nabla}_X Y$ in [6]. The characteristic condition of a Kähler-Norden manifold $(\overline{M}, \overline{J}, \overline{g}, \overline{\overline{g}})$ is

$\Phi = 0$ which implies that the Levi-Civita connections $\overline{\nabla}$ and $\overline{\overline{\nabla}}$ of $\overline{M}$ coincide. Hence the curvature tensors of type (1,3) $\overline{R}$ and $\overline{\overline{R}}$ of $\overline{\nabla}$ and $\overline{\overline{\nabla}}$ coincide, too. The curvature tensors $\overline{R}$ and $\overline{\overline{R}}$ of type (0,4) are given by $\overline{R}(X,Y,Z,W) = \overline{g}(\overline{R}(X,Y,Z),W)$ and

$$\overline{\overline{R}}(X,Y,Z,W) = \overline{\overline{g}}(\overline{\overline{R}}(X,Y,Z),W) = \overline{R}(X,Y,Z,\overline{J}W), \tag{1}$$

where X,Y,Z,W are arbitrary vector fields on $\overline{M}$. From the condition $\overline{\nabla}\overline{J} = \overline{J}\overline{\nabla}$ it follows $\overline{R}(X,Y,\overline{J}Z) = \overline{J}\overline{R}(X,Y,Z)$. The following tensors are essential in the geometry of Kähler-Norden manifolds

$$\begin{aligned} \overline{\pi}_1(X,Y,Z,W) &= \overline{g}(Y,Z)\overline{g}(X,W) - \overline{g}(X,Z)\overline{g}(Y,W), \\ \overline{\pi}_2(X,Y,Z,W) &= \overline{g}(Y,\overline{J}Z)\overline{g}(X,\overline{J}W) - \overline{g}(X,\overline{J}Z)\overline{g}(Y,\overline{J}W), \\ \overline{\pi}_3(X,Y,Z,W) &= -\overline{\pi}_1(X,Y,Z,\overline{J}W) + \overline{\pi}_2(X,Y,Z,\overline{J}W). \end{aligned} \tag{2}$$

We denote by $\overline{\overline{\pi}}_i$ $(i = 1,2,3)$ the corresponding tensors with respect to $\overline{\overline{g}}$. They are related with $\overline{\pi}_i$ $(i = 1,2,3)$ as follows

$$\overline{\overline{\pi}}_1 = \overline{\pi}_2, \quad \overline{\overline{\pi}}_2 = \overline{\pi}_1, \quad \overline{\overline{\pi}}_3 = -\overline{\pi}_3. \tag{3}$$

For every non-degenerate section $\alpha = span\{x,y\}$, $x,y \in T_p\overline{M}$ with respect to $\overline{g}$, the following two sectional curvatures $\overline{\nu}$ and $\overline{\overline{\nu}}$ are defined in [2]

$$\overline{\nu}(\alpha;p) = \frac{\overline{R}(x,y,y,x)}{\overline{\pi}_1(x,y,y,x)}, \quad \overline{\overline{\nu}}(\alpha;p) = \frac{\overline{R}(x,y,y,\overline{J}x)}{\overline{\pi}_1(x,y,y,x)}.$$

Analogously, if $\alpha = span\{x,y\}$ is a non-degenerate section with respect to $\overline{\overline{g}}$, we can define two sectional curvatures $\overline{\nu}'$ and $\overline{\overline{\nu}}'$ given by

$$\overline{\nu}'(\alpha;p) = \frac{\overline{\overline{R}}(x,y,y,x)}{\overline{\overline{\pi}}_1(x,y,y,x)}, \quad \overline{\overline{\nu}}'(\alpha;p) = \frac{\overline{\overline{R}}(x,y,y,\overline{J}x)}{\overline{\overline{\pi}}_1(x,y,y,x)}.$$

A section α is said to be *totally real* with respect to $\overline{g}$ (resp. $\overline{\overline{g}}$) if $\overline{J}\alpha$ is orthogonal to α with respect to $\overline{g}$ (resp. $\overline{\overline{g}}$). The sectional curvatures $\overline{\nu}$, $\overline{\overline{\nu}}$ (resp. $\overline{\nu}'$, $\overline{\overline{\nu}}'$) of a non-degenerate totally real section with respect to $\overline{g}$ (resp. $\overline{\overline{g}}$) are called *totally real sectional curvatures* with respect to $\overline{g}$ (resp. $\overline{\overline{g}}$).

Theorem 2.1 ([2]). *Let $(\overline{M},\overline{J},\overline{g},\overline{\overline{g}})$ be a $2n$-dimensional $(2n \geq 4)$ Kähler-Norden manifold. Then $\overline{M}$ is of constant totally real sectional curvatures $\overline{\nu}$ and $\overline{\overline{\nu}}$ with respect to $\overline{g}$, i.e. $\overline{\nu}(\alpha;p) = \overline{\nu}(p)$, $\overline{\overline{\nu}}(\alpha;p) = \overline{\overline{\nu}}(p)$ $(p \in \overline{M})$ if and only if*

$$\overline{R} = \overline{\nu}[\overline{\pi}_1 - \overline{\pi}_2] + \overline{\overline{\nu}}\overline{\pi}_3. \tag{4}$$

Both functions $\overline{\nu}$ and $\overline{\overline{\nu}}$ are constants if $\overline{M}$ is connected and $2n \geq 6$.

Remark 2.1. Suppose $\overline{M}$ is of constant totally real sectional curvatures $\overline{\nu}$ and $\overline{\overline{\nu}}$ with respect to $\overline{g}$. Then from (4) by using (1), (2) and (3) we obtain $\overline{\overline{R}} = -\overline{\overline{\nu}}[\overline{\overline{\pi}}_1 - \overline{\overline{\pi}}_2] + \overline{\nu}\overline{\overline{\pi}}_3$. Then Theorem 2.1 implies that $\overline{M}$ is of constant totally real sectional curvatures $\overline{\nu}' = -\overline{\overline{\nu}}$ and $\overline{\overline{\nu}}' = \overline{\nu}$ with respect to $\overline{\overline{g}}$.

2.3. *Radical transversal lightlike hypersurfaces of a Kähler-Norden manifold*

Let M be a real hypersurface of an $(m+2)$-dimensional semi-Riemannian manifold $(\overline{M}, \overline{g})$, where $\overline{g}$ is an indefinite metric of index $q \in \{1, \dots, m+1\}$. Denote by T_pM, $T_pM^{\perp}$ and $\operatorname{Rad} T_pM$ the tangent space, the normal space and their intersection at $p \in M$, respectively. As $\overline{g}$ is indefinite, the dimension of $\operatorname{Rad} T_pM$ may not be zero. If a hypersurface M satisfies $\operatorname{Rad} T_pM \neq \{0\}$ for any $p \in M$, it is called a *lightlike hypersurface* of $\overline{M}$. The mapping $\operatorname{Rad} TM : M \ni p \mapsto \operatorname{Rad}T_pM$ defines a smooth distribution on M of rank 1 which is called *a radical distribution* on M. Thus for lightlike hypersurface M the radical distribution $\operatorname{Rad} TM$ coincides with the normal bundle $TM^{\perp}$ and the induced metric g by $\overline{g}$ on M has a constant rank m. There also exists a non-degenerate complementary vector bundle $S(TM)$ of the normal bundle $TM^{\perp}$ in the tangent bundle TM, called *a screen distribution* on M ([3,4]). We have the following decomposition of TM

$$TM = S(TM) \perp TM^{\perp}, \tag{5}$$

where $\perp$ denotes an orthogonal direct sum. Denote by $\mathcal{F}(M)$ the algebra of smooth functions on M and by $\Gamma(E)$ the $\mathcal{F}(M)$-module of smooth sections of a vector bundle E over M. It is well known (Theorem 1.1, p. 79, [3]) that there exists a unique *transversal vector bundle* $\operatorname{tr}(TM)$ of rank 1 over M satisfying the following condition: For any non-zero section ξ of $TM^{\perp}$ on a coordinate neighborhood $U \subset M$, there exists a unique section N of $\operatorname{tr}(TM)$ on U which satisfies

$$\overline{g}(N, \xi) = 1, \qquad \overline{g}(N, N) = \overline{g}(N, W) = 0, \ \forall W \in \Gamma(S(TM)). \tag{6}$$

Hence for any screen distribution $S(TM)$ we have a unique $\operatorname{tr}(TM)$, which is a lightlike complementary vector bundle (but not orthogonal) to TM in $T\overline{M}$ satisfying

$$T\overline{M} = TM \oplus \operatorname{tr}(TM) = S(TM) \perp \left(TM^{\perp} \oplus \operatorname{tr}(TM)\right),$$

where $\oplus$ denotes a non-orthogonal direct sum.

Let $\overline{\nabla}$ be the Levi-Civita connection of $\overline{g}$ on $\overline{M}$ and P be the projection morphism of $\Gamma(TM)$ on $\Gamma(S(TM))$ with respect to the decomposition (5). The local Gauss and Weingarten formulas of M and $S(TM)$ are given by

$$\overline{\nabla}_X Y = \nabla_X Y + B(X,Y)N, \tag{7}$$

$$\overline{\nabla}_X N = -A_N X + \tau(X)N; \tag{8}$$

$$\nabla_X PY = \nabla^*_X PY + C(X,PY)\xi, \tag{9}$$

$$\nabla_X \xi = -A^*_\xi X - \tau(X)\xi, \tag{10}$$

for arbitrary $X, Y \subset \Gamma(TM)$. The induced connections ∇ and ∇^* on TM and $S(TM)$, respectively, are linear connections. We denote by A_N and A^*_ξ the shape operators on TM and $S(TM)$, respectively, and by τ a 1-form on TM. Both local second fundamental forms B and C are related to their shape operators by

$$B(X,Y) = g(A^*_\xi X, Y), \qquad \overline{g}(A^*_\xi X, N) = 0; \tag{11}$$

$$C(X,PY) = g(A_N X, PY), \quad \overline{g}(A_N X, N) = 0. \tag{12}$$

Since $\overline{\nabla}$ is torsion-free, ∇ is also torsion-free and B is symmetric on TM. From (7) we have $B(X,Y) = \overline{g}(\overline{\nabla}_X Y, \xi)$ for all $X, Y \in \Gamma(TM)$. This implies

$$B(X,\xi) = 0, \quad \forall X \in \Gamma(TM). \tag{13}$$

From (11) and (13), it follows that the shape operator A^*_ξ is $S(TM)$-valued, self-adjoint with respect to g and $A^*_\xi \xi = 0$. In general, the induced connection ∇^* on $S(TM)$ is not torsion-free. This fact and (12) show that the shape operator A_N is not self-adjoint and is $S(TM)$-valued. The linear connection ∇^* is a metric connection on $S(TM)$, but ∇ is not a metric connection and satisfies

$$(\nabla_X g)(Y,Z) = B(X,Y)\eta(Z) + B(X,Z)\eta(Y), \tag{14}$$

where η is a 1-form given by $\eta(X) = \overline{g}(X,N)$ for $X \in \Gamma(TM)$.

A lightlike hypersurface M is *totally umbilical* if and only if on each coordinate neighborhood $U \subset M$ there exists a smooth function ρ satisfying $B(X,Y) = \rho g(X,Y)$ for all $X, Y \in \Gamma(TM_{|U})$ (see [3]). In case $\rho = 0$ on U, we see that M is totally geodesic. The screen distribution $S(TM)$ of a lightlike hypersurface M is said to be *totally umbilical* if on each coordinate neighborhood $U \subset M$ there exists a smooth function k satisfying

$$C(X,PY) = kg(X,PY), \quad \forall X, Y \in \Gamma(TM_{|U}) \tag{15}$$

(see [3]). In case $k = 0$ (resp. $k \neq 0$) on U, the screen distribution $S(TM)$ is said to be *totally geodesic* (resp. *proper totally umbilical*) in M.

Denote by $\overline{R}$ and R the curvature tensors of $\overline{\nabla}$ and ∇, respectively. By using (9) and (10) we obtain

$$\begin{aligned} &\overline{g}(\overline{R}(X,Y,PZ),N) \\ &\quad = (\nabla_X C)(Y,PZ) - (\nabla_Y C)(X,PZ) \\ &\qquad + \tau(Y)C(X,PZ) - \tau(X)C(Y,PZ) \end{aligned} \tag{16}$$

for $X, Y, Z \in \Gamma(TM)$, where

$$(\nabla_X C)(Y,PZ) = X(C(Y,PZ)) - C(\nabla_X Y, PZ) - C(Y, \nabla^*_X PZ).$$

In [10] we introduced the class of radical transversal lightlike hypersurfaces of an almost complex manifold with Norden metric. Such a hypersurface does not exist when the ambient manifold is an indefinite almost Hermitian manifold. A lightlike hypersurface $(M, g, S(TM))$ of an almost complex manifold with Norden metric $(\overline{M}, \overline{J}, \overline{g}, \overline{\widetilde{g}})$ is called a *radical transversal lightlike hypersurface* of $\overline{M}$ if $\overline{J}(TM^{\perp}) = \mathrm{tr}(TM)$ holds. We proved in [10] that a radical transversal lightlike hypersurface M of $\overline{M}$ has a unique screen distribution up to a semi-orthogonal transformation and a unique transversal vector bundle. This guarantees that the induced geometrical objects on M are well defined. Let $\{\xi, N\}$ be the pair on a radical transversal lightlike hypersurface M which satisfies (6). From the definition of M we have

$$\overline{J}\xi = bN, \qquad b \in \mathcal{F}(\overline{M}), \quad b \neq 0. \tag{17}$$

Taking into account (5) we have the following decomposition $X = PX + \eta(X)\xi$ for an arbitrary $X \in \Gamma(TM)$. From this decomposition and (17) we obtain

$$\overline{J}X = \overline{J}(PX) + b\eta(X)N. \tag{18}$$

Since $S(TM)$ is holomorphic with respect to $\overline{J}$ (see Theorem 3.3, p. 544, [10]), it follows that $\overline{J}(PX)$ belongs to $S(TM)$. In case $(\overline{M}, \overline{J}, \overline{g}, \overline{\widetilde{g}})$ is a Kähler-Norden manifold, the shape operators A^*_ξ, A_N and the corresponding local second fundamental forms B, C of a radical transversal lightlike hypersurface of $\overline{M}$ are related as follows

$$A^*_\xi X = -b\overline{J}(A_N X), \; B(X,Y) = -bC(X, \overline{J}(PY)), \tag{19}$$

for all $X, Y \in \Gamma(TM)$. Moreover, the 1-form τ is expressed by the function b in the following way

$$\tau(X) = -\frac{1}{2b}X(b). \tag{20}$$

There exist two Norden metrics $\overline{g}$ and $\overline{\overline{g}}$ on an almost complex manifold with Norden metric $(\overline{M}, \overline{J}, \overline{g}, \overline{\overline{g}})$. Therefore we can consider two induced metrics g and $\widetilde{g}$ (by $\overline{g}$ and $\overline{\overline{g}}$, respectively) on a hypersurface M of $\overline{M}$. Now, let (M, g) be a non-degenerate hypersurface of $\overline{M}$ defined by

$$\overline{g}(\overline{N}, \overline{JN}) = 0, \tag{21}$$

where $\overline{N}$ is a space-like (i.e. $\overline{g}(\overline{N}, \overline{N}) = 1$) or time-like (i.e. $\overline{g}(\overline{N}, \overline{N}) = -1$) unit normal to $\overline{M}$.

Theorem 2.2 ([10]). *Let $(\overline{M}, \overline{J}, \overline{g}, \overline{\overline{g}})$ be an almost complex manifold with Norden metric and M be a hypersurface of $\overline{M}$. Then (M, g) is a non-degenerate hypersurface defined by* (21) *if and only if $(M, \widetilde{g})$ is a radical transversal lightlike hypersurface.*

Remark 2.2. Both Norden metrics $\overline{g}$ and $\overline{\overline{g}}$ are of equal worth. Hence, if we interchange the roles of $\overline{g}$ and $\overline{\overline{g}}$ in Theorem 2.2, we find that the corresponding assertion holds.

Further, we study radical transversal lightlike hypersurfaces with proper totally umbilical screen distributions. For convenience, we use the term *screen totally umbilical radical transversal lightlike hypersurfaces* for them.

3. Non-existence of totally umbilical and screen totally umbilical radical transversal lightlike hypersurfaces

In this section, we first prove two theorems about screen totally umbilical radical transversal lightlike hypersurfaces of a Kähler-Norden manifold of constant $\overline{\nu}$ and $\overline{\overline{\nu}}$. Then by using these theorems and some results about totally umbilical radical transversal lightlike hypersurfaces in [11], we find classes of Kähler-Norden manifolds which do not contain hypersurfaces belonging to some of the examined classes of hypersurfaces.

Theorem 3.1. *Let $(M, g, S(TM))$ be a screen totally umbilical radical transversal lightlike hypersurface of a Kähler-Norden manifold $(\overline{M}, \overline{J}, \overline{g}, \overline{\overline{g}})$ $(\dim\overline{M} = 2n \geq 4)$ of constant totally real sectional curvatures $\overline{\nu}$ and $\overline{\overline{\nu}}$ with respect to $\overline{g}$. Then the function k given in* (15) *satisfies the following conditions*:

$$PX(k) - k\tau(PX) = 0, \quad \overline{\nu} = \xi(k) - k\tau(\xi), \tag{22}$$

$$\overline{\overline{\nu}} = -bk^2, \tag{23}$$

where $X \in \Gamma(TM)$ and b is the function given in (17).

Proof. By using (4), (15), (16) and (17) we obtain

$$\begin{aligned} &\overline{\nu}\left[g(Y,PZ)\eta(X) - g(X,PZ)\eta(Y)\right] \\ &\quad + \overline{\overline{\nu}}\left[-g(Y,\overline{J}(PZ))\eta(X) + g(X,\overline{J}(PZ))\eta(Y)\right] \\ &= \left[X(k) - k\tau(X)\right] g(Y,PZ) - \left[Y(k) - k\tau(Y)\right] g(X,PZ) \\ &\quad + k\left[(\nabla_X g)(Y,PZ) - (\nabla_Y g)(X,PZ)\right], \end{aligned} \tag{24}$$

for all $X, Y, Z \in \Gamma(TM)$. If we replace X, Y in (24) by PX, PY, respectively, by taking into account (14), (15), (19) and $\eta(PX) = \eta(PY) = 0$, we find that the equality (24) becomes

$$\left[PX(k) - k\tau(PX)\right] g(PY,PZ) - \left[PY(k) - k\tau(PY)\right] g(PX,PZ) = 0. \tag{25}$$

From (25), by noticing that $S(TM)$ is non-degenerate, we obtain

$$\left[PX(k) - k\tau(PX)\right] PY = \left[PY(k) - k\tau(PY)\right] PX.$$

Suppose that there exists a vector field $X_0 \in \Gamma(TM)$ satisfying $PX_0(k) - k\tau(PX_0) \neq 0$ at a point $p \in M$. Then it follows that all vectors belonging to $S(TM)$ are collinear with $(PX_0)_p$. This is a contradiction because $\dim S(TM) = 2n - 2 \geq 2$. Hence $PX(k) - k\tau(PX) = 0$ at any point $p \in M$. Next we replace X, Y, Z in (24) by PX, ξ, PZ, respectively. Taking into account (13), (14), (15) and (19), we find that the equality (24) becomes

$$\begin{aligned} &-\overline{\nu} g(PX,PZ) + \overline{\overline{\nu}} g(PX,\overline{J}(PZ)) \\ &\quad = -\left[\xi(k) - k\tau(\xi)\right] g(PX,PZ) - bk^2 g(PX,\overline{J}(PZ)). \end{aligned} \tag{26}$$

Because $S(TM)$ is non-degenerate, from (26) we have

$$\left[\overline{\nu} - \xi(k) + k\tau(\xi)\right] PZ + \left[-\overline{\overline{\nu}} - bk^2\right] \overline{J}(PZ) = 0.$$

Finally, by using that PZ and $\overline{J}(PZ)$ are linearly independent, we can establish the second equality in (22) and equality (23). □

Since both $\overline{g}$ and $\overline{\overline{g}}$ are Norden metrics, Theorem 3.1 is valid if we substitute $\overline{\overline{g}}$ for $\overline{g}$ and $\widetilde{g}$ for g in it. Thus by taking into account Remark 2.1 we obtain

Theorem 3.2. *Let $(M, \widetilde{g}, S(TM))$ be a screen totally umbilical radical transversal lightlike hypersurface (i.e. $C(X, PY) = \widetilde{k}\widetilde{g}(X, PY)$) of a Kähler-Norden manifold $(\overline{M}, \overline{J}, \overline{g}, \overline{\widetilde{g}})$ ($\dim \overline{M} = 2n \geq 4$) of constant totally real sectional curvatures $\overline{\nu}$ and $\overline{\widetilde{\nu}}$ with respect to $\overline{g}$. Then the function $\widetilde{k}$ satisfies the following conditions:*

$$PX(\widetilde{k}) - \widetilde{k}\tau(PX) = 0, \quad \overline{\widetilde{\nu}} = -\xi(\widetilde{k}) + \widetilde{k}\tau(\xi), \tag{27}$$

$$\overline{\nu} = -b\widetilde{k}^2, \tag{28}$$

where $X \in \Gamma(TM)$ and b is the function given in (17).

We denote by $(\mathcal{M}, g)$ and $(\mathcal{M}, \widetilde{g})$ the classes of radical transversal lightlike hypersurfaces of an almost complex manifold with Norden metric $(\overline{M}, \overline{J}, \overline{g}, \overline{\widetilde{g}})$ with respect to g and $\widetilde{g}$, respectively. In [11] we studied totally umbilical hypersurfaces belonging to $(\mathcal{M}, g)$ and $(\mathcal{M}, \widetilde{g})$ when the ambient manifold $\overline{M}$ is Kähler-Norden of constant totally real sectional curvatures $\overline{\nu}$ and $\overline{\widetilde{\nu}}$. In Theorem 3.2 of [11] (resp. Theorem 3.3 of [11]) we proved that if a hypersurface M of $\overline{M}$ belongs to $(\mathcal{M}, g)$ (resp. $(\mathcal{M}, \widetilde{g})$) and M is totally umbilical, then $\overline{\nu} = 0$ (resp. $\overline{\widetilde{\nu}} = 0$). As an immediate consequence from these results, (23) and (28) we state

Theorem 3.3. *Let $(\overline{M}, \overline{J}, \overline{g}, \overline{\widetilde{g}})$ be a Kähler-Norden manifold of constant totally real sectional curvatures $\overline{\nu}$ and $\overline{\widetilde{\nu}}$ with respect to $\overline{g}$.*

1) *If $\overline{\nu} = \overline{\widetilde{\nu}} = 0$, then there exist no screen totally umbilical hypersurfaces of $\overline{M}$ belonging to $(\mathcal{M}, g)$ and $(\mathcal{M}, \widetilde{g})$.*
2) *If $\overline{\nu} \neq 0$ and $\overline{\widetilde{\nu}} \neq 0$, then there exist no totally umbilical hypersurfaces of $\overline{M}$ belonging to $(\mathcal{M}, g)$ and $(\mathcal{M}, \widetilde{g})$.*
3) *If $\overline{\nu} \neq 0$ and $\overline{\widetilde{\nu}} = 0$, then there exist no totally umbilical and screen totally umbilical hypersurfaces of $\overline{M}$ belonging to $(\mathcal{M}, g)$.*
4) *If $\overline{\nu} = 0$ and $\overline{\widetilde{\nu}} \neq 0$, then there exist no totally umbilical and screen totally umbilical hypersurfaces of $\overline{M}$ belonging to $(\mathcal{M}, \widetilde{g})$.*

4. Examples of totally umbilical and screen totally umbilical radical transversal lightlike hypersurfaces

In this section we study radical transversal lightlike hypersurfaces in a real 4-dimensional Lie groups which are Kähler-Norden manifolds. We shall start with recalling the construction of an almost complex structure and metrics on a 4-dimensional Lie group given in [8].

Let $\overline{G}$ be a 4-dimensional real connected Lie group and $\{X_1, X_2, X_3, X_4\}$ be a basis of left invariant vector fields of the Lie algebra $\overline{\mathfrak{g}}$ of $\overline{G}$. A bi-invariant almost complex structure $\overline{J}$ and a left invariant metric $\overline{g}$ were defined on $\overline{G}$ by the conditions

$$\begin{aligned} &\overline{J}X_1 = X_3, \quad \overline{J}X_2 = X_4, \quad \overline{J}X_3 = -X_1, \quad \overline{J}X_4 = -X_2; \\ &\overline{g}(X_1, X_1) = \overline{g}(X_2, X_2) = -\overline{g}(X_3, X_3) = -\overline{g}(X_4, X_4) = 1, \\ &\overline{g}(X_i, X_j) = 0, \quad i \neq j. \end{aligned} \tag{29}$$

The non-zero Lie brackets were given by

$$\overline{\mathfrak{g}}: \left| \begin{array}{l} [X_1, X_2] = -[X_3, X_4] = \lambda_1 X_1 + \lambda_2 X_2 + \lambda_3 X_3 + \lambda_4 X_4, \\ [X_2, X_3] = -[X_1, X_4] = \lambda_3 X_1 + \lambda_4 X_2 - \lambda_1 X_3 - \lambda_2 X_4, \end{array} \right. \tag{30}$$

where $\lambda_i \in \mathbb{R}\,(i = 1, 2, 3, 4)$. It was proved in [8] that $(\overline{G}, \overline{J}, \overline{g}, \overline{\widetilde{g}})$ with a Lie algebra $\overline{\mathfrak{g}}$ is a 4-dimensional Kähler-Norden manifold. Moreover $\overline{J}$ is necessarily complex and $\overline{G}$ is of totally real sectional curvatures

$$\overline{\nu} = -(\lambda_1^2 + \lambda_2^2 - \lambda_3^2 - \lambda_4^2), \qquad \overline{\widetilde{\nu}} = 2(\lambda_1\lambda_3 + \lambda_2\lambda_4). \tag{31}$$

Considering λ_i $(i = 1, 2, 3, 4)$ as parameters, we obtain a family of 4-dimensional Kähler-Norden manifolds. Further, we find that each Lie group belonging to this family admits two classes of hypersurfaces, which are the classes of totally umbilical and screen totally umbilical hypersurfaces, that belong to either $(\mathcal{M}, \widetilde{g})$ or $(\mathcal{M}, g)$ for this Lie group.

4.1. *Examples of totally umbilical and screen totally umbilical hypersurfaces belonging to* $(\mathcal{M}, \widetilde{g})$

By using (30) we obtain

Lemma 4.1. *Let $\overline{\mathfrak{g}}$ be the Lie algebra given by* (30). *The following assertions are mutually equivalent to each other:*

i) $\mathfrak{g}_1 = \mathrm{span}\{X_1, X_2, \overline{J}X_2\}$ *is a Lie subalgebra of* $\overline{\mathfrak{g}}$.
ii) $\mathfrak{g}_2 = \mathrm{span}\{\overline{J}X_1, \overline{J}X_2, X_2\}$ *is a Lie subalgebra of* $\overline{\mathfrak{g}}$.
iii) $\lambda_1 = \lambda_3 = 0$ *in the equalities* (30).

Let $\overline{\mathfrak{g}}_1$ be the Lie algebra with non-zero Lie brackets obtained from (30) with $\lambda_1 = \lambda_3 = 0$, i.e.

$$\overline{\mathfrak{g}}_1: \left| \begin{array}{l} [X_1, X_2] = -[\overline{J}X_1, \overline{J}X_2] = \lambda_2 X_2 + \lambda_4 \overline{J}X_2, \\ [X_1, \overline{J}X_2] = -[X_2, \overline{J}X_1] = -\lambda_4 X_2 + \lambda_2 \overline{J}X_2. \end{array} \right. \tag{32}$$

Denote by $(\overline{G}_1, \overline{J}, \overline{g}, \overline{\overline{g}})$ the 4-dimensional Kähler-Norden manifold corresponding to $\overline{\mathfrak{g}}_1$. From Lemma 4.1 we have that $\mathfrak{g}_1$ and $\mathfrak{g}_2$ are Lie subalgebras of $\overline{\mathfrak{g}}_1$. Let G_1 and G_2 be the Lie subgroups of $\overline{G}_1$ with Lie algebras $\mathfrak{g}_1$ and $\mathfrak{g}_2$, respectively. Denote by g_1 and g_2 (resp. $\widetilde{g}_1$ and $\widetilde{g}_2$) the induced metrics by $\overline{g}$ (resp. $\overline{\overline{g}}$) on G_1 and G_2, respectively.

Proposition 4.1. *The hypersurfaces $(G_1, \widetilde{g}_1)$ and $(G_2, \widetilde{g}_2)$ of $\overline{G}_1$ are radical transversal lightlike hypersurfaces of $\overline{G}_1$ for arbitrary $\lambda_2, \lambda_4 \in \mathbb{R}$. Moreover, we have*

$$\xi_1 = -X_1, \quad N_1 = \overline{J}X_1, \quad S(\mathfrak{g}_1) = \operatorname{span}\{X_2, \overline{J}X_2\}, \tag{33}$$

$$\xi_2 = \overline{J}X_1, \quad N_2 = -X_1, \quad S(\mathfrak{g}_2) = \operatorname{span}\{X_2, \overline{J}X_2\}, \tag{34}$$

where $S(\mathfrak{g}_i)$ are the screen distributions of $\mathfrak{g}_i$ $(i = 1, 2)$, and $\{\xi_i, N_i\}$ are the pairs of vector fields of G_i $(i = 1, 2)$ satisfying the following equalities:

$$\overline{\overline{g}}(N_i, \xi_i) = 1, \ \widetilde{g}(\xi_i, \xi_i) = \overline{\overline{g}}(N_i, N_i) = \overline{\overline{g}}(N_i, W_i) = 0,$$

for all $W_i \in S(\mathfrak{g}_i)$, $i = 1, 2$.

Proof. By using (29) we obtain that (G_1, g_1) and (G_2, g_2) are non-degenerate hypersurfaces of $\overline{G}_1$. The vectors $\overline{N}_1 = \overline{J}X_1$ and $\overline{N}_2 = -X_1$ are normal to (G_1, g_1) and (G_2, g_2), respectively. Since $\overline{N}_1$ and $\overline{N}_2$ satisfy (21), from Theorem 2.2 we obtain that $(G_1, \widetilde{g}_1)$ and $(G_2, \widetilde{g}_2)$ are radical transversal lightlike hypersurfaces of $\overline{G}_1$. By using (29) we get

$$\begin{aligned}
&\widetilde{g}_1(X_1, X_1) = \widetilde{g}_1(X_2, X_2) = \widetilde{g}_1(\overline{J}X_2, \overline{J}X_2) = \widetilde{g}_1(X_1, X_2) \\
&\quad = \widetilde{g}_1(X_1, \overline{J}X_2) = \overline{\overline{g}}(\overline{J}X_1, \overline{J}X_1) = \overline{\overline{g}}(\overline{J}X_1, X_2) = \overline{\overline{g}}(\overline{J}X_1, \overline{J}X_2) = 0; \\
&\overline{\overline{g}}(X_1, \overline{J}X_1) - \widetilde{g}_1(X_2, \overline{J}X_2) = -1.
\end{aligned} \tag{35}$$

$$\begin{aligned}
&\widetilde{g}_2(\overline{J}X_1, \overline{J}X_1) = \widetilde{g}_2(\overline{J}X_2, \overline{J}X_2) = \widetilde{g}_2(X_2, X_2) = \widetilde{g}_2(\overline{J}X_1, \overline{J}X_2) \\
&\quad = \widetilde{g}_2(\overline{J}X_1, X_2) = \overline{\overline{g}}(X_1, X_1) = \overline{\overline{g}}(X_1, X_2) = \overline{\overline{g}}(X_1, \overline{J}X_2) = 0; \\
&\overline{\overline{g}}(X_1, \overline{J}X_1) = \widetilde{g}_2(X_2, \overline{J}X_2) = -1.
\end{aligned} \tag{36}$$

The equalities (35) and (36) imply (33) and (34), respectively. □

Remark 4.1. If we take Lie subalgebras of $\overline{\mathfrak{g}}$ as $\mathfrak{g}_1 = \operatorname{span}\{X_1, X_2, \overline{J}X_1\}$ and $\mathfrak{g}_2 = \operatorname{span}\{\overline{J}X_1, \overline{J}X_2, X_1\}$, we can obtain similar results.

Theorem 4.1. *Let $(\overline{M}, \overline{J}, \overline{g}, \overline{\overline{g}})$ be a $2n$-dimensional Kähler-Norden manifold and (M_1, g_1), (M_2, g_2) be radical transversal lightlike hypersurfaces of $\overline{M}$ satisfying the conditions*

$$S(TM_1) = S(TM_2), \qquad \xi_2 = N_1, \qquad N_2 = \xi_1, \tag{37}$$

where $S(TM_i)$ are the screen distributions of M_i $(i = 1, 2)$ and $\{\xi_i, N_i\}$ are the pairs of vector fields of M_i $(i = 1, 2)$ satisfying (6). Then the one of M_1 and M_2 is totally umbilical if and only if the other one is screen totally umbilical.

Proof. Let $\overline{J}\xi_1 = b_1 N_1$ and $\overline{J}\xi_2 = b_2 N_2$, where $b_1, b_2 \in \mathcal{F}(\overline{M})$. Hence $\overline{J}N_1 = -(1/b_1)\,\xi_1$. Then from (37) we obtain $\overline{J}\xi_2 = -(1/b_1)\,N_2$, which means that $b_2 = -1/b_1$. By using (20), for the 1-forms τ_1 and τ_2 of M_1 and M_2, respectively, we have

$$\tau_1(Y) = -\frac{1}{2b_1}Y(b_1), \quad \tau_2(Z) = -\frac{1}{2b_2}Z(b_2) = \frac{1}{2b_1}Z(b_1), \tag{38}$$

for all $Y \in \Gamma(TM_1)$ and $Z \in \Gamma(TM_2)$. From (38), (5) and the condition that $S(TM_1) = S(TM_2)$, we find

$$\tau_1(X) = -\tau_2(X) \tag{39}$$

for each $X \in S\,\big(:= S(TM_1) = S(TM_2)\big)$. By using (7), (8), (10), (13) and (37), for arbitrary $X \in S$ we obtain

$$\begin{aligned} &\overline{\nabla}_X\xi_1 = -A^{1*}_{\xi_1}(X) - \tau_1(X)\xi_1, \\ &\overline{\nabla}_X\xi_1 = \overline{\nabla}_X N_2 = -A^2_{N_2}(X) + \tau_2(X)N_2 = -A^2_{N_2}(X) + \tau_2(X)\xi_1, \end{aligned} \tag{40}$$

where by $A^{i*}_{\xi_i}$ and $A^i_{N_i}$ we denote the screen shape operators and the shape operators of M_i $(i = 1, 2)$, respectively. Taking into account (39), the equalities (40) imply

$$A^{1*}_{\xi_1}(X) = A^2_{N_2}(X), \quad \forall X \in S. \tag{41}$$

Analogously we obtain

$$A^{2*}_{\xi_2}(X) = A^1_{N_1}(X), \quad \forall X \in S. \tag{42}$$

From (19) and $A^{i*}_{\xi_i}(\xi_i) = 0$ it follows that $A^i_{N_i}(\xi_i) = 0$ $(i = 1, 2)$. Hence, (M_i, g_i) is screen totally umbilical if and only if $A^i_{N_i}(X) = kX$ $(i = 1, 2)$, for all $X \in S$. Finally, taking into account that (M_i, g_i) is totally umbilical if and only if $A^{i*}_{\xi_i}(X) = \rho X$ $(i = 1, 2)$ for all $X \in S$, by using (41) and (42), we complete the proof. □

Remark 4.2. Since $\widetilde{\overline{g}}$ is also Norden metric on $\overline{M}$, the assertion in Theorem 4.1 is valid if M_1 and M_2 are radical transversal lightlike hypersurfaces of $\overline{M}$ with respect to $\widetilde{g}$.

Proposition 4.2. *The radical transversal lightlike hypersurface* $(G_1, \widetilde{g}_1)$ *of* $\overline{G}_1$ *is totally umbilical if and only if the non-zero Lie brackets of* $\overline{\mathfrak{g}}_1$ *are given by*

$$\left| \begin{array}{l} [X_1, X_2] = -[\overline{J}X_1, \overline{J}X_2] = \lambda_2 X_2, \\ [X_1, \overline{J}X_2] = -[X_2, \overline{J}X_1] = \lambda_2 \overline{J}X_2, \end{array} \right. \tag{43}$$

with non-zero λ_2.

Proof. From (33) we have $\overline{J}\xi_1 = -N_1$, which means that $b_1 = -1$. Then from (38) it follows that $\tau_1(X) = 0$. By using (7), (10) and (13) we obtain

$$\overline{\overline{\nabla}}_X \xi_1 = \widetilde{\nabla}_X \xi_1 = -\widetilde{A}^{1*}_{\xi_1}(X), \quad X \in \mathfrak{g}_1, \tag{44}$$

where $\widetilde{\nabla}$ is the induced connection on G_1 by $\overline{\overline{\nabla}}$, and $\widetilde{A}^{1*}_{\xi_1}$ is the screen shape operator. Now, we suppose $(G_1, \widetilde{g}_1)$ be totally umbilical. Hence $\widetilde{A}^{1*}_{\xi_1}(X) = \rho PX$ for $X \in \mathfrak{g}_1$. Replacing $-X_1$ by ξ_1 in (44) we get

$$\overline{\overline{\nabla}}_{X_2} X_1 = \rho X_2. \tag{45}$$

Since the metric $\overline{\overline{g}}$ is left invariant, for the Levi-Civita connection $\overline{\overline{\nabla}}$ of $\overline{\overline{g}}$ we have

$$2\overline{\overline{g}}\left(\overline{\overline{\nabla}}_{\overline{X}}\overline{Y}, \overline{Z}\right) = \overline{\overline{g}}\left(\left[\overline{X}, \overline{Y}\right], \overline{Z}\right) + \overline{\overline{g}}\left(\left[\overline{Z}, \overline{X}\right], \overline{Y}\right) + \overline{\overline{g}}\left(\left[\overline{Z}, \overline{Y}\right], \overline{X}\right) \tag{46}$$

for arbitrary $\overline{X}, \overline{Y}, \overline{Z} \in \overline{\mathfrak{g}}_1$. By using (35) and (45) we find

$$\overline{\overline{g}}\left(\overline{\overline{\nabla}}_{X_2} X_1, \overline{J}X_2\right) = -\rho, \quad \overline{\overline{g}}\left(\overline{\overline{\nabla}}_{X_2} X_1, X_2\right) = 0. \tag{47}$$

On the other hand, by using (32), (35) and (46) we obtain

$$\overline{\overline{g}}\left(\overline{\overline{\nabla}}_{X_2} X_1, \overline{J}X_2\right) = \lambda_2, \quad \overline{\overline{g}}\left(\overline{\overline{\nabla}}_{X_2} X_1, X_2\right) - -\lambda_4. \tag{48}$$

The equalities (47) and (48) imply $\lambda_2 = -\rho \neq 0$ and $\lambda_4 = 0$. Putting $\lambda_4 = 0$ in (32) we obtain (43). Conversely, we suppose the non-zero Lie brackets of $\overline{\mathfrak{g}}_1$ are given by (43). According to (11) we see $\widetilde{A}^{1*}_{\xi_1}(X)$ belongs to $S(\mathfrak{g}_1)$. Then replacing $-X_1$ by ξ_1 in (44) we have

$$\begin{array}{ll} \widetilde{A}^{1*}_{\xi_1}(X_2) = \overline{\overline{\nabla}}_{X_2} X_1 = \lambda X_2 + \mu \overline{J}X_2, & \lambda, \mu \in \mathbb{R} \\ \widetilde{A}^{1*}_{\xi_1}(\overline{J}X_2) = \overline{\overline{\nabla}}_{\overline{J}X_2} X_1 = \alpha X_2 + \beta \overline{J}X_2, & \alpha, \beta \in \mathbb{R}. \end{array} \tag{49}$$

From (49), by using (35), (43) and (46) we get $\lambda = \beta = -\lambda_2$ and $\mu = \alpha = 0$. Hence $\widetilde{A}^{1*}_{\xi_1}(X_2) = -\lambda_2 X_2$ and $\widetilde{A}^{1*}_{\xi_1}(\overline{J}X_2) = -\lambda_2 \overline{J}X_2$. If we express $X \in \mathfrak{g}_1$ as $X = x^0\xi_1 + x^1 X_2 + x^2 \overline{J}X_2$ with $x^0, x^1, x^2 \in \mathbb{R}$, then we have $\widetilde{A}^{1*}_{\xi_1}(X) = -\lambda_2(x^1 X_2 + x^2 \overline{J}X_2) = -\lambda_2 PX$, which means that $(G_1, \widetilde{g}_1)$ is totally umbilical. □

Analogously as Proposition 4.2 we obtain

Proposition 4.3. *The radical transversal lightlike hypersurface* $(G_2, \widetilde{g}_2)$ *of* $\overline{G}_1$ *is totally umbilical if and only if the non-zero Lie brackets of* $\overline{\mathfrak{g}}_1$ *are given by*

$$\left| \begin{array}{l} [X_1, X_2] = -[\overline{J}X_1, \overline{J}X_2] = \lambda_4 \overline{J}X_2, \\ [X_1, \overline{J}X_2] = -[X_2, \overline{J}X_1] = -\lambda_4 X_2, \end{array} \right. \tag{50}$$

with non-zero λ_4.

Denote by $\overline{\mathfrak{g}}_{11}$ and $\overline{\mathfrak{g}}_{12}$ the Lie algebras with non-zero Lie brackets given by (43) and (50), respectively. We note that (43) and (50) are obtained from (30) by $\lambda_1 = \lambda_3 = \lambda_4 = 0$, $\lambda_2 \neq 0$ and $\lambda_1 = \lambda_3 = \lambda_2 = 0$, $\lambda_4 \neq 0$, respectively.

Proposition 4.4. *Let* $(\overline{G}_{11}, \overline{J}, \overline{g}, \overline{\overline{g}})$ *and* $(\overline{G}_{12}, \overline{J}, \overline{g}, \overline{\overline{g}})$ *be the 4-dimensional Kähler-Norden manifolds with Lie algebras* $\overline{\mathfrak{g}}_{11}$ *and* $\overline{\mathfrak{g}}_{12}$, *respectively. Then the following hold.*

1) $(G_1, \widetilde{g}_1)$ *and* $(G_2, \widetilde{g}_2)$ *are radical transversal lightlike hypersurfaces of* $\overline{G}_{11}$ *such that* $(G_1, \widetilde{g}_1)$ *is totally umbilical and* $(G_2, \widetilde{g}_2)$ *is screen totally umbilical.*
2) $(G_1, \widetilde{g}_1)$ *and* $(G_2, \widetilde{g}_2)$ *are radical transversal lightlike hypersurfaces of* $\overline{G}_{12}$ *such that* $(G_1, \widetilde{g}_1)$ *is screen totally umbilical and* $(G_2, \widetilde{g}_2)$ *is totally umbilical.*
3) *There exist no totally umbilical and screen totally umbilical hypersurfaces of* $\overline{G}_{11}$ *and* $\overline{G}_{12}$ *belonging to* $(\mathcal{M}, g)$.

Proof. 1), 2): According to Proposition 4.1, $(G_1, \widetilde{g}_1)$ and $(G_2, \widetilde{g}_2)$ are radical transversal lightlike hypersurfaces of $\overline{G}_1$ for arbitrary $\lambda_2, \lambda_4 \in \mathbb{R}$. Hence, the assertion holds for $\overline{G}_{11}$ and $\overline{G}_{12}$. The equalities (33) and (34) show that $(G_1, \widetilde{g}_1)$ and $(G_2, \widetilde{g}_2)$ satisfy the conditions of Theorem 4.1. From Proposition 4.2 and Proposition 4.3 it follows that $(G_1, \widetilde{g}_1)$ and $(G_2, \widetilde{g}_2)$ are totally umbilical radical transversal lightlike hypersurfaces of $\overline{G}_{11}$ and $\overline{G}_{12}$, respectively. Then from Theorem 4.1, taking into account Remark 4.2, we obtain that $(G_2, \widetilde{g}_2)$ and $(G_1, \widetilde{g}_1)$ are screen totally umbilical radical transversal lightlike hypersurfaces of $\overline{G}_{11}$ and $\overline{G}_{12}$, respectively.

We denote by $\overline{\nu}_{11}$, $\overline{\overline{\nu}}_{11}$ and $\overline{\nu}_{12}$, $\overline{\overline{\nu}}_{12}$ the totally real sectional curvatures of $\overline{G}_{11}$ and $\overline{G}_{12}$, respectively. By using (31) we find $\overline{\nu}_{11} = -\lambda_2^2 < 0$, $\overline{\overline{\nu}}_{11} = 0$ and $\overline{\nu}_{12} = \lambda_4^2 > 0$, $\overline{\overline{\nu}}_{12} = 0$. Now the assertion 3) follows from Theorem 3.3. □

We here give a concrete example.

Example 4.1. In Example 5.1 of [11] we consider the Lie subgroup $\overline{G}$ of $SL(2;\mathbb{C})$ given by

$$\overline{G} = \left\{ \begin{pmatrix} z_1 & z_2 \\ 0 & z_1^{-1} \end{pmatrix} : z_1 \in \mathbb{C} \setminus \{0\},\ z_2 \in \mathbb{C} \right\}.$$

The real Lie algebra $\overline{\mathfrak{g}}$ of $\overline{G}$ consists of all (2×2) complex upper triangular traceless matrices, that is,

$$\overline{\mathfrak{g}} = \left\{ \begin{pmatrix} w_1 & w_2 \\ 0 & -w_1 \end{pmatrix} : w_1, w_2 \in \mathbb{C} \right\}.$$

The Lie algebra $\overline{\mathfrak{g}}$ is spanned by the left invariant vector fields $\{X_1, X_2, X_3, X_4\}$, where

$$X_1 = \begin{pmatrix} i & 0 \\ 0 & -i \end{pmatrix},\ X_2 = \begin{pmatrix} 0 & i \\ 0 & 0 \end{pmatrix},\ X_3 = \begin{pmatrix} 1 & 0 \\ 0 & -1 \end{pmatrix},\ X_4 = \begin{pmatrix} 0 & 1 \\ 0 & 0 \end{pmatrix}.$$

The non-zero Lie brackets of the basic vector fields are given by

$$[X_1, X_2] = -[X_3, X_4] = -2X_4\,; \qquad [X_1, X_4] = -[X_2, X_3] = 2X_2\,. \tag{51}$$

A bi-invariant complex structure $\overline{J}$ on $\overline{\mathfrak{g}}$ is defined by $\overline{JX} = -i\overline{X}$, and a left invariant metric $\overline{g}$ by the same equalities as in (29). Then $(\overline{G}, \overline{J}, \overline{g}, \overline{\widetilde{g}})$ is a 4-dimensional Kähler-Norden manifold of constant totally real sectional curvatures $\overline{\nu} = 4$ and $\overline{\widetilde{\nu}} = 0$. Taking into account that $\overline{J}X_1 = X_3$, $\overline{J}X_2 = X_4$ and (51), we conclude that the Lie algebra $\overline{\mathfrak{g}}$ belongs to the 1-parametric family of Lie algebras $\overline{\mathfrak{g}}_{12}$ given by (50) and $\lambda_4 = -2$.

Further, we consider the Lie subalgebra $\mathfrak{g} = \operatorname{span}\{\overline{J}X_1, \overline{J}X_2, X_2\}$ of $\overline{\mathfrak{g}}$ and find the corresponding to $\mathfrak{g}$ Lie subgroup

$$G = \left\{ \begin{pmatrix} a & z \\ 0 & a^{-1} \end{pmatrix} : a \in \mathbb{R} \setminus \{0\},\ z \in \mathbb{C} \right\}.$$

By direct computations we obtained that $(G, \widetilde{g})$ is a totally umbilical radical transversal lightlike hypersurface of $\overline{G}$ with the following properties:

i) the pair satisfying (6) is $\{\xi = -\overline{J}X_1, N = X_1\}$;
ii) the screen distribution $S(\mathfrak{g})$ is spanned by $\{X_2, \overline{J}X_2\}$;
iii) the screen shape operator is given by $A^*_\xi X = A^*_\xi PX = -2PX$ for any $X \in \mathfrak{g}$.

We remark that $(\overline{G}, \overline{J}, \overline{g}, \overline{\widetilde{g}})$ belongs to the family of Kähler-Norden manifolds $(\overline{G}_{12}, \overline{J}, \overline{g}, \overline{\widetilde{g}})$ and the hypersurface $(G, \widetilde{g})$ of $\overline{G}$ is exactly $(G_2, \widetilde{g}_2)$ according to the notations in the present paper.

Now, we take the Lie subalgebra $\mathfrak{g}_1 = \text{span}\{X_1, X_2, \overline{J}X_2\}$ of $\overline{\mathfrak{g}}$. We find that the corresponding to $\mathfrak{g}_1$ Lie subgroup G_1 of $\overline{G}$ is given by

$$G_1 = \left\{ \begin{pmatrix} u & z \\ 0 & u^{-1} \end{pmatrix} : z, u \in \mathbb{C} : |u| = 1 \right\}.$$

By using Proposition 4.4 we obtain that $(G_1, \widetilde{g}_1)$ is a screen totally umbilical radical transversal lightlike hypersurface of $\overline{G}$. We note that the pair $\{\xi_1 = X_1, N_1 = -\overline{J}X_1\}$ satisfies (6) and the screen distribution $S(\mathfrak{g}_1)$ is spanned by $\{X_2, \overline{J}X_2\}$. Hence, $(G, \widetilde{g})$ and $(G_1, \widetilde{g}_1)$ are radical transversal lightlike hypersurfaces of $\overline{G}$ satisfying the conditions (37) from Theorem 4.1. Then from (42) for the shape operator $A^1_{N_1}$ of $(G_1, \widetilde{g}_1)$ we have $A^1_{N_1}X = A^*_\xi X$ for every $X \in S$. Since $A^*_\xi X = -2PX$, it follows that $A^1_{N_1}X = -2PX$ for every $X \in \mathfrak{g}_1$.

4.2. *Examples of totally umbilical and screen totally umbilical hypersurfaces belonging to* $(\mathcal{M}, g)$

The following left invariant vector fields

$$F_1 = \frac{X_1 - \overline{J}X_1}{\sqrt{2}}, \quad F_2 = \frac{X_2 - \overline{J}X_2}{\sqrt{2}}, \quad F_3 = \frac{X_1 + \overline{J}X_1}{\sqrt{2}}, \quad F_4 = \frac{X_2 + \overline{J}X_2}{\sqrt{2}}$$

form a basis of $\overline{\mathfrak{g}}$. By using (29) we obtain

$$\begin{aligned} &\overline{J}F_1 = F_3, \quad \overline{J}F_2 = F_4, \\ &\overline{\overline{g}}(F_1, F_1) = \overline{\overline{g}}(F_2, F_2) = -\overline{\overline{g}}(F_3, F_3) = -\overline{\overline{g}}(F_4, F_4) = 1, \\ &\overline{\overline{g}}(F_i, F_j) = 0, \quad i \neq j, \end{aligned} \tag{52}$$

which means that $\{F_1, F_2, \overline{J}F_1, \overline{J}F_2\}$ is an orthonormal basis of $\overline{\mathfrak{g}}$ with respect to $\overline{\overline{g}}$. By using (30) we find that the non-zero Lie brackets of the Lie algebra $\overline{\mathfrak{g}}$ of $(\overline{G}, \overline{J}, \overline{g}, \overline{\overline{g}})$ with respect to $\{F_1, F_2, \overline{J}F_1, \overline{J}F_2\}$ are as follows:

$$\overline{\mathfrak{g}}: \left| \begin{aligned} &[F_1, F_2] = -[\overline{J}F_1, \overline{J}F_2] = \mu_1 F_1 + \mu_2 F_2 + \mu_3 \overline{J}F_1 + \mu_4 \overline{J}F_2, \\ &[F_1, \overline{J}F_2] = -[F_2, \overline{J}F_1] = -\mu_3 F_1 - \mu_4 F_2 + \mu_1 \overline{J}F_1 + \mu_2 \overline{J}F_2, \end{aligned} \right. \tag{53}$$

where

$$\mu_1 = \frac{\lambda_1 + \lambda_3}{\sqrt{2}}, \quad \mu_2 = \frac{\lambda_2 + \lambda_4}{\sqrt{2}}, \quad \mu_3 = \frac{\lambda_3 - \lambda_1}{\sqrt{2}}, \quad \mu_4 = \frac{\lambda_4 - \lambda_2}{\sqrt{2}}.$$

We verify that the assertions in Lemma 4.1, Proposition 4.1, Proposition 4.2 and Proposition 4.3 are valid if we interchange the roles of $\overline{g}$ and $\overline{\overline{g}}$ with each other. We denote by $\overline{\mathcal{G}}_{11}$ and $\overline{\mathcal{G}}_{12}$ the Lie algebras with non-zero Lie brackets given as follows

$$\overline{\mathcal{G}}_{11}: \left| \begin{array}{l} [F_1, F_2] = -[\overline{J}F_1, \overline{J}F_2] = \mu_2 F_2, \\ [F_1, \overline{J}F_2] = -[F_2, \overline{J}F_1] = \mu_2 \overline{J}F_2, \end{array} \right. \tag{54}$$

$$\overline{\mathcal{G}}_{12}: \left| \begin{array}{l} [F_1, F_2] = -[\overline{J}F_1, \overline{J}F_2] = \mu_4 \overline{J}F_2, \\ [F_1, \overline{J}F_2] = -[F_2, \overline{J}F_1] = -\mu_4 F_2, \end{array} \right. \tag{55}$$

with non-zero μ_2 and non-zero μ_4. We note that (54) and (55) are obtained from (53) by $\mu_1 = \mu_3 = \mu_4 = 0$, $\mu_2 \neq 0$ and $\mu_1 = \mu_3 = \mu_2 = 0$, $\mu_4 \neq 0$, respectively. These conditions are equivalent to the conditions $\lambda_1 = \lambda_3 = 0$, $\lambda_2 = \lambda_4 \neq 0$ and $\lambda_1 = \lambda_3 = 0$, $\lambda_4 = -\lambda_2 \neq 0$, respectively. By using (30), for the non-zero Lie brackets of $\overline{\mathcal{G}}_{11}$ and $\overline{\mathcal{G}}_{12}$ with respect to $\{X_1, X_2, \overline{J}X_1, \overline{J}X_2\}$, we have

$$\overline{\mathcal{G}}_{11}: \left| \begin{array}{l} [X_1, X_2] = -[\overline{J}X_1, \overline{J}X_2] = \lambda_2 (X_2 + \overline{J}X_2) \\ [X_1, \overline{J}X_2] = -[X_2, \overline{J}X_1] = -\lambda_2 (X_2 - \overline{J}X_2), \end{array} \right. \tag{56}$$

$$\overline{\mathcal{G}}_{12}: \left| \begin{array}{l} [X_1, X_2] = -[\overline{J}X_1, \overline{J}X_2] = \lambda_2 (X_2 - \overline{J}X_2) \\ [X_1, \overline{J}X_2] = -[X_2, \overline{J}X_1] = \lambda_2 (X_2 + \overline{J}X_2). \end{array} \right. \tag{57}$$

From (54) and (55) it follows that $\mathcal{G}_1 = \text{span}\{F_1, F_2, \overline{J}F_2\}$ and $\mathcal{G}_2 = \text{span}\{\overline{J}F_1, \overline{J}F_2, F_2\}$ are subalgebras of both of $\overline{\mathcal{G}}_{11}$ and $\overline{\mathcal{G}}_{12}$. If we denote by H_1 and H_2 the Lie subgroups corresponding to $\mathcal{G}_1$ and $\mathcal{G}_2$, respectively, then we obtain the following analogously as Proposition 4.4.

Proposition 4.5. *Let $(\overline{H}_{11}, \overline{J}, \overline{g}, \overline{\widetilde{g}})$ and $(\overline{H}_{12}, \overline{J}, \overline{g}, \overline{\widetilde{g}})$ be the 4-dimensional Kähler-Norden manifolds with Lie algebras $\overline{\mathcal{G}}_{11}$ and $\overline{\mathcal{G}}_{12}$, respectively. Then the following hold.*

1) *(H_1, g_1) and (H_2, g_2) are radical transversal lightlike hypersurfaces of $\overline{H}_{11}$ such that (H_1, g_1) is totally umbilical and (H_2, g_2) is screen totally umbilical.*
2) *(H_1, g_1) and (H_2, g_2) are radical transversal lightlike hypersurfaces of $\overline{H}_{12}$ such that (H_1, g_1) is screen totally umbilical and (H_2, g_2) is totally umbilical.*
3) *There exist no totally umbilical and screen totally umbilical hypersurfaces of $\overline{H}_{11}$ and $\overline{H}_{12}$ belonging to $(\mathcal{M}, \widetilde{g})$.*

Remark 4.3. Denote by $\overline{\nu}_{13}$, $\overline{\widetilde{\nu}}_{13}$ and $\overline{\nu}_{14}$, $\overline{\widetilde{\nu}}_{14}$ the totally real sectional curvatures of $\overline{H}_{11}$ and $\overline{H}_{12}$, respectively. By using (31) we find that $\overline{\nu}_{13} = \overline{\nu}_{14} = 0$, $\overline{\widetilde{\nu}}_{13} = 2\lambda_2^2 > 0$ and $\overline{\widetilde{\nu}}_{14} = -2\lambda_2^2 < 0$.

References

1. D. Blair, *Contact manifolds in Riemannian geometry*, Lecture Notes in Math. 509, Springer Verlag, Berlin, (1976).
2. A. Borisov & G. Ganchev, Curvature properties of Kaehlerian manifolds with B-metric, *Proc. of* 14th *Spring Conf. of UBM*, Sunny Beach, 220–226 (1985).
3. K. L. Duggal & A. Bejancu, *Lightlike Submanifolds of Semi-Riemannian Manifolds and Applications*, Kluwer Academic, (1996).
4. K.L. Duggal & B. Sahin, *Differential Geometry of Lightlike Submanifolds*, Birkhäuser Verlag AG, (2010).
5. G. Ganchev & A. Borisov, Note on almost complex manifolds with Norden metric, *Compt. Rend. Acad. Bulg.* Sci. 39(1986), 31-34.
6. G. Ganchev, K. Gribachev & V. Mihova, B-connections and their conformal invariants on conformally Kähler manifolds with B-metric, *Publ. Inst. Math. Beograd (Nouvelle serie)* **42**, 56, 107–121 (1987).
7. G. Ganchev, V. Mihova & K. Gribachev, Almost contact manifolds with B-metric, *Math. Balkanica* **7**, 262–276 (1993).
8. K. Gribachev & M. Teofilova, Lie groups as four-dimensional complex manifolds with Norden metric, *J. Geom.* **89**, 34-47 (2008).
9. G. Nakova, Some lightlike submanifolds of almost complex manifolds with Norden metric, *J. Geom.* **103**, 293-312 (2012).
10. G. Nakova, Radical transversal lightlike hypersurfaces of almost complex manifolds with Norden metric, *J. Geom.* **104**, 539-556 (2013).
11. G. Nakova, Totally umbilical radical transversal lightlike hypersurfaces of Kähler-Norden manifolds of constant totally real sectional curvatures, arXiv:1407.6924v2[math.DG] 22 Jan 2015.

Received January 25, 2015
Revised May 17, 2015

Current Developments in Differential Geometry
and its Related Fields *pp. 183 – 199*

COMPLEX STATISTICAL MANIFOLDS AND COMPLEX AFFINE IMMERSIONS

Hiroshi MATSUZOE *

Department of Computer Science and Engineering,
Graduate School of Engineering, Nagoya Institute of Technology,
Nagoya, Aichi 466-8555 Japan
E-mail: matsuzoe@nitech.ac.jp

Complex statistical manifolds in complex affine differential geometry are studied in this paper. The H-projective curvature tensor, the dual H-projective curvature tensor and the H-conformal curvature tensor are one of integrability conditions of complex affine immersions. After summarizing a geometry of complex affine hypersurface immersions, it is showed that the Bartlett tensor field of a geometric divergence coincides with a part of the dual H-projective curvature tensor. Geometry of complex centroaffine immersion of codimension two is also summarized. In this case, the Bartlett tensor field of a geometric divergence coincides with a part of the H-projective curvature tensor or of the H-conformal curvature depending on the integrability condition of complex centroaffine immersion of codimension two.

Keywords: Statistical manifold, contrast function, Bartlett tensor field, complex affine immersion, geometric divergence, Norden metric, information geometry.

1. Introduction

A statistical manifold is a semi-Riemannian manifold (M, h) with a torsion free affine connection ∇ with some compatible condition (see §2). The terminology of statistical manifold was originally introduced in information geometry (cf. [1]), a statistical manifold structure also arises naturally in affine differential geometry ([18]). A contrast function is a distance squared like function on a statistical manifold. In particular, a geometric divergence which is naturally constructed from an affine immersion is a typical example of contrast function (see [8] and [11]).

In this paper, we consider a complex generalization of geometry of statistical manifold. It is known that an affine fundamental form in complex affine immersion is always Norden, if it is nondegenerate. Hence we consider complex statistical manifolds with Norden metrics. In addition, the

*This work was partially supported by MEXT KAKENHI (No. 23740047), Japan Society for the Promotion of Science.

dual H-projective curvature tensor is one of integrability conditions of complex affine hypersurface immersions ([6]). Hence we construct geometric divergences whose Bartlett tensor field coincides with a part of the dual H-projective curvature tensor.

Geometric divergences on complex statistical manifolds are originally introduced in complex centroaffine immersions of codimension two ([13]). Hence, at the last section of this paper, we give a brief review of geometry of complex centroaffine immersions of codimension two.

We remark that applications of generalized conformal structures on real statistical manifolds are developing rapidly (see [7,9,14] and [15], for example).

2. Statistical manifolds and contrast functions

In this paper, we assume that all objects are smooth, and a manifold M is simply connected since we consider local geometric properties on a given manifold. Let us review geometry of statistical manifolds and contrast functions.

2.1. *Statistical manifolds*

Let (M, h) be a semi-Riemannian manifold, and ∇ be a torsion free affine connection on M. We say that the triplet (M, ∇, h) is a *statistical manifold* if ∇h is totally symmetric ([8]), that is, the following equation holds:

$$(\nabla_X h)(Y, Z) = (\nabla_Y h)(X, Z),$$

where X, Y and Z are arbitrary vector fields on M. The totally symmetric tensor field $C(X, Y, Z) := (\nabla_X h)(Y, Z)$ is called the *cubic form* of (M, ∇, h).

For a statistical manifold (M, ∇, h), we can define another affine connection ∇^* by

$$Xh(Y, Z) = h(\nabla_X Y, Z) + h(Y, \nabla^*_X Z).$$

We call ∇^* the *dual connection* of ∇ with respect to h. It is easy to check that $(\nabla^*)^* = \nabla$. In addition, ∇^* is torsion free and $\nabla^* h$ is totally symmetric. Hence the triplet (M, ∇^*, h) is a statistical manifold. We call (M, ∇^*, h) the *dual statistical manifold* of (M, ∇, h).

The mean of dual affine connections coincides with the Levi-Civita connection $\nabla^{(0)}$ with respect to h, that is, $\nabla^{(0)} = (\nabla + \nabla^*)/2$. The difference of dual affine connections coincides with the cubic form:

$$C(X, Y, Z) = h(\nabla^*_X Y, Z) - h(\nabla_X Y, Z).$$

Conversely, for a semi-Riemannian manifold (M,h) with a totally symmetric $(0,3)$-tensor field C, we can define mutually dual torsion free affine connections by

$$h(\nabla_X Y, Z) := h(\nabla^{(0)}_X Y, Z) - \frac{1}{2}C(X,Y,Z),$$
$$h(\nabla^*_X Y, Z) := h(\nabla^{(0)}_X Y, Z) + \frac{1}{2}C(X,Y,Z),$$

where $\nabla^{(0)}$ is the Levi-Civita connection with respect to h. In this case, it is known that (M,∇,h) and (M,∇^*,h) are mutually dual statistical manifolds, and the given tensor field C is the cubic form of (M,∇,h).

We remark that the notion of statistical manifold was introduced by Lauritzen [10]. He called the triplet (M,g,C) a statistical manifold, where (M,g) is a Riemannian manifold and C is a totally symmetric $(0,3)$-tensor field. Later, Kurose [8] redefined a statistical manifold from the viewpoint of affine differential geometry.

Suppose that R and R^* are curvature tensors of ∇ and ∇^*, respectively. Then we have

$$h(R(X,Y)Z,V) = -h(Z,R^*(X,Y)V).$$

Hence, for a statistical manifold (M,∇,h), the connection ∇ is flat if and only if its dual connection ∇^* is flat. In this case, we say that the triplet (M,∇,h) is a *flat statistical manifold* ([8]) and a *Hessian manifold* ([21]). We say that the quadruplet (M,h,∇,∇^*) is a *dually flat space* ([1]).

Next, we review geometry of complex statistical manifolds.

Let (M,h) be an n-dimensional complex semi-Riemannian manifold ($\dim_{\mathbf{C}} M = n$) with complex structure J. We say that h is *Hermite* if $h(X,Y) = h(JX,JY)$, and *Norden* if $h(X,Y) = -h(JX,JY)$. From the definition of Norden metrics, h is necessarily indefinite of signature (n,n). We suppose that ∇ is a complex connection on (M,J), that is, ∇ is torsion free and $\nabla J = 0$ everywhere on M. In this case, a quadruplet (M,J,∇,h) is said to be a *complex statistical manifold* if (M,∇,h) is a statistical manifold. In particular, we say that (M,J,∇,h) is a complex statistical manifold of *Norden type* (or *Hermite type*) if h is Norden (or Hermite), respectively. Denote by ∇^* the dual connection of ∇ with respect to h. Then the quadruplet (M,J,∇^*,h) is called the *dual complex statistical manifold* of (M,J,∇,h).

In this paper, we study complex statistical manifolds of Norden type. On the other hand, complex statistical manifolds of Hermite type have been studied in [4], for example. In particular, relations between Kähler structures and symplectic structures on a statistical manifold are studied in

[16]. We remark that the notion of complex statistical manifold has some different definitions (see [13]).

2.2. *Contrast functions*

Next, let us recall geometry of contrast functions.

Let D be a function on $M \times M$. For an arbitrary point $(p,q) \in M \times M$, we identify the tangent space $T_{(p,q)}(M \times M)$ with the direct sum of $T_pM \oplus T_qM$. Then we define a function on M as follows:

$$D[X_1,\ldots,X_i|Y_1,\ldots,Y_j](r) := (X_1)_p \ldots (X_i)_p (Y_1)_q \ldots (Y_j)_q D(p,q)|_{\substack{p=r \\ q=r}},$$

where $X_1,\ldots,X_i$ and $Y_1,\ldots,Y_j$ are arbitrary vector fields on M. We say that a function D on $M \times M$ is a *contrast function* if it satisfies the following conditions ([3,11]):

1) $D[\ |\ ](p) = 0$,
2) $D[X|\ ](p) = D[\ |X](p) = 0$,
3) $h(X,Y) := -D[X|Y]$ is a semi-Riemannian metric on M.

For a contrast function D, we define affine connections ∇ and ∇^* by

$$h(\nabla_X Y, Z) := -D[XY|Z],$$
$$h(Y, \nabla^*_X Z) := -D[Y|XZ],$$

respectively. For example, the Leibniz rule for ∇ is obtained by

$$\begin{aligned} h(\nabla_X(fY), Z) &= -D[X(fY)|Z] = -(Xf)D[Y|Z] - fD[XY|Z] \\ &= h((Xf)Y + f\nabla_X Y, Z). \end{aligned}$$

Differentiating the equation $h(X,Y) = -D[X|Y]$, we can show that ∇ and ∇^* are mutually dual with respect to h. In addition, ∇ and ∇^* are torsion free because

$$h(\nabla_X Y, Z) - h(\nabla_Y X, Z) = -D[XY - YX|Z] = h([X,Y], Z).$$

Since mutually dual affine connections ∇ and ∇^* are torsion free, ∇h and $\nabla^* h$ are totally symmetric, respectively. Hence (M,∇,h) and (M,∇^*,h) are statistical manifolds. We say that (M,∇,h) and (M,∇^*,h) are *induced statistical manifolds* from a contrast function D.

We define $(1,3)$-tensor fields B and B^* on M by the following formulas:

$$h(B(X,Y)Z, V) := -D[XYZ - \nabla_X \nabla_Y Z|V],$$
$$h(V, B^*(X,Y)Z) := -D[V|XYZ - \nabla^*_X \nabla^*_Y Z].$$

We call B the *Bartlett tensor field* of contrast function D and B^* the *dual Bartlett tensor field*, respectively. It is known that the anti-symmetric part of B gives the curvature tensor R of ∇, and B^* gives R^*, respectively.

Proposition 2.1 (cf. [3,11]). *Let D be a contrast function on $M \times M$. Suppose that B and B^* are the Bartlett tensor fields of D, respectively. Then, the following equations hold:*

$$R(X,Y)Z = B(Y,X)Z - B(X,Y)Z,$$
$$R^*(X,Y)Z = B^*(Y,X)Z - B^*(X,Y)Z.$$

3. Holomorphic curvature tensors

In this section, we recall the definitions of H-projective, dual H-projective, and H-conformal curvature tensors. For more details, see [5,6] and [12].

3.1. *H-projective curvature tensors*

Let (M, J) be a complex manifold with a torsion free affine connection ∇. The Ricci tensor field of ∇ is defined by

$$\mathrm{Ric}(X,Y) := \mathrm{tr}\{X \mapsto R(X,Y)Z\},$$

where R is the curvature tensor field of ∇. We suppose that the Ricci tensor is symmetric and pure, that is,

$$\mathrm{Ric}(X,Y) = \mathrm{Ric}(Y,X),$$
$$\mathrm{Ric}(X,JY) = \mathrm{Ric}(JX,Y).$$

Under the above conditions, we define the *holomorphic projective curvature tensor* (or *H-projective curvature tensor*) W_P of ∇ by

$$W_{HP}(X,Y)Z := R(X,Y)Z + S_{HP}(X,Y)Z - S_{HP}(Y,X)Z,$$

where the tensor field S_{HP} is defined by

$$S_{HP}(X,Y)Z := -\frac{1}{2(n-1)}\{\mathrm{Ric}(Y,Z)X - \mathrm{Ric}(Y,JZ)JX\}. \qquad (1)$$

If $\dim_{\mathbf{C}} M \geq 3$, the connection ∇ is said to be *holomorphically projectively flat* (or *H-projectively flat*) if the H-projective curvature tensor of ∇ vanishes everywhere on M. If ∇ is H-projectively flat, the following equation holds:

$$(\nabla_X \mathrm{Ric})(Y,Z) = (\nabla_Y \mathrm{Ric})(X,Z).$$

3.2. *Dual H-projective curvature tensors*

Let (M, J, ∇, h) be a complex statistical manifold of Norden type. Denote by $\mathrm{tr}_h(\mathrm{Ric})$ the trace of the Ricci tensor with respect to h, and by $\mathrm{Ric}^{\#}(X)$ the Ricci operator $h(\mathrm{Ric}^{\#}(X), Y) := \mathrm{Ric}(X, Y)$. Set a $(1,1)$-tensor field L by

$$L(X) := -\frac{1}{2}\mathrm{Ric}^{\#}(X) + \frac{1}{4(n-1)}\{\mathrm{tr}_h(\mathrm{Ric})X - \mathrm{tr}_h(\mathrm{Ric} \circ J)JX\}.$$

We define the *dual holomorphic projective curvature tensor* (or *dual H-projective curvature tensor*) W_{DHP} of ∇ by

$$W_{DHP}(X, Y)Z := R(X, Y)Z + S_{DHP}(X, Y)Z - S_{DHP}(Y, X)Z,$$

where the tensor field S_{DHP} is defined by

$$S_{DHP}(X, Y)Z := -h(Y, Z)LX + h(Y, JZ)JLX. \tag{2}$$

If $\dim_{\mathbf{C}} M \geq 3$, the connection ∇ is said to be *dual holomorphically projectively flat* (or *dual H-projectively flat*) if the dual H-projective curvature tensor of ∇ vanishes everywhere on M (cf. [5]).

3.3. *H-conformal curvature tensors*

Let (M, J, h) be a complex manifold with a Norden metric. Suppose that $\nabla^{(0)}$ is the Levi-Civita connection with respect to h. Denote by $R^{(0)}$ the curvature tensor of $\nabla^{(0)}$, and by $\mathrm{Ric}^{(0)}$ its Ricci tensor. Set a $(1,1)$-tensor field $L^{(0)}$ by

$$\begin{aligned} L^{(0)}(X) := {} & \frac{1}{2(n-2)}\mathrm{Ric}^{(0)\#}(X) \\ & -\frac{1}{8(n-1)(n-2)}\{\mathrm{tr}_h(\mathrm{Ric}^{(0)})X - \mathrm{tr}_h(\mathrm{Ric}^{(0)} \circ J)JX\}, \end{aligned}$$

where $\mathrm{tr}_h(\mathrm{Ric}^{(0)})$ is the trace of the Ricci tensor with respect to h and $\mathrm{Ric}^{(0)\#}(X)$ is the Ricci operator.

For (M, J, h), we define *holomorphic conformal curvature tensor* (or *H-conformal curvature tensor*) W_{HC} of h by

$$W_{HC}(X, Y)Z := R^{(0)}(X, Y)Z + S_{HC}(X, Y)Z - S_{HC}(Y, X)Z,$$

where the tensor field S_{HC} is defined by

$$\begin{aligned} S_{HC}(X, Y)Z := {} & -h(Y, Z)LX + h(Y, JZ)JLX \\ & -h(LY, Z)X + h(LY, JZ)JX. \end{aligned} \tag{3}$$

If $\dim_{\mathbf{C}} M \geq 4$, the Norden metric h is said to be *holomorphically conformally flat* (or *H-conformally flat*) if the H-conformal curvature tensor of h vanishes everywhere on M.

We remark that the H-conformal curvature tensor has some different definitions. For another definition, see [22] for example.

4. Complex affine immersions

In this section, we summerize geometry of complex affine immersions, and construct a contrast function from the viewpoint of complex affine differential geometry. For foundations of affine differential geometry, see [18]. For geometry of complex affine immersions, see [2,6] and [17].

Let (M, J) be a complex manifold of $\dim_{\mathbf{C}} M = n$ (≥ 2). Denote by $\mathbf{C}^{n+1} \cong (\mathbf{R}^{n+1}, \widetilde{J})$ the affine space $\mathbf{R}^{2n+2}$ with the standard complex structure $\widetilde{J}$, and by $\widetilde{\nabla}$ the standard flat affine connection of $\mathbf{C}^{n+1}$. Suppose that $f : (M, J) \to \mathbf{C}^{n+1}$ is a holomorphic immersion, i.e., $f_*JX = \widetilde{J}f_*X$, and ξ is a vector field along f. We say that the pair $\{f, \xi\} : (M, J) \to \mathbf{C}^{n+1}$ is a *complex affine immersion* if the tangent space $T_{f(p)}\mathbf{C}^{n+1} (p \in M)$ is decomposed as follows:

$$T_{f(p)}\mathbf{C}^{n+1} = f_*(T_pM) \oplus \mathbf{R}\{\xi_p\} \oplus \mathbf{R}\{\widetilde{J}\xi_p\}, \tag{4}$$

where $\mathbf{R}\{\xi_p\}$ and $\mathbf{R}\{\widetilde{J}\xi_p\}$ are the real one dimensional subspaces spanned by ξ_p and $\widetilde{J}\xi_p$, respectively. We call ξ a *transversal vector field* of f. When $\xi = -f$ (or $\xi = f$), we say that $\{f, -f\}$ is a *centroaffine immersion.*

From the decomposition of tangent space (4), Gauss and Weingarten formulas are given as follows:

$$\begin{aligned} \widetilde{\nabla}_X f_*Y &= f_*\nabla_X Y + h(X, Y)\xi + \bar{h}(X, Y)\widetilde{J}\xi, \\ \widetilde{\nabla}_X \xi &= -f_*SX + \tau(X)\xi + \bar{\tau}(X)\widetilde{J}\xi. \end{aligned}$$

We say that ∇, h, S and τ are the *induced connection*, the *affine fundamental form*, the *affine shape operator*, and the *transversal connection form* of $\{f, \xi\}$, respectively.

Since the connection $\widetilde{\nabla}$ is compatible with $\widetilde{J}$ and f is holomorphic, then the induced connection ∇ is a complex connection on M. We say that a transversal vector field ξ is *holomorphic* if $\bar{\tau}(X) = -\tau(JX)$ and $SJ = JS$, and ξ is *anti-holomorphic* if $\bar{\tau}(X) = \tau(JX)$ and $SJ = -JS$.

By straightforward calculations, we have the following proposition.

Proposition 4.1. *Let $\{f, \xi\}$ be a complex affine immersion from (M, J)*

to $\mathbf{C}^{n+1}$. *Then we obtain*

$$h(JX, Y) = h(X, JY) = -\bar{h}(X, Y).$$

Proposition 4.2. *Let* $\{f, \xi\}$ *be a complex affine immersion from* (M, J) *to* $\mathbf{C}^{n+1}$. *We change a transversal vector field to*

$$\xi' = \phi\xi + \psi\widetilde{J}\xi + f_* V,$$

where ϕ *and* ψ *are functions on* M *nowhere vanishing simultaneously, and* V *is a vector field on* M. *Denote by* $\nabla', h', \bar{h}', S', \tau'$ *and* $\bar{\tau}'$ *the induced objects from* $\{f, \xi'\}$. *Then we have*

$$\begin{aligned}
&\text{(i)} && \nabla_X Y = \nabla'_X Y + h'(X,Y)V + \bar{h}'(X,Y)JV, \\
&\text{(ii)} && h(X,Y) = \phi h'(X,Y) - \psi\bar{h}'(X,Y), && (5)\\
& && \bar{h}(X,Y) = \psi h'(X,Y) + \phi\bar{h}'(X,Y), && (6)\\
&\text{(iii)} && \phi SX + \psi JSX - \nabla_X V = S'X - \tau'(X)V - \bar{\tau}'(X)JV, \\
&\text{(iv)} && X\phi + \phi\tau(X) - \psi\bar{\tau}(X) + h(X,V) = \phi\tau'(X) - \psi\bar{\tau}'(X), \\
& && X\psi + \psi\tau(X) - \phi\bar{\tau}(X) + h(X,V) = \psi\tau'(X) + \phi\bar{\tau}'(X).
\end{aligned}$$

Equations (5) and (6) imply

$$h' = \frac{1}{\phi^2 + \psi^2}(\phi h + \psi k), \quad k' = \frac{1}{\phi^2 + \psi^2}(\phi k - \psi h).$$

From the above equations and Proposition 4.1, nondegeneracy of h is independent of the choice of ξ. We say that f is *nondegenerate* if h is nondegenerate. In this case, we can choose ξ such that $\tau = \kappa = 0$. We say that $\{f, \xi\}$ is *equiaffine* if τ vanishes everywhere on M.

Since the connection $\widetilde{\nabla}$ is flat, from the decomposition of the tangent space and Proposition 4.1, we have the fundamental equations of complex affine immersions.

$$\begin{aligned}
&\text{Gauss:} && R(X,Y)Z \\
& && \quad = h(Y,Z)SX - h(JY,Z)JSX - h(X,Z)SY + h(JX,Z)JSY, \\
&\text{Codazzi:} && (\nabla_X h)(Y,Z) + \tau(X)h(Y,Z) + \bar{\tau}(X)h(JY,Z) \\
& && \quad = (\nabla_Y h)(X,Z) + \tau(Y)h(X,Z) + \bar{\tau}(Y)h(JY,Z), \qquad (7)\\
& && (\nabla_X S)(Y) + \tau(Y)SX + \bar{\tau}(Y)JSX \\
& && \quad = (\nabla_Y S)(X) + \tau(X)SY + \bar{\tau}(X)JSY, \\
&\text{Ricci:} && h(SX,Y) - h(X,SY) = d\tau(X,Y), \\
& && h(SX,JY) - h(JX,SY) = d\bar{\tau}(X,Y).
\end{aligned}$$

Lemma 4.1. *Let $\{f, \xi\}$ be a nondegenerate complex affine immersion. Then the affine shape operator is given by*

$$SX = -\frac{1}{2}\mathrm{Ric}^{\#}(X) + \frac{1}{4(n-1)}\left\{\mathrm{tr}_h(\mathrm{Ric})X - \mathrm{tr}_h(J \circ \mathrm{Ric})JX\right\},$$

where $\mathrm{Ric}^{\#}$ *is the Ricci operator, and* $\mathrm{tr}_h(\mathrm{Ric})$ *is the trace of Ricci tensor with respect to* h.

Proof. Contracting the Gauss equation, we have

$$\mathrm{Ric}(Y, Z) = \mathrm{tr}\, S\, h(Y, Z) - \mathrm{tr}(J \circ S)h(JY, Z) - 2h(SY, Z). \tag{8}$$

Contracting the above equation with respect to h, we have

$$\mathrm{tr}_h\mathrm{Ric} = (2n-2)\mathrm{tr}S. \tag{9}$$

Substituting (9) to (8), we obtain the result. □

From Codazzi equation (7) for h, if a complex affine immersion $\{f, \xi\}$ is nondegenerate and equiaffine, then the quadruplet (M, J, ∇, h) is a statistical manifold of Norden type. Moreover, we have the fundamental theorem. for complex affine immersions (cf. [6 ,17 ,19] and [20]).

Theorem 4.1. *Let $\{f, \xi\}$ be a complex affine immersion from (M, J) to $\mathbf{C}^{n+1}$. Suppose that $\{f, \xi\}$ is holomorphic nondegenerate equiaffine immersion. Then (M, J, ∇, h) is a statistical manifold of Norden type with a dual H-projectively flat affine connection.*

Conversely, let (M, J, ∇, h) be a simply connected n-dimensional complex statistical manifold of Norden type. Suppose that the given connection ∇ is dual H-projectively flat and its Ricci tensor is symmetric and pure. Then there exists a complex affine immersion from (M, J) to $\mathbf{C}^{n+1}$ such that $\{f, \xi\}$ induces the given statistical manifold (M, J, ∇, h).

Let us construct a contrast function on a complex statistical manifold (M, J, ∇, h) in terms of complex affine immersions.

Let $\mathbf{C}^{n+1} \cong (\mathbf{R}^{2n+2}, \widetilde{J})$ be a complex affine space. Denote by $\mathbf{R}_{2n+2}$ the dual vector space of $\mathbf{R}^{2n+2}$. We define a map $\widetilde{J}^* : \mathbf{R}_{2n+2} \to \mathbf{R}_{2n+2}$ by

$$\langle \widetilde{J}^*\alpha, x\rangle := (\widetilde{J}^*\alpha)(x) := \alpha(\widetilde{J}x),$$

where $\alpha \in \mathbf{R}_{2n+2}$, $x \in \mathbf{R}^{2n+2}$ and $\langle *, *\rangle$ is the canonical pairing of $\mathbf{R}_{2n+2}$ and $\mathbf{R}^{2n+2}$. Then $\mathbf{C}_{n+1}$ is the dual complex vector space of $\mathbf{C}^{n+1}$.

Let $\{f, \xi\}$ be a complex affine immersion. We define a map v from (M, J) to $\mathbf{C}_{n+1}$ by

$$\langle v, \xi\rangle = 1, \quad \langle v, \widetilde{J}\xi\rangle = 0, \quad \langle v, f_*X\rangle = 0.$$

We call v the *conormal map* of $\{f,\xi\}$. From the definition of conormal map, using Gauss, Weingarten equations and Proposition 4.1, we have the following.

Proposition 4.3. *Let $\{f,\xi\}$ be a complex affine immersion, and v the conormal map of $\{f,\xi\}$. Then we obtain*

$$\begin{aligned} \langle v_*X, \xi\rangle &= -\tau(X), & \langle v_*X, f_*Y\rangle &= -h(X,Y),\\ \langle v_*X, \widetilde{J}\xi\rangle &= \bar{\tau}(X), & \langle v_*JX, f_*Y\rangle &= -h(X,JY),\\ \langle v_*JX, \xi\rangle &= -\tau(X), & \langle v_*X, f_*JY\rangle &= -h(X,JY). \end{aligned}$$

Proof. By differentiating $\langle v,\xi\rangle = 1$, we have

$$\begin{aligned} 0 &= X\langle v,\xi\rangle\\ &= \langle v_*X,\xi\rangle + \langle v,\widetilde{\nabla}_X\xi\rangle\\ &= \langle v_*X,\xi\rangle + \langle v, -f_*SX + \tau(X)\xi + \bar{\tau}(X)\widetilde{J}\xi\rangle\\ &= \langle v_*X,\xi\rangle + \tau(X). \end{aligned}$$

Similarly, by differentiating $\langle v, f_*Y\rangle = 0$, we have

$$\begin{aligned} 0 &= X\langle v, f_*Y\rangle\\ &= \langle v_*X, f_*Y\rangle + \langle v, \tilde{\nabla}_X f_*Y\rangle\\ &= \langle v_*X, f_*Y\rangle + \langle v, f_*\nabla_X Y + h(X,Y)\xi + \bar{h}(X,Y)\tilde{J}\xi\rangle\\ &= \langle v_*X, f_*Y\rangle + h(X,Y). \end{aligned}$$

In the same manner, we obtain the other formulas. □

From Proposition 4.3, we have the following result.

Proposition 4.4. *Let $\{f,\xi\}$ be a complex affine immersion, and v the conormal map of $\{f,\xi\}$.*

(1) *ξ is holomorphic if and only if v is holomorphic, i.e., $\widetilde{J}^*\circ v_* = v_*\circ J$.*
(2) *ξ is anti-holomorphic if and only if v is anti-holomorphic, i.e., $\widetilde{J}^*\circ v_* = -v_*\circ J$.*

If $\{f,\xi\}$ is nondegenerate, the affine fundamental form h is always a Norden metric in both cases holomorphic and anti-holomorphic. Hence we assume that ξ is holomorphic from now on.

Proposition 4.5. *Let $\{f,\xi\}$ be a nondegenerate complex affine immersion, and v the conormal map of $\{f,\xi\}$. Suppose that ξ is holomorphic. Then $\{v,-v\}$ is a complex centroaffine immersion from (M,J) to $\mathbf{C}_{n+1}$.*

Proof. From Propositions 4.3 and 4.4, v is a holomorphic immersion from (M, J) to $\mathbf{C}_{n+1}$ since h is nondegenerate. From the definition of the conormal map, v is transversal to v itself. Hence $\{v, -v\}$ is a complex centroaffine immersion. □

Denote by $\widetilde{\nabla}$ the standard flat affine connection on $\mathbf{C}_{n+1}$. The Gauss equation for $\{v, -v\}$ is given by

$$\widetilde{\nabla}_X v_* Y = v_*(\nabla^*_X Y) - h^*(X, Y) - \bar{h}^*(X, Y)\widetilde{J}^* v.$$

Then the relations of induced objects between $\{f, \xi\}$ and $\{v, -v\}$ are given as follows.

$$h^*(X, Y) - h(SX, Y) = (\nabla^*_X \tau)Y - \tau(X)\tau(Y) + \bar{\tau}(X)\bar{\tau}(Y),$$
$$Xh(Y, Z) = h(\nabla_X Y, Z) + h(Y, \nabla^*_X Z) + \tau(Z)h(X, Y) - \bar{\tau}(Z)\bar{h}(X, Y).$$

Next, let us construct a contrast function from the viewpoint of complex affine immersions. Let $\{f, \xi\}$ be a nondegenerate equiaffine complex affine immersion, and v the conormal map of $\{f, \xi\}$. Suppose that (M, J, ∇, h) the complex statistical manifold induced from $\{f, \xi\}$. We define a function D on $M \times M$ by

$$D(p, q) := \langle v(q), f(p) - f(q) \rangle, \qquad (p, q \in M). \tag{10}$$

We call D the *geometric divergence* of (M, J, ∇, h). If M is simply connected, D is independent from the realization of (M, J, ∇, h) into $\mathbf{C}^{n+1}$.

Proposition 4.6. *Let D be a geometric divergence of (M, J, ∇, h). Then D is a contrast function on $M \times M$.*

Proof. From the definitions of geometric divergence (10) and the conormal map, we have

$$D[\,|\,](r) = \langle v(q), f(p) - f(q) \rangle|_{\substack{p=r\\q=r}} = 0,$$
$$D[X|\,](r) = X_p \langle v(q), f(p) - f(q) \rangle|_{\substack{p=r\\q=r}} = \langle v(q), f_* X_p \rangle|_{\substack{p=r\\q=r}} = 0,$$
$$D[X|Y](r) = Y_q \langle v(q), f_* X_p \rangle|_{\substack{p=r\\q=r}} = \langle v_* Y_q, f_* X_p \rangle|_{\substack{p=r\\q=r}} = -h(Y, X).$$

These equations imply that D is a contrast function on $M \times M$. □

We remark that a geometric divergence D induces the induced connection ∇. In fact, since $\{f, \xi\}$ is equiaffine, we obtain

$$\begin{aligned} D[XY|Z](r) &= X_p \langle v_* Z_q, f_* Y_p \rangle|_{\substack{p=r\\q=r}} \\ &= \langle v_* Z_q, f_*(\nabla_X Y)_p + h_p(X, Y)\xi_p + \bar{h}_p(X, Y)\tilde{J}\xi_p \rangle|_{\substack{p=r\\q=r}} \\ &= -h_r(\nabla_X Y, Z). \end{aligned}$$

Theorem 4.2. *Suppose that (M, J, ∇, h) is a simply connected dual H-projectively flat complex statistical manifold of Norden type. Then there exits a geometric divergence of (M, J, ∇, h) such that its Bartlett tensor field coincides with the tensor field S_{DHP}, i.e., $B(X,Y)Z = S_{DHP}(X,Y)Z$.*

Proof. From Theorem 4.1, there exits a complex affine immersion $\{f, \xi\}$ from (M, J) to $\mathbf{C}^{n+1}$ such that $\{f, \xi\}$ induces the given complex statistical manifold (M, J, ∇, h). In this case, the complex affine immersion $\{f, \xi\}$ is holomorphic nondegenerate equiaffine.

By differentiating four times the geometric divergence $D(p, q)$, we have

$$\begin{aligned}
D[XYZ|V](r) &= X_p Y_p Z_p V_q D(p,q)|_{\substack{p=r\\q=r}} \\
&= X_p Y_p Z_p V_q \langle v(q), f(p) - f(q)\rangle|_{\substack{p=r\\q=r}} \\
&= X_p Y_p \langle v_* V_q, f_* Z_p\rangle|_{\substack{p=r\\q=r}} \\
&= X_p \langle v_* V_q, f_*(\nabla_Y Z)_p + h(Y,Z)\xi_p + \bar{h}(Y,Z)\widetilde{J}\xi_p\rangle|_{\substack{p=r\\q=r}} \\
&= -h(\nabla_X \nabla_Y Z, V) \\
&\qquad + h(Y,Z)h(SX,V) - h(Y,JZ)h(JSX,V).
\end{aligned}$$

Hence the Bartlett tensor field is given by

$$B(X,Y)Z = -h(Y,Z)SX + h(Y,JZ)JSX.$$

From Lemma 4.1, since ξ is holomorphic, we have the affine fundamental form by

$$SX = -\frac{1}{2}\mathrm{Ric}^{\#}(X) + \frac{1}{4(n-1)}\left\{\mathrm{tr}_h(\mathrm{Ric})X - \mathrm{tr}_h(\mathrm{Ric}\circ J)JX\right\}.$$

Since (M, J, ∇, h) is dual H-projectively flat, then the Bartlett tensor field coincides with the tensor field S_{DHP}, which is defined by Equation (2). □

5. Complex centroaffine immersions of codimension two

In the previous section, we construct a contrast function from the viewpoint of complex affine immersion. This kind of results was originally obtained in complex centroaffine immersions of codimension two ([12]).

Let (M, J) be a complex manifold of $\dim_{\mathbf{C}} M = n (\geq 2)$. Denote by $\mathbf{C}^{n+2}$ the vector space $\mathbf{R}^{2n+4}$ with the standard complex structure $\widetilde{J}$. Let $f : M \to \mathbf{C}^{n+2}$ be a holomorphic immersion, and ξ a holomorphic vector field along f. We say that the pair $\{f, \xi\} : (M, J) \to \mathbf{C}^{n+2}$ is a *holomorphic centroaffine immersion of codimension two* if the tangent space $T_{f(p)}\mathbf{C}^{n+2}$ is decomposed as

$$T_{f(p)}\mathbf{C}^{n+2} = f_*(T_pM) \oplus \mathbf{R}\{\xi_p\} \oplus \mathbf{R}\{\widetilde{J}\xi_p\} \oplus \mathbf{R}\{f(p)\} \oplus \mathbf{R}\{\widetilde{J}f(p)\},$$

where $\mathbf{R}\{\xi_p\}, \mathbf{R}\{\xi_p\}, \mathbf{R}\{f(p)\}$ and $\mathbf{R}\{\widetilde{J}f(p)\}$ show the real one dimensional subspaces spanned by $\xi_p, \widetilde{J}\xi_p, f(p)$ and $\widetilde{J}f(p)$, respectively. We call ξ a *transversal vector field* of f.

We remark that the transversal vector field is assumed to be holomorphic. Geometry of complex centroaffine immersions of codimension two with anti-holomorphic transversal vector fields is a quite open problem.

Denote by $\widetilde{\nabla}$ the standard flat affine connection on $\mathbf{C}^{n+2}$. From the decomposition of tangent space, Gauss and Weingarten formulas are given as follows:

$$\begin{aligned}\widetilde{\nabla}_X f_*Y &= f_*\nabla_X Y + h(X,Y)\xi + \bar{h}(X,Y)\widetilde{J}\xi + T(X,Y)f + \bar{T}(X,Y)\widetilde{J}f,\\ \widetilde{\nabla}_X\xi &= -f_*SX + \tau(X)\xi + \bar{\tau}(X)\widetilde{J}\xi + \mu(X)f + \bar{\mu}(X)\widetilde{J}f.\end{aligned}$$

We call ∇, h, S and τ the *induced connection*, the *affine fundamental form*, the *affine shape operator*, and the *transversal connection form* of $\{f,\xi\}$, respectively. Since f is a holomorphic immersion and $\widetilde{\nabla}\widetilde{J} = 0$, the induced connection ∇ is a complex connection.

Proposition 5.1. *Suppose that $\{f,\xi\} : (M,J) \to \mathbf{C}^{n+2}$ is a holomorphic centroaffine immersion of codimension two. Then the induced objects have the following formulas:*

$$\begin{aligned}&h(JX,Y) = h(X,JY) = -\bar{h}(X,Y),\\ &T(JX,Y) = T(X,JY) = -\bar{T}(X,Y),\\ &\tau(X) = -\bar{\tau}(JX), \qquad \mu(X) = -\bar{\mu}(JX), \qquad SJ = JS.\end{aligned}$$

We say that f is *nondegenerate* if the affine fundamental form is non-degenerate everywhere on M. This condition is independent of choice of transversal vector fields. If f is non-degenerate, we can take a transversal vector field ξ such that the transversal connection form τ vanishes. In this case, we say that $\{f,\xi\}$ is *equiaffine*.

Since the connection $\widetilde{\nabla}$ on $\mathbf{C}^{n+2}$ is flat, from the decomposition of the tangent space and the relations in Proposition 5.1, we have the fundamental equations of holomorphic centroaffine immersions of codimension two.

$$\begin{aligned}\text{Gauss:}\quad & R(X,Y)Z\\ &= h(Y,Z)SX - h(JY,Z)JSX - T(Y,Z)X + T(JY,Z)JX\\ &\quad -h(X,Z)SY + h(JX,Z)JSY + T(X,Z)Y - T(JX,Z)JY,\end{aligned}$$

$$\begin{aligned}
\text{Codazzi:}\quad & (\nabla_X h)(Y,Z) + \tau(X)h(Y,Z) - \tau(JX)h(JY,Z) \\
&\quad = (\nabla_Y h)(X,Z) + \tau(Y)h(X,Z) - \tau(JY)h(JY,Z), \qquad (11) \\
& (\nabla_X T)(Y,Z) + \mu(X)h(Y,Z) - \mu(JX)h(JY,Z) \\
&\quad = (\nabla_Y T)(X,Z) + \mu(Y)h(X,Z) - \mu(JY)h(JXZ), \\
& (\nabla_X S)(Y) + \tau(Y)SX - \tau(JY)JSX + \mu(Y)X - \mu(JY)JX \\
&\quad = (\nabla_Y S)(X) + \tau(X)SY - \tau(JX)JSY + \mu(X)Y - \mu(JX)JY, \\
\text{Ricci:}\quad & h(SX,Y) - h(X,SY) = d\tau(X,Y), \\
& T(SX,Y) - T(X,SY) = d\mu(X,Y).
\end{aligned}$$

From the Codazzi equation (11) for h, if a complex centroaffine immersion $\{f,\xi\}$ is nondegenerate and equiaffine, then the quadruplet (M,J,∇,h) is a statistical manifold of Norden type. Fundamental theorems of complex affine immersions of codimension two is known in [12].

Theorem 5.1. *Let (M,J,∇,h) be a simply connected complex statistical manifold of Norden type. Suppose that ∇ is a complex connection and its Ricci tensor is symmetric and pure. Then the following hold.*

(1) *Suppose that* $\dim_{\mathbf{C}} M \geq 3$. *Then ∇ is H-projectively flat if and only if there exists a complex centroaffine immersion of codimension two $\{f,\xi\}$ such that it induces the given statistical manifold (M,J,∇,h) and ξ is a parallel vector field.*
(2) *Suppose that* $\dim_{\mathbf{C}} M \geq 3$. *Then ∇ is dual H-projectively flat if and only if there exists a complex centroaffine immersion of codimension two $\{f,\xi\}$ such that it induces the given statistical manifold (M,J,∇,h) and the image $f(M)$ lies on some complex affine hyperplane of $\mathbf{C}^{n+2}$.*
(3) *Suppose that* $\dim_{\mathbf{C}} M \geq 4$. *Then h is H-conformally flat if and only if there exists a complex centroaffine immersion of codimension two $\{f,\xi\}$ such that it induces a statistical manifold $(M,J,\nabla^{(0)},h)$, where $\nabla^{(0)}$ is the Levi-Civita connection with respect to h.*

Let us construct a contrast function on a complex statistical manifold from the viewpoint of holomorphic centroaffine immersions of codimension two.

Let $\mathbf{C}^{n+2}$ be a complex vector space. Denote by $\mathbf{C}_{n+2}$ the dual complex vector space of $\mathbf{C}^{n+2}$. Suppose that $\{f,\xi\}$ is a holomorphic centroaffine immersion of codimension two. We define the *dual map* $\{v,w\}$ of $\{f,\xi\}$ by

$$\langle v,\xi\rangle = 1, \qquad \langle v,f\rangle = 0, \qquad \langle w,\xi\rangle = 0, \qquad \langle w,f\rangle = 1,$$
$$\langle v,\widetilde{J}\xi\rangle = 0, \quad \langle v,\widetilde{J}f\rangle = 0, \qquad \langle w,\widetilde{J}\xi\rangle = 0, \quad \langle w,\widetilde{J}f\rangle = 0,$$
$$\langle v,f_*X\rangle = 0, \qquad \langle w,f_*X\rangle = 0.$$

If f is nondegenerate, v is a holomorphic immersion from (M,J) to $\mathbf{C}_{n+2}$. Since w is transversal to v, the pair $\{v,w\}$ is a centroaffine immersion of codimension two from (M,J) to $\mathbf{C}_{n+2}$. Similar calculations of Proposition 4.3, we have the following.

Proposition 5.2. *Let $\{f,\xi\}$ be a holomorphic centroaffine immersion of codimension two, and $\{v,w\}$ the dual map of $\{f,\xi\}$. Then we obtain*

$$\langle v_*X,f_*Y\rangle = -h(X,Y), \quad \langle v_*JX,f_*Y\rangle = \langle v_*X,f_*JY\rangle = -h(X,JY).$$

Suppose that $\{f,\xi\}$ is nondegenerate equiaffine and $\widetilde{\nabla}$ is the standard flat affine connection on $\mathbf{C}_{n+2}$. The Gauss equation for $\{v,w\}$ is given by

$$\begin{aligned}\widetilde{\nabla}_X v_*Y = v_*(\nabla^*_X Y) + h^*(X,Y) + \bar{h}^*(X,Y)\widetilde{J}^*v \\ +T^*(X,Y)w + \bar{T}^*(X,Y)\widetilde{J}^*w.\end{aligned}$$

Then the induced objects satisfy

$$\begin{aligned}&h(X,Y) - h^*(X,Y), \quad T(X,Y) = T^*(X,Y), \\ &Xh(Y,Z) = h(\nabla_X Y,Z) + h(Y,\nabla^*_X Z).\end{aligned} \tag{12}$$

From equation (12), a complex statistical manifold (M,J,∇,h) is induced from $\{f,\xi\}$ if and only if the dual statistical manifold (M,J,∇^*,h) is induced from $\{v,w\}$.

Next, let us construct a contrast function. Let $\{f,\xi\}$ be a holomorphic nondegenerate equiaffine centroaffine immersion of codimension two. Suppose that $\{v,w\}$ is the dual map of $\{f,\xi\}$ and (M,J,∇,h) is the complex statistical manifold induced from $\{f,\xi\}$. We define the *geometric divergence* D of (M,J,∇,h) by

$$D(p,q) := \langle v(q), f(p) - f(q)\rangle, \qquad (p,q \in M).$$

If M is simply connected, the geometric divergence D is independent from the realization of (M,J,∇,h) into $\mathbf{C}^{n+1}$. The geometric divergence of (M,J,∇,h) is a contrast function on $M \times M$.

Theorem 5.2. *Let (M,J,∇,h) be a simply connected statistical manifold of Norden type. Suppose that ∇ is a complex connection and its Ricci tensor is symmetric and pure. Then the following hold.*

(1) *Suppose that* $\dim_{\mathbf{C}} M \geq 3$ *and* ∇ *is H-projectively flat. Then there exists a geometric divergence of* (M, J, ∇, h) *such that its Bartlett tensor field coincides with the tensor field* S_{HP} *defined by Equation* (1).

(2) *Suppose that* $\dim_{\mathbf{C}} M \geq 4$, $\nabla = \nabla^{(0)}$ *is the Levi-Civita connection with respect to* h *and* h *is H-conformally flat. Then there exists a geometric divergence of* (M, J, ∇, h) *such that its Bartlett tensor field coincides with the tensor field* S_{HC} *defined by Equation* (3).

References

1. S. Amari and H. Nagaoka, *Method of Information Geometry*, Amer. Math. Soc., Providence, Oxford University Press, Oxford, 2000.
2. F. Dillen and L. Vrancken, Complex affine hypersurfaces of $\mathbf{C}^{n+1}$, I, II, *Bull. Soc. Math. Belg. Ser. B*, **22**, 245–271 (1992), **23**, 1–27 (1993).
3. S. Eguchi, Geometry of minimum contrast, *Hiroshima Math. J.*, **22**, 631–647 (1992).
4. H. Furuhata, Hypersurfaces in statistical manifolds, *Diff. Geom. Appl.*, **27**, 420–429 (2009).
5. S. Ivanov, On dual holomorphically projectively flat affine connections, *J. Geom.*, **59**, 67–76 (1997).
6. S. Ivanov, On the fundamental theorem for non-degenerate complex affine hypersurface immersions, *Monatsh. Math.*, **123**, 321–336 (1997).
7. M. Kumon, A. Takemura, and K. Takeuchi, Conformal geometry of statistical manifold with application to sequential estimation, *Sequential Anal.*, **30**, 308-337 (2011).
8. T. Kurose, On the divergences of 1-conformally flat statistical manifolds, *Tôhoku Math. J.*, **46**, 427–433 (1994).
9. T. Kurose, Conformal-projective geometry of statistical manifolds, *Interdiscip. Inform. Sci.*, **8**, 89–100 (2002).
10. S.L. Lauritzen, Statistical manifolds, *Differential geometry in statistical inferences*, IMS Lecture Notes Monograph Series 10, Institute of Mathematical Statistics, Hayward California, 96–163 (1987).
11. H. Matsuzoe, Geometry of contrast functions and conformal geometry, *Hiroshima Math. J.*, **29**, 175–191 (1999).
12. H. Matsuzoe, Contrast functions on statistical manifolds with Norden metric, *JP J. Geom. Topol.*, **2**, 97–116 (2002).
13. H. Matsuzoe and J. Inoguchi, Statistical structures on tangent bundles, *Appl. Sci.*, **5**, 55–75 (2003).
14. H. Matsuzoe and M. Henmi, Hessian structures and divergence func-

tions on deformed exponential families, *Geometric Theory of Information*, F. Nielsen ed., Springer, Heiderberg, 57-80 (2014).
15. H. Matsuzoe and A. Ohara, Geometry for q-exponential families, *Recent progress in Differential Geometry and its Related Fields*, T. Adachi, H. Hashimoto & M. Hristov eds., World Sci. Publ., Hackensack, 55-71 (2011).
16. T. Noda, Symplectic structures on statistical manifolds, *J. Aust. Math. Soc.*, **90**, 374-384 (2011).
17. K. Nomizu, U. Pinkall and F. Podesta, On the geometry of affine Kähler immersions, *Nagoya Math. J.*, **120**, 205–222 (1990).
18. K. Nomizu and T. Sasaki, *Affine differential geometry – Geometry of Affine Immersions –*, Cambridge University Press, Cambridge, 1994.
19. B. Opozda, Fundamental theorems for complex hypersurfaces, *Kobe J. Math.*, **10**, 133–146 (1993).
20. B. Opozda, Equivalence theorems for complex affine hypersurfaces, *Results Math.*, **27**, 316–327 (1995).
21. H. Shima, *The Geometry of Hessian Structures*, World Scientific, 2007.
22. A.L. Singh and S.D. Singh, H-curvature tensors in an almost complex manifold with Norden metric, *Acta Cienc. Indica Math.*, **20**, 183–187 (1994).

Received April 8, 2015
Revised May 9, 2015

Current Developments in Differential Geometry
and its Related Fields pp. 201 – 213

A METHOD OF DETERMINING THE $SO(7)$-INVARIANTS FOR CURVES IN $\mathrm{Im}\,\mathbf{O}$ BY THEIR G_2-INVARIANTS

Misa OHASHI

Department of Mathematics, Nagoya Institute of Technology,
Nagoya 466-8555, Japan
E-mail: ohashi.misa@nitech.ac.jp

We show how to compute the $SO(7)$-invariants from the G_2-invariant functions of a curve in the purely imaginary octonions.

Keywords: G_2, Frenet-Serret formula, $SO(7)$-invariants, G_2-invariants.

1. Introduction

Two regular curves with arclength parameters in 7 dimensional Euclidean space $\mathbf{R}^7$ are said to be $SO(7)$-congruent to each other in the strong sense if they coincide with each other under the action of the isometry group $\mathbf{R}^7 \rtimes SO(7)$. Each curve in $\mathbf{R}^7$ has six curvatures $\{k_i\}_{i\in\{1,\ldots,6\}}$. Fixing an interval I ($\subset \mathbf{R}$) we denote by $[\gamma]_{SO(7)}$ the $SO(7)$-congruence class in the strong sense containing a curve $\gamma : I \to \mathbf{R}^7$, that is, we set

$$[\gamma]_{SO(7)} = \left\{\tilde{\gamma} : I \to \mathbf{R}^7 \mid \tilde{\gamma} \text{ is } SO(7)\text{-congruent to } \gamma \text{ in the strong sense}\right\}.$$

Since two curves which have the same series of curvatures are $SO(7)$-congruent to each other, therefore the class $[\gamma]_{SO(7)}$ of $\gamma : I \to \mathbf{R}^7$ can be identify with a series of six curvatures;

$$[\gamma]_{SO(7)} \longleftrightarrow (k_1, k_2, \ldots, k_6).$$

Let $\mathbf{O}$ be the octonions with the algebraic structure. The purely imaginary part $\mathrm{Im}\,\mathbf{O}$ of $\mathbf{O}$ is isomorphic to $\mathbf{R}^7$. The exceptional simple Lie group G_2 is defined as the group of automorphisms of $\mathbf{O}$ and is contained in $SO(7)$. Two curves in $\mathrm{Im}\,\mathbf{O} \cong \mathbf{R}^7$ are said to be G_2-congruent to each other in the strong sense if they coincide with each other under the action of $\mathrm{Im}\,\mathbf{O} \rtimes G_2$. If we denote the G_2-congruence class in the strong sense containing a curve $\gamma : I \to \mathrm{Im}\,\mathbf{O}$ by

$$[\gamma]_{G_2} = \{\tilde{\gamma} : I \to \mathrm{Im}\,\mathbf{O} \mid \tilde{\gamma} \text{ is } G_2\text{-congruent with } \gamma \text{ in the strong sense}\},$$

then it can be identified with the series of its six G_2-invariants $(k_1, \rho_1, \kappa_2, \alpha, \rho_3, \beta_1)$ (see [2]). Since we reduce the group of action from

$\mathbf{R}^7 \rtimes SO(7)$ to $\mathrm{Im}\,\mathbf{O} \rtimes G_2$, we have

$$[\gamma]_{SO(7)} = \bigcup_{\gamma' \in [\gamma]_{SO(7)}} [\gamma']_{G_2} \quad \text{(disjoint union)}.$$

The purpose of this paper is to show how to compute the curvatures ($SO(7)$-invariants) from the G_2-invariants of each curve and give concrete example of curves in 4-dimensional Euclidean space $V^4 \subset \mathrm{Im}\,\mathbf{O}$.

2. Preliminaries

Let $\mathbf{H}$ be the quaternions having the basis $\{1, i, j, k\}$ satisfying the following rules of multiplication:

$$i^2 = j^2 = k^2 = -1, \; ij = -ji = k, \; jk = -kj = i, \; ki = -ik = j.$$

We take an element $\varepsilon = (0, 1)$ of $\mathbf{H} \oplus \mathbf{H}$. The octonions $\mathbf{O}$ can be considered as the direct sum $\mathbf{H} \oplus \mathbf{H}\varepsilon$ with the following multiplication:

$$(a + b\varepsilon)(c + d\varepsilon) = (ac - \bar{d}b) + (da + b\bar{c})\varepsilon,$$

where $a, b, c, d \in \mathbf{H}$, the symbol " $^-$ " denotes the quaternionic conjugation. The octonions is a non-commutative, non-associative, alternative division algebra. For arbitrary $u, \; v \in \mathbf{O}$, we have

$$\langle uv, \; uv \rangle = \langle u, \; u \rangle \langle v, \; v \rangle , \tag{1}$$

where $\langle \; , \; \rangle$ is the canonical inner product of $\mathbf{R}^8 \cong \mathbf{O}$. The algebra which satisfy condition (1) is called the "normed algebra". The exterior product of $\mathbf{O}$ is given by $u \times v = (\bar{v}u - \bar{u}v)/2$, where the conjugation $\bar{v}$ of v is given by $\bar{v} = 2\langle v, \; 1\rangle - v$, for any $u, \; v \in \mathbf{O}$. We see that $u \times v \in \mathrm{Im}\,\mathbf{O}$ for any $u, \; v \in \mathbf{O}$. Here, $\mathrm{Im}\,\mathbf{O}$ is the purely imaginary part of $\mathbf{O}$ given as $\mathrm{Im}\,\mathbf{O} = \{u \in \mathbf{O} \mid \langle u, \; 1\rangle = 0\}$. If $u, \; v \in \mathrm{Im}\,\mathbf{O}$ and $\langle u, \; v\rangle = 0$, then $u \times v = uv$. A 3-dimensional vector spce V is called an associative 3-plane, if it satisfies associative law under the multiplication of $\mathbf{O}$. Its orthogonal complement $V^{\perp}$ of $\mathrm{Im}\,\mathbf{O}$ is called a coassociative 4-plane. The vector space $\mathrm{span}_{\mathbf{R}} \{u, \; v, \; u \times v\}$ is an associative 3-plane.

The exceptional simple Lie group G_2 is defined as the group of automorphisms of $\mathbf{O}$

$$G_2 = \{g \in SO(8) \mid g(uv) = g(u)g(v) \;\; \text{for } {}^{\forall}u, v \in \mathbf{O}\} = Aut(\mathbf{O}).$$

Since $g(1) = 1$, we have $G_2 \subset SO(7)$. By the definition of G_2, the action of G_2 preserves the decomposition of an associative 3-plane and a coassociative 4-plane. That is, for an arbitrary $g \in G_2$, we find that $g(\mathrm{Im}\,\mathbf{H})$ is an associative 3-plane and $g(\mathbf{H}\varepsilon)$ is also a coassociative 4-plane of $\operatorname{Im}\mathbf{O}$. We note that the Grassmannian manifold of all associative 3-planes coincides with $G_2/SO(4)$ (which is isomorphic to the Grassmannian manifold of all coassociative 4-planes by duality).

3. $SO(7)$-invariants and G_2-invariants of curves

In this section, we give the definitions of the G_2-frame field and the G_2-invariants along a curve in $\operatorname{Im}\mathbf{O}$, and recall a fundamental theorem of curves.

First we recall the fundamental theorem on curves in $\mathbf{R}^7$. Let $\gamma : I \to \operatorname{Im}\mathbf{O}$ and $\tilde{\gamma} : \tilde{I} \to \operatorname{Im}\mathbf{O}$ be two regular curves in $\operatorname{Im}\mathbf{O} \cong \mathbf{R}^7$ which are parameterized by their arclengths $s(\in I)$ and $\tilde{s}(\in \tilde{I})$, respectively. They are said to be G_2 (resp. $SO(7)$)-congruent to each other, if there exist $g \in G_2$ (resp. $SO(7)$), $a \in \operatorname{Im}\mathbf{O} \cong \mathbf{R}^7$ and $s_0 \in \mathbf{R}$ satisfying

$$g \circ \gamma(s) + a = \tilde{\gamma}(s + s_0),$$

for all $s \in I$. In particular, their domain-intervals coincide with each other by parallel displacement in a real line. In order to avoid redundant complexity, in this paper we do not make clear the domains of curves, and mix the congruency and the congruency in the strong sense on curves.

Let $\gamma : I \to \mathbf{R}^7$ be a C^∞-curve in $\mathbf{R}^7$ and $(v_1, v_2, \ldots, v_7)$ be the Frenet frame field along γ, which is defined by

$$v_1 = \gamma' \left(= \frac{d\gamma}{ds} \right), \quad v_2 = \frac{v_1'}{k_1}, \quad k_1 = \|v_1'\|,$$

$$v_i = \frac{1}{k_{i-1}} \left(v_{i-1}' - \langle v_{i-1}', v_{i-2} \rangle v_{i-2} \right),$$

$$k_{i-1} = \left(\|v_{i-1}'\|^2 - \langle v_{i-1}', v_{i-2} \rangle^2 \right)^{1/2},$$

for $i \in \{3, 4, 5, 6\}$. Here we assume the functions $k_1, \ldots, k_5$ to be everywhere-positive on I. We define a unit vector field v_7 along γ so that it is perpendicular to all the vector field $v_1, \ldots, v_6$ and satisfies $\det(v_1 \ldots v_6\, v_7) = 1$. We set $k_6 = \langle v_6', v_7 \rangle$. The function k_i is called the i-th curvature and these 6 functions are $SO(7)$-invariants. Then the Frenet-Serret formula of this curve in $\mathbf{R}^7$ is given by

$$\frac{d}{ds}(v_1\, v_2\, \dots\, v_7) = (v_1\, v_2\, \dots\, v_7)\begin{pmatrix} 0 & -k_1 & 0 & \cdots & 0 \\ k_1 & 0 & -k_2 & & \vdots \\ 0 & k_2 & 0 & \ddots & 0 \\ \vdots & & \ddots & \ddots & -k_6 \\ 0 & \cdots & 0 & k_6 & 0 \end{pmatrix}.$$

From this formula, we obtain the following classical $SO(7)$-congruence theorem.

Fact. *Let γ, $\tilde{\gamma} : I \to \boldsymbol{R}^7$ be two curves in $\boldsymbol{R}^7$ with same arclength parameter $s \in I$. They are $SO(7)$-congruent to each other if and only if their series of curvatures $\{k_i\}_{i\in\{1,\dots,6\}}$ and $\{\tilde{k}_i\}_{i\in\{1,\dots,6\}}$ coincide with each other (i.e. $k_i = \tilde{k}_i$ all $i \in \{1, \dots, 6\}$).*

By this Fact, we find that the $SO(7)$-congruence class of a curve $\gamma : I \to \mathbf{R}^7$ can be identified with $SO(7)$-invariants (curvatures)

$$\begin{aligned}[\gamma]_{SO(7)} &= \{\tilde{\gamma} : I \to \mathbf{R}^7 \mid \tilde{\gamma} \text{ is } SO(7)\text{-congruent with } \gamma\} \\ &\longleftrightarrow (k_1, k_2, \dots, k_6).\end{aligned}$$

Next we explain the G_2-congruence theorem of curves in $\operatorname{Im}\mathbf{O}$. We here give the construction method of G_2-frame field along the curve $\gamma : I \to \operatorname{Im}\mathbf{O}$. The G_2-frame field $(e_4\ e_1\ e_2\ e_3\ e_5\ e_6\ e_7)$ along γ is defined by

$$\begin{aligned} &e_4 = v_1, \quad e_1 = v_2, \quad e_5 = e_1 \times e_4, \\ &e_2 = \frac{1}{\kappa_2}\left(e_1' - \langle e_1', e_4\rangle e_4 - \langle e_1', e_5\rangle e_5\right), \\ &e_3 = e_1 \times e_2, \quad e_6 = e_2 \times e_4, \quad e_7 = e_3 \times e_4, \end{aligned}$$

where $\kappa_2 = \sqrt{\|e_1'\|^2 - \langle e_1', e_4\rangle^2 - \langle e_1', e_5\rangle^2}$. We can construct the above G_2-frame field along γ under the assumption that κ_2 is a everywhere-positive function, then . Note that a curve with $\kappa_2 \equiv 0$ lies on some associative 3-plane. From this construction of G_2-frame field, we have the following Frenet-Serret formula of G_2 type

$$\frac{d}{ds}(e_4\, e_1\, e_2\, e_3\, e_5\, e_6\, e_7) = (e_4\, e_1\, e_2\, e_3\, e_5\, e_6\, e_7)\left(\begin{array}{c|c|c} 0 & -{}^t\mu & 0_{1\times 3} \\ \hline \mu & A & -{}^tB \\ \hline 0_{3\times 1} & B & A \end{array}\right),$$

where

$$\mu = \begin{pmatrix} k_1 \\ 0 \\ 0 \end{pmatrix}, \quad A = \begin{pmatrix} 0 & -\kappa_2 & 0 \\ \kappa_2 & 0 & -\alpha \\ 0 & \alpha & 0 \end{pmatrix}, \quad B = \begin{pmatrix} \rho_1 & 0 & 0 \\ 0 & \rho_2 & \beta_2 \\ 0 & \beta_1 & \rho_3 \end{pmatrix}.$$

For these functions of entries, we have the following two relations

$$\rho_1 + \rho_2 + \rho_3 = 0, \quad \beta_1 - \beta_2 - k_1 = 0.$$

We call the above six functions $(k_1, \rho_1, \kappa_2, \alpha, \rho_3, \beta_1)$ the G_2-(complete) invariants of γ. In fact, we have the following G_2-congruence theorem.

Theorem 3.1 ([2]). *Let γ, $\tilde{\gamma} : I \to \mathrm{Im}\,\boldsymbol{O}$ be two curves in $\mathrm{Im}\,\boldsymbol{O}$ with same arclength parameter $s \in I$. They are G_2-congruent to each other if and only if their series of G_2-(complete) invariants $(k_1, \rho_1, \kappa_2, \alpha, \rho_3, \beta_1)$ and $(\tilde{k}_1, \tilde{\rho}_1, \tilde{\kappa}_2, \tilde{\alpha}, \tilde{\rho}_3, \tilde{\beta}_1)$ coincide with each other, that is,*

$$k_1 = \tilde{k}_1, \quad \rho_1 = \tilde{\rho}_1, \quad \kappa_2 = \tilde{\kappa}_2, \quad \alpha = \tilde{\alpha}, \quad \rho_3 = \tilde{\rho}_3, \quad \beta_1 = \tilde{\beta}_1.$$

By this result we find that the G_2-congruence class of a curve $\gamma : I \to \mathrm{Im}\,\mathbf{O}$ can be identified with the series of six G_2-(complete) invariants

$$\begin{aligned} [\gamma]_{G_2} &= \{\tilde{\gamma} : I \to \mathrm{Im}\,\mathbf{O} \mid \tilde{\gamma} \text{ is } G_2\text{-congruent with } \gamma\} \\ &\longleftrightarrow (k_1, \rho_1, \kappa_2, \alpha, \rho_3, \beta_1). \end{aligned}$$

Since $G_2 \subset SO(7)$, if γ and $\tilde{\gamma}$ are G_2-congruent to each other, then they are $SO(7)$-congruent to each other. Hence

$$[\gamma]_{SO(7)} = \bigcup_{\gamma' \in [\gamma]_{SO(7)}} [\gamma']_{G_2} \quad \text{(disjoint union)}.$$

4. Computation method of $SO(7)$-invariants from G_2-invariants of curves

Let $(v_1, v_2, \ldots, v_7)$ be the Frenet-frame field and $(e_4\; e_1\, e_2\, e_3\; e_5\, e_6\, e_7)$ be the G_2-frame field along a curve $\gamma : I \to \mathrm{Im}\,\mathbf{O}$ with the arclength parameter $s \in I$. We assume that $k_6 \neq 0$ and $\kappa_2 > 0$. From the construction, the

G_2-frame field can be written by the $SO(7)$-Frenet-frame field as follows

$$\begin{cases} e_4 = v_1, \quad e_1 = v_2, \quad e_5 = v_2 \times v_1, \\ e_2 = \dfrac{1}{\kappa_2}\left(k_2 v_3 - \rho_1 v_2 \times v_1\right), \\ e_3 = \dfrac{1}{\kappa_2}\left(k_2 v_2 \times v_3 + \rho_1 v_1\right), \\ e_6 = \dfrac{1}{\kappa_2}\left(k_2 v_3 \times v_1 + \rho_1 v_2\right), \\ e_7 = \dfrac{k_2}{\kappa_2}\left(v_2 \times v_3\right) v_1. \end{cases}$$

If we put $\cos\sigma = \langle v_3,\ v_2 \times v_1 \rangle$, then we have

$$\cos\sigma = \frac{\rho_1}{k_2}, \quad |\sin\sigma| = \frac{\kappa_2}{k_2}.$$

We set the functions a_{ij} as follows

$$\begin{aligned} a_{i1} &= \langle v_{i+3},\ v_2 \times v_1 \rangle, \quad a_{i2} = \langle v_{i+3},\ v_3 \times v_2 \rangle, \\ a_{i3} &= \langle v_{i+3},\ v_3 \times v_1 \rangle, \quad a_{i4} = \langle v_{i+3},\ (v_3 \times v_2) \times v_1 \rangle, \end{aligned} \tag{2}$$

for each $i \in \{1, 2, 3, 4\}$. Then the G_2- frame field and the G_2-(complete) invariants are written by

$$(e_4\ e_1\ e_2\ e_3\ e_5\ e_6\ e_7) = (v_1\ v_2\ v_3\ v_4\ v_5\ v_6\ v_7)M,$$

with a function M which takes values in $S^1 \times SO(4)/\mathbf{Z}_2$ and is given as

$$M = M(s) = \left(\begin{array}{c|ccc|ccc} 1 & 0 & 0 & 0 & 0 & 0 & 0 \\ \hline 0 & 1 & 0 & 0 & 0 & 0 & 0 \\ 0 & 0 & |\sin\sigma| & 0 & \cos\sigma & 0 & 0 \\ 0 & 0 & -\frac{\cos\sigma}{|\sin\sigma|}a_{11} & -\frac{a_{12}}{|\sin\sigma|} & a_{11} & \frac{a_{13}}{|\sin\sigma|} & -\frac{a_{14}}{|\sin\sigma|} \\ \hline 0 & 0 & -\frac{\cos\sigma}{|\sin\sigma|}a_{21} & -\frac{a_{22}}{|\sin\sigma|} & a_{21} & \frac{a_{23}}{|\sin\sigma|} & -\frac{a_{24}}{|\sin\sigma|} \\ 0 & 0 & -\frac{\cos\sigma}{|\sin\sigma|}a_{31} & -\frac{a_{32}}{|\sin\sigma|} & a_{31} & \frac{a_{33}}{|\sin\sigma|} & -\frac{a_{34}}{|\sin\sigma|} \\ 0 & 0 & -\frac{\cos\sigma}{|\sin\sigma|}a_{41} & -\frac{a_{42}}{|\sin\sigma|} & a_{41} & \frac{a_{43}}{|\sin\sigma|} & -\frac{a_{44}}{|\sin\sigma|} \end{array}\right), \tag{3}$$

and

$$\begin{cases} k_1 = \|e_4{}'\|, \\ \rho_1 = \langle e_1', \, e_5\rangle = k_2 \cos\sigma, \\ \kappa_2 = \sqrt{\|e_1'\|^2 - \langle e_1', \, e_4\rangle^2 - \langle e_1', e_5\rangle^2} = k_2|\sin\sigma|, \\ \alpha = \langle e_2', \, e_3\rangle = -\dfrac{k_3 a_{12}}{\sin^2\sigma}, \\ \rho_3 = \langle e_3', \, e_7\rangle = -\dfrac{k_3 a_{13}}{\sin^2\sigma}, \\ \beta_1 = \langle e_2', \, e_7\rangle = -\dfrac{k_3 a_{14}}{\sin^2\sigma}, \end{cases} \tag{4}$$

respectively. We note that the condition $\kappa_2 > 0$ is equivalent to the condition $|\sin\sigma| > 0$. Since $(v_1, v_2, \ldots, v_7)$ is an $SO(7)$-valued function, we find that

$$\mathscr{A}_\sigma(s) = \frac{1}{|\sin\sigma|}\begin{pmatrix} a_{11} & \cdots & a_{14} \\ \vdots & \ddots & \vdots \\ a_{41} & \cdots & a_{44} \end{pmatrix},$$

is an $SO(4)$-valued function.

Remark 4.1 ([1]). Through the identification

$$S^1 \times SO(4)/\mathbf{Z}_2 \cong SO(7)/G_2 = \mathbf{RP}^7,$$

we can consider the matrix-valued function $M(s)$ as a $\mathbf{RP}^7$-valued function.

We give the method of computation of $SO(7)$-invariants of curves from G_2 ones.

Remark 4.2. We study the case of $k_6 = 0$.

(1) In the case that the curve lies on a 3-dimensional vector space $\mathbf{R}^3 \subset \operatorname{Im}\mathbf{O}$, that is the case that $k_3 = \cdots = k_6 = 0$, if we put $M_1(s) = \big(I_3 \; O_{3\times 4}\big)\, M(s)$, then we have $(v_1\ v_2\ v_3) = (e_4\ e_1\, e_2\, e_3\ e_5\, e_6\, e_7)^t M_1$. In this case, M_1 can be considered as an S^1-valued function.

(2) In the case that the curve lies on a 4-dimensional vector space $\mathbf{R}^4 \subset \operatorname{Im}\mathbf{O}$, that is the case that $k_4 = \cdots = k_6 = 0$, if we put $M_2(s) = \big(I_4 \; O_{4\times 3}\big)\, M(s)$, then we have $(v_1\ v_2\ v_3\ v_4) = (e_4\ e_1\, e_2\, e_3\ e_5\, e_6\, e_7)^t M_2$. In this case, M_2 can be considered as an S^3-valued function.

(3) In the case that the curve lies on a 5-dimensional vector space $\mathbf{R}^5 \subset \operatorname{Im}\mathbf{O}$, that is the case that $k_5 = k_6 = 0$, if we put $M_3(s) = \left(I_5\ O_{5\times 2}\right) M(s)$, then we have $(v_1\ v_2\ v_3\ v_4\ v_5) = (e_4\ e_1\ e_2\ e_3\ e_5\ e_6\ e_7)^t M_3$. In this case, M_3 can be considered as a $\mathbf{CP}^3$-valued function.

(4) In the case that the curve lies on a 6-dimensional vector space $\mathbf{R}^6 \subset \operatorname{Im}\mathbf{O}$, that is the case that $k_6 = 0$, we can take a unit vector $v_7 \in (\mathbf{R}^6)^{\perp}$ as a constant vector.

A curve γ in $\operatorname{Im}\mathbf{O}$ is said to be fully contained in some $\mathbf{R}^n$ $\left(\subset \operatorname{Im}\mathbf{O}\right)$ $(n \in \{3, \ldots, 6\})$ if it lies on this $\mathbf{R}^n$ but is not contained in $\mathbf{R}^m$ of lower dimension m. Here, we just call it an n-full curve. For an n-full curve, we have $k_i > 0$ $(1 \leqq i \leqq n-2)$ and $k_{n-1} \neq 0$. We hence put $k_n = \cdots = k_6 = 0$ as a convention.

Theorem 4.1. *Each $SO(7)$-invariants (curvatures) k_i of a curve in $\boldsymbol{R}^7$ can be computated by its six G_2-(complete) invariants $(k_1, \rho_1, \kappa_2, \alpha, \rho_3, \beta_1)$.*

$$k_i = k_i(k_1, \rho_1, \kappa_2, \alpha, \rho_3, \beta_1) \qquad i \in \{1, \ldots, 6\}.$$

Proof. First, by (4), we see that

$$k_2 = \sqrt{{\kappa_2}^2 + {\rho_1}^2}.$$

Therefore we have

$$\cos\sigma = \frac{\rho_1}{\sqrt{{\kappa_2}^2 + {\rho_1}^2}}, \qquad \sin\sigma = \frac{\kappa_2}{\sqrt{{\kappa_2}^2 + {\rho_1}^2}}.$$

Secondly, we compute the 3rd curvature k_3. From Frenet-Serret formula, we get

$$(\cos\sigma)' = \langle v_3,\ v_2 \times v_1\rangle' = k_3 \langle v_4,\ v_2 \times v_1\rangle = k_3 a_{11}.$$

Since $\sum_{j=1}^{4} {a_{1j}}^2 = \sin^2\sigma$, we have

$$k_3 = \frac{\sqrt{({\kappa_2}'\rho_1 - \kappa_2{\rho_1}')^2 + {\kappa_2}^2({\rho_1}^2 + {\kappa_2}^2)({\rho_3}^2 + \alpha^2 + {\beta_1}^2)}}{{\rho_1}^2 + {\kappa_2}^2},$$

$$a_{11} = \frac{1}{k_3}\frac{\kappa_2(\kappa_2{\rho_1}' - \rho_1{\kappa_2}')}{({\rho_1}^2 + {\kappa_2}^2)^{3/2}}, \qquad a_{12} = -\frac{1}{k_3}\frac{\alpha{\kappa_2}^2}{{\rho_1}^2 + {\kappa_2}^2},$$

$$a_{13} = -\frac{1}{k_3}\frac{\rho_3{\kappa_2}^2}{{\rho_1}^2 + {\kappa_2}^2}, \qquad a_{14} = -\frac{1}{k_3}\frac{\beta_1{\kappa_2}^2}{{\rho_1}^2 + {\kappa_2}^2}.$$

Thirdly, we will compute the 4th curvature k_4. By (2), the derivative of a_{11} is given by

$$\begin{aligned}(a_{11})' &= \langle v_4,\ v_2 \times v_1 \rangle' \\ &= k_4 \langle v_5,\ v_2 \times v_1 \rangle - k_3 \langle v_3,\ v_2 \times v_1 \rangle + k_2 \langle v_4,\ v_3 \times v_1 \rangle \\ &= k_4 a_{21} - k_3 \cos\sigma + k_2 a_{13}. \end{aligned} \tag{5}$$

In the similar way, by (2), we have

$$\begin{aligned}(a_{12})' &= k_4 a_{22} - k_1 a_{13}, \\ (a_{13})' &= k_4 a_{23} - k_2 a_{11} + k_1 a_{12}, \\ (a_{14})' &= k_4 a_{24}. \end{aligned} \tag{6}$$

Since $\sum_{j=1}^{4} {a_{2j}}^2 = \sin^2\sigma$, we see that

$$\begin{aligned} k_4 = \frac{{\kappa_2}^2 + {\rho_1}^2}{{\kappa_2}^2} \Big\{ &\big((a_{11})' + k_3 \cos\sigma - k_2 a_{13}\big)^2 + \big((a_{12})' + k_1 a_{13}\big)^2 \\ &+ \big((a_{13})' + k_2 a_{11} - k_1 a_{12}\big)^2 + \big((a_{14})'\big)^2 \Big\}^{1/2}. \end{aligned} \tag{7}$$

By (5), (6), (7), we can show that the functions a_{2j} ($j \in \{1, 2, 3, 4\}$) are written by the G_2-(complete) invariants

$$a_{21} = \frac{(a_{11})' + k_3 \cos\sigma - k_2 a_{13}}{k_4}, \qquad a_{22} = \frac{(a_{12})' + k_1 a_{13}}{k_4},$$

$$a_{23} - \frac{(a_{13})' + k_2 a_{11} - k_1 (a_{12})}{k_4}, \qquad a_{24} = \frac{(a_{14})'}{k_4}.$$

Fourthly, to compute the 5th curvature k_5, we take the derivative of a_{2j}. Especially, the derivative of a_{22} is given by

$$(a_{22})' = k_5 a_{32} - k_4 a_{12} - k_1 a_{23} + k_3 \langle v_5,\ v_4 \times v_2 \rangle .$$

By (3), wc have

$$\begin{aligned} v_2 &= e_1, \\ v_4 &= \frac{1}{\sin\sigma} \left(-a_{11} \cos\sigma e_2 - a_{12} e_3 + a_{11} \sin\sigma e_5 + a_{13} e_6 - a_{14} e_7\right), \end{aligned}$$

and

$$v_5 = \frac{1}{\sin\sigma} \left(-a_{21} \cos\sigma e_2 - a_{22} e_3 + a_{21} \sin\sigma e_5 + a_{23} e_6 - a_{24} e_7\right).$$

Hence we get

$$v_4 \times v_2 = \frac{1}{\sin\sigma}\left(-a_{21}\cos\sigma e_3 - a_{22}e_2 + a_{21}\sin\sigma e_4 + a_{23}e_7 + a_{24}e_6\right).$$

So, we obtain

$$\langle v_5,\ v_4 \times v_2\rangle = -\frac{1}{\sin^2\sigma}\left(\cos\sigma\begin{vmatrix} a_{11} & a_{12} \\ a_{21} & a_{22}\end{vmatrix} + \begin{vmatrix} a_{13} & a_{14} \\ a_{23} & a_{24}\end{vmatrix}\right).$$

Therefore the 5th curvature k_5 is written as

$$\begin{aligned} k_5 = \frac{{\kappa_2}^2 + {\rho_1}^2}{{\kappa_2}^2}\Big\{&\big((a_{21})' + k_4 a_{11} - k_2 a_{23}\big)^2 \\ &+ \big((a_{22})' + k_4 a_{12} + k_1 a_{23} - k_3\langle v_5, v_4 \times v_2\rangle\big)^2 \\ &+ \big((a_{23})' + k_4 a_{13} + k_2 a_{21} + k_1 a_{22} - k_3\langle v_5, v_4 \times v_1\rangle\big)^2 \\ &+ \big((a_{24})' + k_4 a_{14} - k_3\langle v_5, (v_4 \times v_2) \times v_1\rangle\big)^2\Big\}^{1/2}, \end{aligned}$$

where

$$\langle v_5,\ v_4 \times v_1\rangle = \frac{1}{\sin^2\sigma}\left(-\cos\sigma\begin{vmatrix} a_{11} & a_{13} \\ a_{21} & a_{23}\end{vmatrix} + \begin{vmatrix} a_{12} & a_{14} \\ a_{22} & a_{24}\end{vmatrix}\right),$$

and

$$\langle v_5,\ (v_4 \times v_2) \times v_1\rangle = -\frac{1}{\sin^2\sigma}\left(\cos\sigma\begin{vmatrix} a_{11} & a_{14} \\ a_{21} & a_{24}\end{vmatrix} + \begin{vmatrix} a_{12} & a_{13} \\ a_{22} & a_{23}\end{vmatrix}\right).$$

Similarly, the 6th curvature k_6 is given by

$$\begin{aligned} k_6 = \frac{{\kappa_2}^2 + {\rho_1}^2}{{\kappa_2}^2}\Big\{&\big((a_{31})' + k_5 a_{21} - k_2 a_{33}\big)^2 \\ &+ \big((a_{32})' + k_5 a_{22} + k_1 a_{33} - k_3\langle v_6, v_4 \times v_2\rangle\big)^2 \\ &+ \big((a_{33})' + k_5 a_{23} + k_2 a_{31} + k_1 a_{32} - k_3\langle v_6, v_4 \times v_1\rangle\big)^2 \\ &+ \big((a_{34})' + k_5 a_{24} - k_3\langle v_6, (v_4 \times v_2) \times v_1\rangle\big)^2\Big\}^{1/2}, \end{aligned}$$

where

$$\begin{aligned} a_{31} &= \frac{(a_{21})' + k_4 a_{11} - k_2 a_{23}}{k_5}, \\ a_{32} &= \frac{(a_{22})' + k_1 a_{23} + k_3\langle v_5, v_4 \times v_2\rangle}{k_5}, \\ a_{33} &= \frac{(a_{23})' + k_4 a_{13} - k_2 a_{21} + k_1 a_{22} + k_3\langle v_5, v_4 \times v_1\rangle}{k_4}, \\ a_{34} &= \frac{(a_{24})' + k_4 a_{14} - k_3\langle v_5, (v_4 \times v_2) \times v_1\rangle}{k_4}, \end{aligned}$$

and

$$\langle v_6, v_4 \times v_2 \rangle = -\frac{1}{\sin^2 \sigma}\left(\cos\sigma \begin{vmatrix} a_{11} & a_{12} \\ a_{31} & a_{32} \end{vmatrix} + \begin{vmatrix} a_{13} & a_{14} \\ a_{33} & a_{34} \end{vmatrix}\right),$$

$$\langle v_6, v_4 \times v_1 \rangle = \frac{1}{\sin^2 \sigma}\left(-\cos\sigma \begin{vmatrix} a_{11} & a_{13} \\ a_{31} & a_{33} \end{vmatrix} + \begin{vmatrix} a_{12} & a_{14} \\ a_{32} & a_{34} \end{vmatrix}\right),$$

$$\langle v_6, (v_4 \times v_2) \times v_1 \rangle = -\frac{1}{\sin^2 \sigma}\left(\cos\sigma \begin{vmatrix} a_{11} & a_{14} \\ a_{31} & a_{34} \end{vmatrix} + \begin{vmatrix} a_{12} & a_{13} \\ a_{32} & a_{33} \end{vmatrix}\right).$$

From the above, we obtain the computation method of the $SO(7)$-invariants from the G_2-invariants. □

We give an example related to Theorem 4.1.

Example 4.1. Let V^4 be a following 4-dimensional vector space in $\mathrm{Im}\,\mathbf{O}$.

$$V^4 = \mathrm{span}_{\mathbf{R}}\{i,\ j,\ \cos\theta\, k + \sin\theta\,\varepsilon,\ \alpha_1(\sin\theta\, k + \cos\theta\,\varepsilon) + \alpha_2 i\varepsilon + \alpha_3 j\varepsilon + \alpha_4 k\varepsilon\},$$

where

$$(\theta,\ \alpha_1,\ \alpha_2,\ \alpha_3,\ \alpha_4) \in [0,\pi] \times S^3/_\sim \cong S^4\left(\sum_{i=1}^{4} \alpha_i^2 = 1\right).$$

Let $\mathbf{R}^4$ be a 4-dimensional vector space in $\mathrm{Im}\,\mathbf{O}$ spanned by an orthonormal frame $(u_1,\ u_2,\ u_3,\ u_4)$, and $\tilde{\gamma} : I \to \mathbf{R}^4 \subset \mathrm{Im}\,\mathbf{O}$ be an arbitrary helix. Here, a helix is the curve given as an orbit of 1-parameter subgroup of $\mathbf{R}^4$. Then this curve $\tilde{\gamma}$ is G_2- congruent to the following helix $\gamma : \mathbf{R} \to V^4$ given by

$$\begin{aligned}\gamma(t) &= r_1\cos(p_1 t)\cdot i + r_1 \sin(p_1 t)\cdot j \\ &\quad + r_2\cos(p_2 t)\cdot(\cos\theta\, k + \sin\theta\,\varepsilon) \\ &\quad + r_2 \sin(p_2 t)\cdot\{\alpha_1(\sin\theta\, k + \cos\theta\,\varepsilon) + \alpha_2 i\varepsilon + \alpha_3 j\varepsilon + \alpha_4 k\varepsilon\},\end{aligned}$$

where $r_i,\ p_i,\ \theta,\ \alpha_i$ are constants satisfying $r_1,\ r_2 > 0$, $p_1p_2(p_1+p_2)(p_1 - p_2) \neq 0$. In fact, let $V_4(\mathbf{R}^7)$ be the Stiefel manifold of orthonormal 4-frames in $\mathbf{R}^7$. Since $V_4(\mathbf{R}^7)/G_2 \cong S^4$ (see [1]), there exists $g \in G_2$ satisfying

$$\begin{aligned}&g(u_1) = i,\quad g(u_2) = j,\quad g(u_3) = \cos\theta\, k + \sin\theta\,\varepsilon, \\ &g(u_4) = \alpha_1(\sin\theta\, k + \cos\theta\,\varepsilon) + \alpha_2 i\varepsilon + \alpha_3 j\varepsilon + \alpha_4 k\varepsilon.\end{aligned}$$

From the existence of such an element $g \in G_2$, we see that γ and $\tilde{\gamma}$ are G_2-congruent to each other.

If we put

$$C_1 = \sqrt{r_1^2p_1^2 + r_2^2p_2^2},\quad C_2 = \sqrt{r_1^2p_1^4 + r_2^2p_2^4}$$

and set four vectors $(\xi_1, \xi_2, \xi_3, \xi_4)$ as

$$\xi_1 = i, \quad \xi_2 = j, \quad \xi_3 = \cos\theta\, k + \sin\theta\, \varepsilon,$$
$$\xi_4 = \alpha_1(\sin\theta\, k + \cos\theta\, \varepsilon) + \alpha_2 i\varepsilon + \alpha_3 j\varepsilon + \alpha_4 k\varepsilon,$$

then the Frenet-frame field (v_1, v_2, v_3, v_4) along γ is given by

$$(v_1, v_2, v_3, v_4) = (\xi_1\, \xi_2\, \xi_3\, \xi_4)F(t),$$

where the matrix-valued function $F(t)$ is given by

$$C_1C_2F(t)$$
$$= \begin{pmatrix} -C_2r_1p_1\sin(p_1t) & -C_1r_1p_1^2\cos(p_1t) & C_2r_2p_2\sin(p_1t) & C_1r_2p_2^2\cos(p_1t) \\ C_2r_1p_1\cos(p_1t) & -C_1r_1p_1^2\sin(p_1t) & -C_2r_2p_2\cos(p_1t) & C_1r_2p_2^2\sin(p_1t) \\ -C_2r_2p_2\sin(p_2t) & -C_1r_2p_2^2\cos(p_2t) & -C_2r_1p_1\sin(p_2t) & -C_1r_1p_1^2\cos(p_2t) \\ C_2r_2p_2\cos(p_2t) & -C_1r_2p_2^2\sin(p_2t) & C_2r_1p_1\cos(p_2t) & -C_1r_1p_1^2\cos(p_2t) \end{pmatrix}.$$

If we set

$$\mu_1 = \sqrt{\alpha_2^2 + \alpha_3^2}\,\sin\theta, \quad \mu_2 = \sqrt{\cos^2\theta + \alpha_1^2\sin^2\theta},$$
$$\tau_1 = -\arctan\frac{\alpha_2}{\alpha_3}, \qquad \tau_2 = \arctan(\alpha_1\tan\theta),$$

then G_2-(complete) invariants are written by

$$\begin{cases} k_1 = \dfrac{C_2}{C_1^2}, \\ \rho_1 = \dfrac{r_1r_2p_1p_2(p_1^2 - p_2^2)}{C_1^2C_2^2}\left(r_2p_2^2\mu_1\sin(p_1t+\tau_1) + r_1p_1^2\mu_2\sin(p_2t+\tau_2)\right), \\ \kappa_2 = \dfrac{r_1r_2p_1p_2(p_1^2 - p_2^2)}{C_1^2C_2^2} \\ \qquad \times\sqrt{C_2^2 - (r_2p_2^2\mu_1\sin(p_1t+\tau_1) + r_1p_1^2\mu_2\sin(p_2t+\tau_2))^2}, \\ \alpha = -\dfrac{p_1p_2}{C_1}\cdot\dfrac{-r_1p_1\mu_1\cos(p_1t+\tau_1) + r_2p_2\mu_2\cos(p_2t+\tau_2)}{C_2^2 - (r_2p_2^2\mu_1\sin(p_1t+\tau_1) + r_1p_1^2\mu_2\sin(p_2t+\tau_2))^2}, \\ \rho_3 = -p_1p_2\cdot\dfrac{-r_1p_1^2\mu_1\sin(p_1t+\tau_1) + r_2p_2^2\mu_2\sin(p_2t+\tau_2)}{C_2^2 - (r_2p_2^2\mu_1\sin(p_1t+\tau_1) + r_1p_1^2\mu_2\sin(p_2t+\tau_2))^2}, \\ \beta_1 = \dfrac{p_1p_2\alpha_4\sin\theta}{C_2^2 - (r_2p_2^2\mu_1\sin(p_1t+\tau_1) + r_1p_1^2\mu_2\sin(p_2t+\tau_2))^2}. \end{cases}$$

By the above expressions of G_2-invariants and by Theorem 4.1, we can show that the curvatures k_1, k_2, k_3 are given by

$$\begin{cases} k_1 = C_2/C_1^2, \\ k_2 = \sqrt{{\rho_1}^2 + {\kappa_2}^2} = r_1 r_2 p_1 p_2 (p_1 + p_2)(p_1 p_2)/(C_1^2 C_2), \\ k_3 = \dfrac{\sqrt{({\kappa_2}'\rho_1 - \kappa_2 {\rho_1}')^2 + {\kappa_2}^2({\rho_1}^2 + {\kappa_2}^2)({\rho_3}^2 + \alpha^2 + {\beta_1}^2)}}{{\rho_1}^2 + {\kappa_2}^2} \\ \quad = p_1 p_2 / C_2. \end{cases}$$

(Recall our convention on curvatures.) We obtain the relationship between $SO(4)$-invariants and G_2-invariants of curves in $V^4 \subset \operatorname{Im} \mathbf{O}$. In this case, the other invariants are given by

$$\begin{cases} \cos\sigma &= \rho_1 \Big/ \sqrt{{\kappa_2}^2 + {\rho_1}^2} \\ &= \left\{ r_2 p_2^2 \mu_1 \sin(p_1 t + \tau_1) + r_1 p_1^2 \mu_2 \sin(p_2 t + \tau_2) \right\} / C_2, \\ a_{11} &= \dfrac{1}{k_3} \dfrac{\kappa_2(\kappa_2 {\rho_1}' - \rho_1 {\kappa_2}')}{({\rho_1}^2 + {\kappa_2}^2)^{3/2}} \\ &= \left\{ r_2 p_2 \mu_1 \cos(p_1 t + \tau_1) + r_1 p_1 \mu_2 \cos(p_2 t + \tau_2) \right\} / C_1, \\ a_{12} &= -\dfrac{1}{k_3} \dfrac{\alpha {\kappa_2}^2}{{\rho_1}^2 + {\kappa_2}^2} \\ &= \left\{ -r_1 p_1 \mu_1 \cos(p_1 t + \tau_1) + r_2 p_2 \mu_2 \cos(p_2 t + \tau_2) \right\} / C_1, \\ a_{13} &= -\dfrac{1}{k_3} \dfrac{\rho_3 {\kappa_2}^2}{{\rho_1}^2 + {\kappa_2}^2} \\ &= \left\{ -r_1 p_1^2 \mu_1 \sin(p_1 t + \tau_1) + r_2 p_2^2 \mu_2 \sin(p_2 t + \tau_2) \right\} / C_2, \\ a_{14} &= -\alpha_4 \sin\theta. \end{cases}$$

References

1. H. Hashimoto & M. Ohashi, On fibre bundle structures of Stiefel manifolds related to the octonions, To appear in *Topology and its Applications.*
2. M. Ohashi, G_2-congruence theorem for curves in purely imaginary octonions and its application, *Geometriae Dedicata* **163**, 1–17 (2013).

Received February 13, 2015
Revised May 19, 2015

Current Developments in Differential Geometry
and its Related Fields *pp. 215 – 224*

MAGNETIC JACOBI FIELDS FOR SURFACE MAGNETIC FIELDS

Qingsong SHI

Division of Mathematics and Mathematical Science,
Nagoya Institute of Technology
Gokiso, Nagoya, 466-8555, Japan
E-mail: sqs120012@yahoo.co.jp

A surface magnetic field is a closed 2-form on an orientable Riemann surface. In order to study variations of trajectories for this magnetic field, we take magnetic Jacobi fields corresponding to these variations and study them by comparing magnetic Jacobi fields for uniform magnetic fields on Riemann surface of constant curvature. By applying this study we give estimates on areas of trajectory disks.

Keywords: Surface magnetic fields, trajectories, Jacobi fields, comparison theorems, trajectory disks.

1. Introduction

When we study Riemannian manifolds, it is needless to say that geodesics play quite an important role. In some sense global study of Riemannian manifolds is developed with studies of geodesics: For examples, we have Rauch's and Toponogov's comparison theorems, Hopf-Rinow's theorem and so on. From curve-theoretic point of view, it is natural to consider that if we take a family of curves which are associated with geometric structures on Riemannian manifolds we would be able to get more information on underlying manifolds by properties of these curves. In this context S. Maeda and his coauthors [5 ,6 ,10 ,11] studied submanifolds by using circles and their generalizations. Though circles are nice curves in the sense of Frenet formula, as they have velocity vectors and normal vectors, it is not so easy to treat them and the family of all circles is a bit too big.

Another family of curves which was used to study Kähler manifolds is the family of trajectories for Kähler magnetic fields. A closed 2-form on a Riemannian manifold M is said to be a magnetic field. On a Kähler manifold we have a natural closed 2-form which is called the Kähler form. A constant multiple of this form is said to be a Kähler magnetic field. Its trajectories are circles of special kind and have many similar properties as of geodesics. In his papers [1 ,3], T. Adachi generalized some classical results on geodesics to those on trajectories for Kähler magnetic fields.

If we take an orientable Riemann surface M, we can regard it a Kähler manifold with complex structure J. Since every 2-form $\mathbb{B}_\varphi = \varphi\, dvol_M$ on M with a smooth function φ is closed, we shall call it a surface magnetic field on M. A smooth curve parameterized by its arclength is said to be a trajectory for $\mathbb{B}_\varphi$ if it satisfies $\nabla_{\dot\gamma}\dot\gamma = \varphi(\gamma)J\dot\gamma$. As a first step of our study of surface magnetic fields we study their magnetic Jacobi fields which are associated with variations of trajectories. In this paper we give results corresponding to Rauch's comparison theorem. Our results are closely related with the result in [8], but our way is much more direct. As applications we give estimates on volumes of trajectory balls for surface magnetic fields, which corresponds to Bisop's comparison theorem ([7]) and the results for Kähler magnetic fields in [4] and [13].

2. Variations of trajectories

Let M be an orientable Riemann surface. Regarding this as a 1-dimensional complex manifold we denote by J is complex structure. Given a smooth function $\varphi \in C^\infty(M)$ we take a surface magnetic field $\mathbb{B}_\varphi = \varphi\, dvol_M$. A smooth curve γ parameterized by its arclength is said to be a *trajectory* for $\mathbb{B}_\varphi$ if it satisfies $\nabla_{\dot\gamma}\dot\gamma(t) = \varphi\big(\gamma(t)\big)J\dot\gamma(t)$. We call a smooth map $\alpha : I \times (-\epsilon, \epsilon) \to M$ a variation of trajectories for $\mathbb{B}_\varphi$ if for each $s \in (-\epsilon,\epsilon)$ the map $\alpha_s = \alpha(\ ,s) : I \to M$ is a trajectory for $\mathbb{B}_\varphi$. Since α_s is parameterized by its arc length, we have $\big\langle \nabla_{\frac{\partial\alpha}{\partial t}} \frac{\partial\alpha}{\partial s}, \frac{\partial\alpha}{\partial t}\big\rangle = 0$. Also we have

$$\nabla_{\frac{\partial\alpha}{\partial t}}\nabla_{\frac{\partial\alpha}{\partial t}}\frac{\partial\alpha}{\partial s} + R\Big(\frac{\partial\alpha}{\partial s},\frac{\partial\alpha}{\partial t}\Big)\frac{\partial\alpha}{\partial t} = \frac{\partial}{\partial s}\varphi\big(\alpha(t,s)\big)J\frac{\partial\alpha}{\partial s} + \varphi\big(\alpha(t,s)\big)J\nabla_{\frac{\partial\alpha}{\partial t}}\frac{\partial\alpha}{\partial s}.$$

We hence say a vector field Y along a trajectory γ for $\mathbb{B}_\varphi$ to be a *magnetic Jacobi field* for $\mathbb{B}_\varphi$ if it satisfies

$$\begin{cases} \nabla_{\dot\gamma}\nabla_{\dot\gamma}Y + R(Y,\dot\gamma)\dot\gamma - (Y\varphi)J\dot\gamma - \varphi(\gamma)J\nabla_{\dot\gamma}Y = 0, \\ \langle \nabla_{\dot\gamma}Y, \dot\gamma\rangle \equiv 0. \end{cases} \tag{2.1}$$

Since M is 2-dimensional, we denote as $Y = f_Y\dot\gamma + g_Y J\dot\gamma$ with smooth functions f_Y, g_Y along γ. We then find that (2.1) turns to

$$\begin{cases} g_Y'' + g_Y\{K(\gamma) + \varphi(\gamma)^2 - (J\dot\gamma)\varphi\} = 0, \\ f_Y' = \varphi(\gamma)g_Y, \end{cases}$$

where $K\big(\gamma(t)\big)$ denotes the sectional curvature at $\gamma(t)$. We set $\mathcal{J}(\gamma)$ the set of all magnetic Jacobi fields along γ, which is a 3-dimensional vector space.

For a vector field $X = g_X J\dot{\gamma}$ along a trajectory γ for $\mathbb{B}_\varphi$ which is orthogonal to $\dot{\gamma}$ and a positive T, we set

$$Ind_0^T(X) = \int_0^T \Big\{ {g'_X}^2 + g_X^2\big((J\dot{\gamma})\varphi - K(\gamma) - \varphi^2(\gamma)\big)\Big\} dt.$$

For a vector field Y along γ, we denote by $Y^\perp$ its component orthogonal to $\dot{\gamma}$, that is, we set $Y^\perp = Y - \langle Y, \dot{\gamma}\rangle\dot{\gamma}$ $(= g_Y J\dot{\gamma})$. Along the same way as for ordinary Jacobi fields we have the following.

Lemma 2.1. *When Y is a magnetic Jacobi field along γ we have*

$$Ind_0^T(Y^\perp) = \langle \nabla_{\dot{\gamma}} Y^\perp(T), Y^\perp(T)\rangle - \langle \nabla_{\dot{\gamma}} Y^\perp(0), Y^\perp(0)\rangle.$$

Proof. Since $Y^\perp = g_Y J\dot{\gamma}$ we get $\nabla_{\dot{\gamma}} Y^\perp = -\varphi(\gamma) g_Y \dot{\gamma} + g'_Y J\dot{\gamma}$. Thus we have

$$\begin{aligned}
&\langle \nabla_{\dot{\gamma}} Y^\perp(T), Y^\perp(T)\rangle - \langle \nabla_{\dot{\gamma}} Y^\perp(0), Y^\perp(0)\rangle \\
&\quad = g'_Y(T) g_Y(T) - g'_Y(0) g_Y(0) = \int_0^T \{g'_Y(t) g(t)\}'\, dt \\
&\quad = \int_0^T \{g''_Y(t) g(t) + g'^2_Y\}\, dt = Ind_0^T(Y^\perp),
\end{aligned}$$

which is the conclusion. □

Let γ be a trajectory for a surface magnetic field $\mathbb{B}_\varphi$. A point $\gamma(t_0)$ is said to be a *magnetic conjugate point* of $\gamma(0)$ along γ if there is a non-trivial magnetic Jacobi field for $\mathbb{B}_\varphi$ along γ with $Y(0) = 0$ and $Y^\perp(t_0) = 0$. In this case we call t_0 a magnetic conjugate value. Putting $p = \gamma(0)$ we denote by ${t_c}^\varphi(p;\gamma)$ the minimal positive magnetic conjugate value of p along γ. When there are no magnetic conjugate points of p on $\gamma((0,\infty))$, we set ${t_c}^\varphi(p;\gamma) = \infty$.

Proposition 2.1. *Let γ be a trajectory for $\mathbb{B}_\varphi$ and Y be a magnetic Jacobi field along γ satisfying $Y(0) = 0$. Suppose there are no magnetic conjugate points of $\gamma(0)$ on $\gamma((0,T])$. We take a vector field X along γ which is orthogonal to $\dot{\gamma}$. If it satisfies $X(0) = 0$ and $X(T) = Y^\perp(T)$, then we have $Ind_0^T(X) \geq Ind_0^T(Y^\perp)$. The equality holds if and only if $X \equiv Y^\perp$.*

Proof. Since we have magnetic conjugate points of $\gamma(0)$ on $\gamma((0,\infty))$, we have a non-vanishing smooth function h satisfying $X(t) = h(t)Y^\perp(t)$. We denote as $X = g_X J\dot{\gamma}$ and $Y^\perp = g_Y J\dot{\gamma}$. We then have $g_X = h g_Y$ and

$h(T)=1$. As Y is a magnetic Jacobi field with $g_Y(0)=0$, we obtain

$$\begin{aligned}
Ind_0^T(X) &= \int_0^T \Big\{(g_Y'h+g_Yh')^2 + h^2g_Y^2\big((J\dot\gamma)\varphi - K(\gamma)-\varphi^2(\gamma)\big)\Big\}dt \\
&= \int_0^T h^2\Big\{g_Y'^2 + g_Y^2\big((J\dot\gamma)\varphi - K(\gamma)-\varphi^2(\gamma)\big)\Big\}dt \\
&\quad + \int_0^T \{g_Yg_Y'(h^2)' + g_Y^2h'^2)\,dt \\
&= \int_0^T h^2\Big\{g_Y'^2 + g_Y^2\big((J\dot\gamma)\varphi - K(\gamma)-\varphi^2(\gamma)\big)\Big\}dt \\
&\quad + g_Y(T)g_Y'(T) - \int_0^T (g_Y'^2h^2 + g_Yg_Y''h^2)\,dt + \int_0^T g_Y^2h'^2\,dt \\
&= g_Y(T)g_Y'(T) + \int_0^T g_Y^2h'^2\,dt \\
&= Ind_0^T(Y^\perp) + \int_0^T g_Y^2h'^2\,dt \geq Ind_0^T(Y^\perp).
\end{aligned}$$

One can easily see that the equality holds if and only if h is a constant function, which means that $h\equiv 1$. Thus we get the conclusion. □

3. Comparison theorems on magnetic Jacobi fields

In this section we give estimates of norms of magnetic Jacobi fields. We first recall magnetic Jacobi fields for Kähler magnetic fields on a 2-dimensional real space form $\mathbb{R}M^2(c)$ of constant sectional curvature c, which is a standard sphere $S^2(c)$, a Euclidean space $\mathbb{R}^2$ and a real hyperbolic space $H^2(c)$ according that c is positive, zero and negative (see [2]). We define two functions $\hat{\mathfrak{s}}(t;\kappa,c)$ and $\hat{\mathfrak{c}}(t;\kappa,c)$ by

$$\hat{\mathfrak{s}}(t;\kappa,c) = \begin{cases} \big(1/\sqrt{\kappa^2+c}\big)\ \sin\big(\sqrt{\kappa^2+c}\,t\big), & \text{if } \kappa^2+c>0, \\ t, & \text{if } \kappa^2+c=0, \\ \big(1/\sqrt{|c|-\kappa^2}\big)\ \sinh\big(\sqrt{|c|-\kappa^2}\,t\big), & \text{if } \kappa^2+c<0, \end{cases}$$

$$\hat{\mathfrak{c}}(t;\kappa,c) = \begin{cases} \cos\big(\sqrt{\kappa^2+c}\,t\big), & \text{if } \kappa^2+c>0, \\ 1, & \text{if } \kappa^2+c=0, \\ \cosh\big(\sqrt{|c|-\kappa^2}\,t\big), & \text{if } \kappa^2+c<0. \end{cases}$$

Given a trajectory γ for a Kähler magnetic field $\mathbb{B}_\kappa$ on $\mathbb{R}M^2(c)$, the vertical component $Y^\perp$ of a magnetic Jacobi field Y for $\mathbb{B}_\kappa$ along γ is given as

$$Y^\perp(t) = \Big\{\langle Y(0), J\dot\gamma(0)\rangle \hat{\mathfrak{c}}(t;\kappa,c) \\ + \big\{\big\langle\big(\nabla_{\dot\gamma}Y\big)(0), J\dot\gamma(0)\big\rangle - \kappa\langle Y(0),\dot\gamma(0)\rangle\big\}\hat{\mathfrak{s}}(t;\kappa,c)\Big\}J\dot\gamma(t).$$

In particular, when $Y(0)=0$ we have

$$Y^\perp(t) = \big\langle\big(\nabla_{\dot\gamma}Y\big)(0), J\dot\gamma(0)\big\rangle\,\hat{\mathfrak{s}}(t;\kappa,c)J\dot\gamma(t).$$

We here give estimates on norms of vertical components of magnetic Jacobi fields for a surface magnetic field by comparing those for Kähler magnetic fields. First, we give an estimate from above.

Theorem 3.1. *Let M and $\widehat{M}$ be two complete orientable Riemann surfaces and φ be a smooth function on M. We suppose $\beta := \inf_{p\in M}(\varphi^2 - \|\nabla\varphi\|) \geq 0$. Putting $\kappa = \sqrt{\beta}$, we take a trajectory γ for $\mathbb{B}_\varphi$ on M and a trajectory $\hat\gamma$ for $\mathbb{B}_\kappa$ on $\widehat{M}$. Let Y be a magnetic Jacobi field for $\mathbb{B}_\varphi$ along γ and $\widehat{Y}$ be a magnetic Jacobi field for $\mathbb{B}_\kappa$ along $\hat\gamma$ which satisfy $Y^\perp(0) = \widehat{Y}^\perp(0) = 0$, $\|\nabla_{\dot\gamma}Y^\perp(0)\| = \|\nabla_{\dot{\hat\gamma}}\widehat{Y}^\perp(0)\| (\neq 0)$.*

If sectional curvatures satisfy $K\big(\gamma(t)\big) \geq K\big(\hat\gamma(t)\big)$ for $0 \leq t \leq t_c{}^\varphi(\gamma(0),\gamma)$, then we have the following.

1. *$t_c{}^\varphi(\gamma(0);\gamma) \leq t_c{}^\kappa(\hat\gamma(0);\hat\gamma)$;*
2. *The function $\|\widehat{Y}^\perp(t)\|/\|Y^\perp(t)\|$ is monotonic increase for $0 \leq t \leq t_c{}^\varphi(\gamma(0);\gamma)$;*
3. $\dfrac{\langle\nabla_{\dot{\hat\gamma}}\widehat{Y}^\perp(t), \widehat{Y}^\perp(t)\rangle}{\langle\widehat{Y}^\perp(t), \widehat{Y}^\perp(t)\rangle} \geq \dfrac{\langle\nabla_{\dot\gamma}Y^\perp(t), Y^\perp(t)\rangle}{\langle Y^\perp(t), Y^\perp(t)\rangle}$ *for $0 \leq t \leq t_c{}^\varphi(\gamma(0);\gamma)$;*
4. *$\|\widehat{Y}^\perp(t)\| \geq \|Y^\perp(t)\|$ for $0 \leq t \leq t_c{}^\varphi(\gamma(0);\gamma)$.*
5. $|\langle Y(t),\dot\gamma(t)\rangle| \leq \big(\sup_{p\in M}|\varphi|\big)\displaystyle\int_0^t \|\widehat{Y}^\perp(s)\|\,ds.$

Proof. We denote Y and $\widehat{Y}$ as $Y = f\dot\gamma + gJ\dot\gamma$ and $\widehat{Y} = \hat f\dot{\hat\gamma} + \hat gJ\dot{\hat\gamma}$, respectively. Since $Y^\perp = gJ\dot\gamma$ and $\widehat{Y}^\perp = \hat gJ\dot{\hat\gamma}$, we see by the assumption on Y and $\widehat{Y}$ that $g(0) = \hat g(0) = 0$ and $|g'(0)| = |\hat g'(0)|$. Applying de l'Hôspital's rule, we find that the function $F(t) = \hat g(t)^2/g(t)^2$ satisfies

$$\lim_{t\downarrow 0} F(t) = \lim_{t\downarrow 0}\frac{\hat g'(t)\hat g(t)}{g'(t)g(t)} = \lim_{t\downarrow 0}\frac{\hat g''(t)\hat g(t) + \hat g'(t)^2}{g''(t)g(t) + g'(t)^2} = 1.$$

Moreover, we have

$$\frac{d}{dt}F(t) = \frac{d}{dt}\Big(\frac{\hat g(t)^2}{g(t)^2}\Big) = 2\,\frac{\hat g(t)^2}{g(t)^2}\left(\frac{\hat g'(t)}{\hat g(t)} - \frac{g'(t)}{g(t)}\right).$$

If we can show that $F'(t) \geq 0$ for $0 < t \leq \min\{t_c{}^{\varphi}(\gamma(0);\gamma), t_c{}^{\kappa}(\hat{\gamma}(0);\hat{\gamma})\}$, we find $\hat{g}(t) \geq g(t)$ on this interval and get the assertions. So we need to show $\hat{g}'/g \geq g'/g$ on this interval.

We take an arbitrary positive T which satisfies $0 < T < \min\{t_c{}^{\varphi}(\gamma(0);\gamma), t_c{}^{\kappa}(\gamma(0);\gamma)\}$, and set $\widehat{G}_T(t) = \hat{g}(t)/\hat{g}(T)$ and $G_T(t) = g(t)/g(T)$. Clearly we have $\hat{g}'(T)/\hat{g}(T) = \widehat{G}'_T(T)\widehat{G}_T(T)$ and $g'(T)/g(T) = G'_T(T)G_T(T)$. If we set $\widehat{Y}_T(t) = \widehat{Y}(t)/\hat{g}(T)$ and $Y_T(t) = Y(t)/g(T)$, they are magnetic Jacobi fields for $\mathbb{B}_\kappa$ along $\hat{\gamma}$ and for $\mathbb{B}_\varphi$ along γ, respectively. We take a vector field X along γ given by $X(t) = \widehat{G}_T(t)J\dot{\gamma}(t)$. This satisfies $X(0) = 0$ and $X(T) = Y_T^{\perp}(T)$. Since $|(J\dot{\gamma})\varphi| \leq \|\nabla\varphi\|$, we have by Proposition 2.1 that

$$\begin{aligned}
\widehat{G}'_T(T)\widehat{G}_T(T) &= \widehat{G}'_T(T)\widehat{G}_T(T) - \widehat{G}'_T(0)\widehat{G}_T(0) = Ind_0^T(\widehat{Y}_T^{\perp}) \\
&= \int_0^T \Big\{\widehat{G}'_T(t)^2 + \widehat{G}_T(t)^2\{-K(\hat{\gamma}(t)) - \kappa^2\}\Big\}dt \\
&\geq \int_0^T \Big\{\widehat{G}'_T(t)^2 + \widehat{G}_T(t)^2\{\|(\nabla\varphi)\big(\gamma(t)\big)\| - K(\gamma(t)) - \varphi\big(\gamma(t)\big)^2\}\Big\}dt \\
&\geq \int_0^T \Big\{\widehat{G}'_T(t)^2 + \widehat{G}_T(t)^2\{(J\dot{\gamma})\varphi(t) - K(\gamma(t)) - \varphi\big(\gamma(t)\big)^2\}\Big\}dt \\
&= Ind_0^T(X) \geq Ind_0^T(Y_T^{\perp}) = G'_T(T)G_T(T).
\end{aligned}$$

We hence get $\hat{g}'(T)/\hat{g}(T) \geq g'(T)/g(T)$. As T is an arbitrary point on the interval $0 < T < \min\{t_c{}^{\varphi}(\gamma(0);\gamma), t_c{}^{k}(\gamma(0);\gamma)\}$, we obtain the conclusions on g. Since we have $f' = (\varphi \circ \gamma) \cdot g$, we get the last assertion. □

Next we give an estimate from below.

Theorem 3.2. *Let M and $\widehat{M}$ be two complete orientable Riemann surfaces and φ be a smooth function on M. We suppose $\rho := \sup_{p\in M}(\|\nabla\varphi\| + \varphi^2) < \infty$. Putting $\kappa = \sqrt{\rho}$, we take a trajectory γ for $\mathbb{B}_\varphi$ on M and a trajectory $\hat{\gamma}$ for $\mathbb{B}_\kappa$ on $\widehat{M}$. Let Y be a magnetic Jacobi field for $\mathbb{B}_\varphi$ along γ and $\widehat{Y}$ be a magnetic Jacobi field for $\mathbb{B}_\kappa$ along $\hat{\gamma}$ which satisfy $Y^{\perp}(0) = \widehat{Y}^{\perp}(0) = 0$, $\|\nabla_{\dot{\gamma}}Y^{\perp}(0)\| = \|\nabla_{\dot{\hat{\gamma}}}\widehat{Y}^{\perp}(0)\| (\neq 0)$.*

If sectional curvatures satisfy $K\big(\gamma(t)\big) \leq K\big(\hat{\gamma}(t)\big)$ for $0 \leq t \leq t_c{}^{\varphi}(\gamma(0);\gamma)$, then we have the following:

(1) *$t_c{}^{\kappa}(\hat{\gamma}(0);\hat{\gamma}) \leq t_c{}^{\varphi}(\gamma(0);\gamma)$;*
(2) *The function $\|\widehat{Y}^{\perp}(t)\|/\|Y^{\perp}(t)\|$ is monotonic decrease for $0 \leq t \leq t_c{}^{\kappa}(\gamma(0);\gamma)$;*

(3) $\dfrac{\langle \nabla_{\dot{\hat{\gamma}}} \widehat{Y}^{\perp}(t), \widehat{Y}^{\perp}(t)\rangle}{\langle \widehat{Y}^{\perp}(t), \widehat{Y}^{\perp}(t)\rangle} \le \dfrac{\langle \nabla_{\dot{\gamma}} Y^{\perp}(t), Y^{\perp}(t)\rangle}{\langle Y^{\perp}(t), Y^{\perp}(t)\rangle}$ *for* $0 \le t \le t_c{}^{\kappa}(\gamma(0);\gamma)$;

(4) $\|\widehat{Y}^{\perp}(t)\| \le \|Y^{\perp}(t)\|$ *for* $0 \le t \le t_c{}^{\kappa}(\gamma(0);\gamma)$.

Proof. We use the same notations as in the proof of Theorem 3.1. We have only to show that $G'_T(T)G_T(T) \ge \widehat{G}'_T(T)\widehat{G}_T(T)$ for an arbitrary T with $0 < T \le \min\{t_c{}^{\varphi}(\gamma(0);\gamma), t_c{}^{\kappa}(\gamma(0);\gamma)\}$. We take a vector field $\widehat{X}$ along $\hat{\gamma}$ defined by $\widehat{X}(t) = G_T(t)J\dot{\hat{\gamma}}(t)$. This vector field satisfies $\widehat{X}(0) = 0$ and $\widehat{X}(T) = \widehat{Y}_T^{\perp}(T)$. We therefore have

$$\begin{aligned}
G'_T(T)G_T(T) &= G'_T(T)G_T(T) - G'_T(0)G_T(0) = Ind_0^T(Y_T^{\perp}) \\
&= \int_0^T \Big\{G'_T(t)^2 + G_T(t)^2\big\{(J\dot{\gamma})\varphi(t) - K(\gamma(t)) - \varphi\big(\gamma(t)\big)^2\big\}\Big\}dt \\
&\ge \int_0^T \Big\{G'_T(t)^2 + G_T(t)^2\big\{-\|(\nabla\varphi)\big(\gamma(t)\big)\| - K(\gamma(t)) - \varphi\big(\gamma(t)\big)^2\big\}\Big\}dt \\
&\ge \int_0^T \Big\{G'_T(t)^2 + G_T(t)^2\big\{-K(\hat{\gamma}(t)) - \kappa^2\big\}\Big\}\,dt \\
&= Ind_0^T(X) \ge Ind_0^T(\widehat{Y}_T^{\perp}) = \widehat{G}'_T(T)\widehat{G}_T(T),
\end{aligned}$$

and get the conclusion. □

As a consequence of these theorems, we obtain the following estimates on the first magnetic conjugate values.

Corollary 3.1. *Let M be an orientable Riemann surface whose sectional curvatures satisfy $K_M \le c$ for some constant c. If a smooth function φ on M satisfies $\rho := \sup_{p\in M}(\|\nabla\varphi\| + \varphi^2) < \infty$, then for an arbitrary trajectory for $\mathbb{B}_{\varphi}$ we have $t_c{}^{\varphi}(\gamma(0);\gamma) \ge \pi/\sqrt{\rho + c}$. In particular, if $\rho + c \le 0$, then there are no magnetic conjugate points for $\mathbb{B}_{\varphi}$.*

Corollary 3.2. *Let M be an orientable Riemann surface whose sectional curvatures satisfy $K_M \ge c$ for some constant c. If a smooth function φ on M satisfies $\beta := \inf_{p\in M}(\|\nabla\varphi\| - \varphi^2) \ge 0$ and $\beta + c > 0$, then for an arbitrary trajectory for $\mathbb{B}_{\varphi}$ we have $t_c{}^{\varphi}(\gamma(0);\gamma) \le \pi/\sqrt{\beta + c}$.*

4. Areas of trajectory disks

We here give an application. Let M be an orientable Riemann surface and φ be a smooth function on M. Given a point $p \in M$, we define the magnetic exponential map at p $\mathbb{B}_\varphi\exp_p : T_pM \to M$ for a surface magnetic field $\mathbb{B}_\varphi$ of the tangent space T_pM at p by

$$\mathbb{B}_\varphi\exp_p(v) = \begin{cases} \gamma_{v/\|v\|}(\|v\|), & \text{when } v \neq 0_p, \\ p, & \text{when } v = 0_p, \end{cases}$$

where, γ_u for a unit tangent vector u denotes the trajectory for $\mathbb{B}_\varphi$ satisfying $\dot\gamma(0) = u$. When $\varphi \equiv 0$, this is the ordinary exponential map at p. Clearly, its differential $(d\mathbb{B}_\varphi\exp_p)_{ru} : T_{ru}(T_pM) \to T_{\gamma_u(r)}M$ is singular if and only if r is a magnetic conjugate values of p along γ_u.

We set $B_r^\varphi(p) = \mathbb{B}_\varphi\exp_p\big(\{v \in T_pM \mid \|v\| \le r\}\big)$, and call this image of a Euclidean r-disk a *trajectory ball* or a *trajectory disk* of radius r centered at p. When $\varphi \equiv 0$, this disk is a geodesic ball centered at p. By applying Theorems 3.1 and 3.2, we give estimates of the volume $\mathrm{Vol}(B_r^\varphi(p))$ of this disk. We set $t_c{}^\varphi(p) = \min\{t_c{}^\varphi(p;u) \mid u \in U_pM\}$ and define a map $\Theta_{\varphi,p} : \big(0, t_c{}^\varphi(p)\big) \times U_pM \to M$ by $\Theta_{\varphi,p}(t,u) = \mathbb{B}_\varphi\exp_p(tu)$. Given $u \in U_pM$ we take a magnetic Jacobi field Y_u along a trajectory γ_u so that it satisfies $\big(\nabla_{\dot\gamma_u}Y\big)(0) = Ju$ by regarding M as a 1-dimensional Kähler manifold. We then have

$$\dot\gamma_u(t) = \big(d\mathbb{B}_\varphi\exp_p\big)_{tu}(u) \quad \text{and} \quad Y_u(t) = \big(d\mathbb{B}_\varphi\exp_p\big)_{tu}(tJu).$$

Therefore, by considering the Jacobian when we take the polar coordinate, we fing that the pull back $\Theta_{\varphi,p}^* dvol_M$ of the volume element of M at $\gamma_u(t)$ is given as $\det(\dot\gamma_u, Y_u)\, dtd\omega = \|Y_u^\perp(t)\|\, dtd\omega$, where $d\omega$ denotes the line element of a unit circle U_pM. We here set the *magnetic injectivity radius* at p by $\iota_\varphi(p) = \sup\{r \mid \mathbb{B}_\varphi|_{B_r(0_p)} \text{ is injective}\}$. Hence if $r \le \min\{\iota_\varphi(p), t_c{}^\varphi(p)\}$, we obtain

$$\mathrm{Vol}(B_r^\varphi(p)) = \int_0^r dt \int \|Y_u^\perp(t)\|\, d\omega.$$

For a Kähler magnetic field $\mathbb{B}_\kappa$ on a real space form $\mathbb{R}M^2(c)$ of constant sectional curvature c, we have

$$\mathrm{Vol}(B_r^\kappa(p)) = 2\pi \int_0^r \hat{\mathfrak{s}}(t;\kappa,c)\, dt = 2\pi\hat{\mathfrak{d}}(r;\kappa,c),$$

where

$$\hat{\mathfrak{d}}(t;\kappa,c) = \begin{cases} \big\{1 - \cos\big(\sqrt{\kappa^2+c}\, r\big)\big\}/(\kappa^2+c), & \text{if } \kappa^2+c>0, \\ r^2/2, & \text{if } \kappa^2+c=0, \\ \big\{\cosh\big(\sqrt{|c|-\kappa^2}\, r\big) - 1\big\}/(\kappa^2+c), & \text{if } \kappa^2+c<0. \end{cases}$$

We now apply Theorems 3.1 and 3.2 to get estimates of areas of trajectory disks for an arbitrary surface magnetic field on an orientable Riemann surface M. Given a smooth function φ on M we set

$$\rho_r(p) := \max\big\{ \|\nabla\varphi\| + \varphi^2 \bigm| q \in B_r^{\varphi}(p)\big\}.$$
$$\beta_r(p) := \min\big\{ \|\nabla\varphi\| - \varphi^2 \bigm| q \in B_r^{\varphi}(p)\big\}.$$

Since a trajectory disk $B_r^{\varphi}(p)$ is contained in a geodesic disk $B_r^0(p)$, we have

$$\rho_r(p) \le \max\big\{ \|\nabla\varphi\| + \varphi^2 \bigm| q \in B_r^0(p)\big\} \le \rho,$$
$$\beta_r(p) \ge \min\big\{ \|\nabla\varphi\| - \varphi^2 \bigm| q \in B_r^0(p)\big\} \ge \beta.$$

Here, ρ and β are the constants given in §3. In view of the proof of Theorem 3.2, if the sectional curvatures of M is not greater than c on $B_r^{\varphi}(p)$, then a magnetic Jacobi field Y with $Y^{\perp}(0) = 0$ along a trajectory γ for $\mathbb{B}_{\varphi}$ emanating from p satisfies

$$\|Y^{\perp}(t)\| \ge \|\nabla_{\dot\gamma} Y^{\perp}(0)\|\, \hat{\mathfrak{s}}\big(t; \sqrt{\rho_r(p)}, c\big)$$

for $0 \le t \le \min\big\{r, t_c{}^{\sqrt{\rho_r(p)}}(p)\big\}$. Thus we have the following.

Theorem 4.1. *Let φ be a smooth function on an orientable Riemann surface M whose sectional curvatures satisfy* Riem $\le c$. *If r with $r \le \iota_{\varphi}(p)$ satisfies $r \le t_c{}^{\sqrt{\rho_r(p)}}(p)$, then the volume of a trajectory disc $B_r^{\varphi}(p)$ for a surface magnetic field $\mathbb{B}_{\varphi}$ is estimated from below as*

$$\mathrm{Vol}(B_r^{\varphi}(p)) \ge 2\pi\hat{\mathfrak{d}}(r; \sqrt{\rho_r(p)}, c).$$

In particular, if $\rho := \sup_{p\in M}(\|\nabla\varphi\| + \varphi^2) < \infty$, then for r with $r \le \min\big\{\iota_{\varphi}(p), t_c{}^{\sqrt{\rho}}(p)\big\}$ we have $\mathrm{Vol}(B_r^{\varphi}(p)) \ge 2\pi\hat{\mathfrak{d}}(r; \sqrt{\rho}, c)$.

Similary, in view of the proof of Theorem 3.1, if the sectional curvatures of M is not smaller than c on $B_r^{\varphi}(p)$ and if $\beta_r(p) \ge 0$, then a magnetic Jacobi field Y with $Y^{\perp}(0) = 0$ along a trajectory γ for $\mathbb{B}_{\varphi}$ emanating from p satisfies

$$\|Y^{\perp}(t)\| \le \|\nabla_{\dot\gamma} Y^{\perp}(0)\|\, \hat{\mathfrak{s}}\big(t; \sqrt{\beta_r(p)}, c\big)$$

for $0 \le t \le \min\{r, t_c{}^{\varphi}(p;\gamma)\}$. Hence we have the following.

Theorem 4.2. *Let φ be a smooth function on an orientable Riemann surface M whose sectional curvatures satisfy* Riem $\ge c$. *If $\beta_r(p) \ge 0$ and $r \le 2\pi/\sqrt{\beta_r(p) + c}$, then the volume of a trajectory disc $B_r^{\varphi}(p)$ for $\mathbb{B}_{\varphi}$ is estimated from above as*

$$\mathrm{Vol}(B_r^{\varphi}(p)) \le 2\pi\hat{\mathfrak{d}}(r; \sqrt{\beta_r(p)}, c).$$

In particular, if $\beta := \inf_{p\in M}\big(\|\nabla\varphi\| - \varphi^2\big) \geq 0$*, then for* r *with* $r \leq 2\pi/\sqrt{\beta+c}$ *we have* $\mathrm{Vol}(B_r^{\varphi}(p)) \leq 2\pi\hat{\mathfrak{d}}(r;\sqrt{\beta},c)$.

We shall discuss the relation between the injectivity radius and first conjugate values in the forthcoming paper.

References

1. T. Adachi, Kähler magnetic flows for a manifold of constant holomorphic sectional curvature, *Tokyo Math. J.* **18**, 473–483 (1995).
2. ———, Magnetic Jacobi fields for Kähler magnetic fields, *Recent Progress in Differential Geometry and its Related Fields*, 41–53 T. Adachi, H. Hashimoto & M. Hristov eds., World Scientific, Singapole (2011).
3. ———, A theorem of Hadamard-Cartan type for Kähler magnetic fields, *J. Math. Soc. Japan* **64**, 1–21 (2012).
4. T. Adachi & P. Bai, Volumes of trajectory-balls for Kähler magnetic fields, *J. Geom.* **105**, 369–389 (2014).
5. T. Adachi & S. Maeda, Characterizations of space forms by circles on their geodesic spheres, *Proc. Japan Acad. Ser. A Math. Sci.* **78**, 143–147 (2002).
6. ———, Characterization of totally umbilic hypersurfaces in a space form by circles, *Czechoslovak Math. J.* **55**, 201–207 (2005).
7. R.L. Bishop & R. Critenden, *Geometry of manifolds*, Academic Press, New York (1964).
8. N. Goda, Magnetic flows of Anosov type, *Tôhoku Math. J.* **49**, 165–183 (1997).
9. ———, The theorem of E. Hopf under uniform magnetic fields, *J. Math. Soc. Japan* **50**, 767–779 (1998).
10. S. Maeda, A characterization of constant isotropic immersions by circles, *Arch. Math.* (*Basel*) **81**, 90–95 (2003).
11. S. Maeda & H. Tanabe, Totally geodesic immersions of Kähler manifolds and Kähler Frenet curves, *Math. Z.* 252, 787–795 (2006).
12. T. Sakai, *Riemannian Geometry*, Shokabo, Tokyo (1992) (in Japanese) and Translations of Math. Monographs vol. 149, A.M.S. (1996).
13. Q. Shi & T. Adachi, An estimate on volumes of trajectory-balls for Kähler magnetic fields, preprint (2014).

Received May 12, 2015
Revised May 20, 2015

Current Developments in Differential Geometry and its Related Fields pp. 225 –233

A GEOMETRIC STUDY ON LAPLACE TRANSFORMED CURVES

Milen J. HRISTOV *

Department of Algebra and Geometry, Faculty Mathematics and Informatics, "St. Cyril and St. Methodius" University of Veliko Tarnovo, 5000 Veliko Tarnovo, Bulgaria
E-mail: m.hristov@uni-vt.bg

In this paper we define Laplace transformation of smooth regular curves in a Euclidean space. We express the Frenet frame, curvature and torsion of the Laplace transformed curve in terms of Laplace transformations of the position vector of an original curve.

Keywords: Laplace integral transformation, smooth regular curve, Frenet frame, curvatures and torsions.

1. Introduction

A continuous function $f : [0, \infty) \to \mathbb{R}$ is said to be of exponential order λ if $|f(t)|e^{-\lambda t}$ is bounded for $t > 0$. We denote by $\mathcal{O}$ the set of all continuous functions defined on the interval $[0, \infty)$ which are of exponential order for some constant λ. For such a function $f \in \mathcal{O}$ of exponential order λ, we can define a function $\mathcal{L}[f] : (\lambda, \infty) \to \mathbb{R}$ by

$$\mathcal{L}[f](s) = \int_0^\infty e^{-st} f(t)\, dt.$$

This function is called the Laplace transformed function of f (see [1, 2] for more detail). Usually this function is defined on $\{s \in \mathbb{C} \mid \mathrm{Re}(z) > \lambda\}$, but we here consider only on real s. In the field of analysis, this Laplace transformation is useful; to solve differential equations and so on.

In this paper we attempt to introduce this analytic operation in the field of Euclidean geometry. Given a smooth curve in a Euclidean space which is parameterized by its arclength we define its Laplace transformed curve in this space. Since differentials are translated into algebraic multiplications through the Laplace transformation, we give some fundamental calculations on Laplace transformed curves from the viewpoint of Frenet-Serre formula

*Supported by Scientific researches fund of "St.Cyril and St. Methodius" University of Veliko Tarnovo under contract RD-09-422-13/09.04.2014.

by using this translation. As we find that our Laplace transformation on curves does not preserve geometric invariants, we show how some elementary curves are transformed as for examples. In order to study geometric properties of curves which do not depend on parameters of curves, in this paper we restrict ourselves to curves parameterized by their arclength. In [3] we treat some analytic aspects of Laplace transformed curves.

2. Laplace transformed images of smooth regular curves

Let $c : \vec{r}(t) = \vec{r}_c(t) = (x_1(t), x_2(t), x_3(t)) : [0, \infty) \to \mathbb{R}^3$ be a smooth regular curve in an affine Euclidean 3-space $\mathbb{R}^3$ which is parametrized by its arc-length. In this paper we treat smooth curves which are defined on some open interval containing $[0, \infty)$. We denote its Frenet frame by $\{\vec{t}(t), \vec{n}(t), \vec{b}(t)\}$, and its curvature and torsion by $\kappa_c(t)$ and $\tau_c(t)$, respectively. That is, we set

$$\vec{t}(t) = \vec{r}\,'(t), \quad \vec{n}(t) = \frac{\vec{r}\,''(t)}{|\vec{r}\,''(t)|}, \quad \vec{b}(t) = \vec{t}(t) \times \vec{n}(t),$$

$$\kappa_c(t) = |\vec{r}\,''(t)|, \quad \tau_c(t) = \det\big(\vec{r}\,'(t), \vec{r}\,''(t), \vec{r}\,'''(t)\big)\{\kappa_c(t)\}^{-2}.$$

We hence have the following Frenet formula:

$$\begin{pmatrix} \vec{t}\,'(t) \\ \vec{n}\,'(t) \\ \vec{b}\,'(t) \end{pmatrix} = \mathbb{F}(t) \begin{pmatrix} \vec{t}(t) \\ \vec{n}(t) \\ \vec{b}(t) \end{pmatrix}, \qquad \mathbb{F}(t) = \begin{pmatrix} 0 & \kappa_c(t) & 0 \\ -\kappa_c(t) & 0 & \tau_c(t) \\ 0 & -\tau_c(t) & 0 \end{pmatrix}. \tag{2.1}$$

We now define Laplace transformed curves. Let $c : \vec{r}_c(t)$ be a smooth regular curve defined on the interval $[0, \infty)$ parameterized by its arc-length. We suppose all its coordinate functions $x_i(t)$, $i = 1, 2, 3$ belong to $\mathcal{O}$. This condition is clearly equivalent to the condition $\|\vec{r}_c\| \in \mathcal{O}$. We define its Laplace transformed curve $\mathcal{L}[c]$ by

$$\mathcal{L}[c] = \mathcal{C} : \vec{R}(s) = \mathcal{L}[\vec{r}_c](s) = \big(\mathcal{L}[x_1](s), \mathcal{L}[x_2](s), \mathcal{L}[x_3](s)\big).$$

If we set $\lambda_c = \inf\{\lambda \mid \|\vec{r}_c\| \text{ is of exponential order } \lambda\}$, the Laplace transformed curve $\mathcal{L}[c]$ is defined on the interval (λ_c, ∞). Since the Laplace transformation is quite an analytic operation we should note that taking the parameter of arc-length is important. That is, even if a smooth curve c is expressed as $\big(x_1(u), x_2(u), x_3(u)\big)$ with non-arc-length parameter u, we can define a curve $\big(\mathcal{L}[x_1(u)](s), \mathcal{L}[x_2(u)](s), \mathcal{L}[x_3(u)](s)\big)$, but its image does not coincide with the image of $\mathcal{L}[c]$, in general.

Example 2.1. Let $c : \vec{r}(t) = \vec{q} + t\vec{p} : [0, \infty) \to \mathbb{R}^3$ be a ray of initial point $Q(q_1, q_2, q_3)$ and of initial unit vector $\vec{p} = (p_1, p_2, p_3)$. Then its Laplace transformed curve is

$$\mathcal{L}[c] : \vec{R}(s) = \frac{1}{s}\vec{q} + \frac{1}{s^2}\vec{p} = \Big(\frac{q_1}{s} + \frac{p_1}{s^2}, \frac{q_2}{s} + \frac{p_2}{s^2}, \frac{q_3}{s} + \frac{p_3}{s^2}\Big).$$

When Q is the origin, the images of c and $\mathcal{L}[c]$ coincide and their directions are opposite. When Q is not the origin, the curve $\mathcal{L}[c] = (X_1, X_2, X_3)$ is not a ray, in general. For example, we consider the case $p_3 = q_3 = 0$. When $p_1q_2 - p_2q_1 = 0$, it is a ray satisfying $p_2X_1 - p_1X_2 = 0$. But when $p_1q_2 - p_2q_1 \neq 0$ and $q_1q_2 \neq 0$, it is a quadratic curve satisfying $(q_2X_1 - q_1X_2)^2 = (p_1q_2 - p_2q_1)(p_2X_1 - p_1X_2)$.

Moreover, as we have $\mathcal{L}[q + pu^n](s) = (qs^n + n!p)/s^{n+1}$, one can understand the importance of taking the parameter of arc-length.

The above example shows that even if two curves c_1, c_2 are congruent to each other by some parallel translation there Laplace transformed curves $\mathcal{L}[c_1], \mathcal{L}[c_2]$ are not congruent to each other in general. On the other hand, as the Laplace transformation acting on functions is linear, we have the following.

Lemma 2.1. *Each linear congruence transformation φ on $\mathbb{R}^3$ is commutative to the Laplace transformation. More precisely, if we take the curve $\varphi(c)$ for a given smooth regular curve c, we find $\mathcal{L}[\varphi(c)] = \varphi(\mathcal{L}[c])$.*

Thus, when we study Laplace transformation acting on curves in $\mathbb{R}^3$, we only consider the congruency by the action of orthogonal transformations and do not consider the congruency by the action of parallel translations.

Example 2.2. Let $\vec{v}, \vec{w}$ be an orthonormal pair of vectors. We take a circle of positive curvature κ:

$$c : \vec{r}\,(t) = \vec{q} + \frac{1}{\kappa}\cos\kappa t\,\vec{v} + \frac{1}{\kappa}\sin\kappa t\,\vec{w} : \ [0,\infty) \to \mathbb{R}^3.$$

Its Laplace transformed curve is

$$\mathcal{L}[c] : \vec{R}(s) = \frac{1}{s}\,\vec{q} + \frac{s}{\kappa(s^2+\kappa^2)}\,\vec{v} + \frac{1}{s^2+\kappa^2}\,\vec{w}.$$

When $\vec{v}, \vec{w}$ are orthogonal to $\vec{q}$, this transformed curve lies on a cylinder: If we take $\vec{v} = (1,0,0)$, $\vec{w} = (0,1,0)$, $\vec{q} = (0,0,d)$, then it lies on the cylinder $X^2 + Y^2 = Y/\kappa^2$. In particular, when $\vec{q} = \vec{0}$, its image is a circle-arc.

Example 2.3. Let $\vec{u}, \vec{v}, \vec{w}$ be an orthonormal triplet of vectors. We take a helix of curvature κ and torsion τ:

$$c : \vec{r}\,(t) = \vec{q} + \frac{t}{\sqrt{2}}\,\vec{u} + \frac{1}{\sqrt{\kappa^2+\tau^2}}\left\{\cos\sqrt{\frac{\kappa^2+\tau^2}{2}}\,t\,\vec{v} + \sin\sqrt{\frac{\kappa^2+\tau^2}{2}}\,t\,\vec{w}\right\}$$
$$: [0,\infty) \to \mathbb{R}^3.$$

Its Laplace transformed curve is

$$\mathcal{L}[c] : \vec{R}(s) = \frac{1}{s}\,\vec{q} + \frac{1}{\sqrt{2}\,s^2}\,\vec{u} + \frac{1}{2s^2+\kappa^2+\tau^2}\left\{\frac{2s}{\sqrt{\kappa^2+\tau^2}}\vec{v} + \sqrt{2}\,\vec{w}\right\}.$$

Example 2.4. Let $\vec{v}, \vec{w}$ be an orthonormal pair of vectors. We take a catenary with positive a:

$$c : \vec{r}\,(t) = \vec{q} + a\log\left\{a^{-1}\left(t+\sqrt{t^2+a^2}\right)\right\}\vec{v} + \sqrt{t^2+a^2}\,\vec{w} : \ [0,\infty) \to \mathbb{R}^3.$$

Its Laplace transformed curve is given as

$$\mathcal{L}[c] : \vec{R}(s) = \frac{1}{s}\,\vec{q} + \left(\frac{a}{s}\int_0^\infty \frac{e^{-st}}{\sqrt{t^2+a^2}}\,dt\right)\vec{v} + \left(\int_0^\infty e^{-st}\sqrt{t^2+a^2}\,dt\right)\vec{w}.$$

Though we use arc-length parameter to define Laplace transformed curve, we can apply the Laplace transformation to arbitrary vector-valued functions and matrix-valued functions. For a vector-valued function $F(u) = \big(f_1(u),\dots,f_n(u)\big)$ satisfying $f_i \in \mathcal{O}$ $(i = 1,\dots,n)$, we define a vector-valued function $\mathcal{L}[F]$ by $\mathcal{L}[F](s) = \big(\mathcal{L}[f_1](s),\dots,\mathcal{L}[f_n](s)\big)$. In this sense we can define Laplace transformation of curves in $\mathbb{R}^n$. When a smooth curve c parameterized by its arc-length is expressed as $\vec{r}_c(t) = \big(x_1(t),\dots,x_n(t)\big)$, we define its Laplace transformed curve by regarding c as a vector-valued function.

We come back to the transformation of curves in $\mathbb{R}^3$. By the Laplace transformation we have $\mathcal{L}[f'](s) = s\mathcal{L}[f](s) - f(0)$. Therefore we have

$$\begin{aligned} \mathcal{L}[\vec{t}\,](s) &= \mathcal{L}[\vec{r}_c\,'](s) = s\mathcal{L}[\vec{r}_c](s) - \vec{r}\,(0) = s\vec{R}(s) - \vec{r}\,(0), \\ \mathcal{L}[\vec{n}](s) &= \mathcal{L}\Big[\frac{1}{\kappa}\,\vec{r}_c\,''\Big](s) = \mathcal{L}\Big[\frac{1}{\kappa}\,\vec{t}\,'\Big](s). \end{aligned} \tag{2.2}$$

If we restrict ourselves to curves having constant curvature, the second equality in (2.2) shows that

$$\mathcal{L}[\vec{n}](s) = \frac{1}{\kappa}\mathcal{L}[\vec{t}\,'](s) = \frac{s}{\kappa}\mathcal{L}[\vec{t}\,](s) - \frac{1}{\kappa}\vec{t}\,(0).$$

Hence $\mathcal{L}[\vec{t}\,](s)$ and $\mathcal{L}[\vec{n}](s)$ are not orthogonal, in general, and we have

$$\mathcal{L}[\vec{t}\,](s) \times \mathcal{L}[\vec{n}](s) = -\frac{1}{\kappa}\mathcal{L}[\vec{t}\,](s) \times \vec{t}\,(0).$$

We here show a classical result by using the Laplace transformation and the Laplace inverse transformation $\mathcal{L}^{-1}$.

Proposition 2.1. *Let $c : \vec{r}_c = \vec{r}_c(t)$ be a helix of curvature κ and torsion τ which is parameterized by its arc-length. Then its Frenet frame at an arbitrary point $\vec{r}_c(t)$ is expressed as*

$$\begin{pmatrix} \vec{t}\,(t) \\ \vec{n}\,(t) \\ \vec{b}\,(t) \end{pmatrix} = A(t) \begin{pmatrix} \vec{t}\,(0) \\ \vec{n}\,(0) \\ \vec{b}\,(0) \end{pmatrix}$$

with

$$A(t) = \frac{e^t}{\kappa^2+\tau^2} \begin{pmatrix} \tau^2 & 0 & \kappa\tau \\ 0 & 0 & 0 \\ \kappa\tau & 0 & \kappa^2 \end{pmatrix} + \frac{\cos\sqrt{\kappa^2+\tau^2}\,t}{\kappa^2+\tau^2} \begin{pmatrix} \kappa^2 & 0 & 0 \\ 0 & \kappa^2+\tau^2 & 0 \\ 0 & 0 & \tau^2 \end{pmatrix}$$

$$+ \frac{\sin\sqrt{\kappa^2+\tau^2}\,t}{(\kappa^2+\tau^2)^{3/2}} \begin{pmatrix} 0 & \kappa(\kappa^2+\tau^2) & \kappa\tau \\ -\kappa(\kappa^2+\tau^2) & 0 & \tau(\kappa^2+\tau^2) \\ \kappa\tau & -\tau(\kappa^2+\tau^2) & 0 \end{pmatrix}.$$

Proof. We take the Laplace transformation of the Frenet formula given by (2.1). Since the curvature and the torsion of a helix is constant, the linearity of the Laplace transformation shows that

$$\mathcal{L}\left[\begin{pmatrix} \vec{t}\,' \\ \vec{n}\,' \\ \vec{b}\,' \end{pmatrix}\right] = \begin{pmatrix} 0 & \kappa & 0 \\ -\kappa & 0 & \tau \\ 0 & -\tau & 0 \end{pmatrix} \mathcal{L}\left[\begin{pmatrix} \vec{t} \\ \vec{n} \\ \vec{b} \end{pmatrix}\right].$$

As we have

$$\mathcal{L}\left[\begin{pmatrix} \vec{t}\,' \\ \vec{n}\,' \\ \vec{b}\,' \end{pmatrix}\right](s) = \begin{pmatrix} s\mathcal{L}[\vec{t}\,](s) - \vec{t}\,(0) \\ s\mathcal{L}[\vec{n}\,](s) - \vec{n}\,(0) \\ s\mathcal{L}[\vec{b}\,](s) - \vec{b}\,(0) \end{pmatrix} = s\mathcal{L}\left[\begin{pmatrix} \vec{t} \\ \vec{n} \\ \vec{b} \end{pmatrix}\right](s) - \begin{pmatrix} \vec{t}\,(0) \\ \vec{n}\,(0) \\ \vec{b}\,(0) \end{pmatrix},$$

we obtain

$$\begin{pmatrix} s & -\kappa & 0 \\ \kappa & s & -\tau \\ 0 & \tau & s \end{pmatrix} \mathcal{L}\left[\begin{pmatrix} \vec{t} \\ \vec{n} \\ \vec{b} \end{pmatrix}\right](s) = \begin{pmatrix} \vec{t}\,(0) \\ \vec{n}\,(0) \\ \vec{b}\,(0) \end{pmatrix},$$

hence have

$$\mathcal{L}\left[\begin{pmatrix} \vec{t} \\ \vec{n} \\ \vec{b} \end{pmatrix}\right](s) = \frac{1}{s(s^2+\kappa^2+\tau^2)} \begin{pmatrix} s^2+\tau^2 & \kappa s & \kappa\tau \\ -\kappa s & s^2 & \tau s \\ \kappa\tau & -\tau s & s^2+\kappa^2 \end{pmatrix} \begin{pmatrix} \vec{t}\,(0) \\ \vec{n}\,(0) \\ \vec{b}\,(0) \end{pmatrix}.$$

Since the Laplace inverse transformation is linear, we have

$$\begin{pmatrix} \vec{t}\,(t) \\ \vec{n}\,(t) \\ \vec{b}\,(t) \end{pmatrix} = \mathcal{L}^{-1}\left[\frac{1}{s(s^2+\kappa^2+\tau^2)}\begin{pmatrix} s^2+\tau^2 & \kappa s & \kappa\tau \\ -\kappa s & s^2 & \tau s \\ \kappa\tau & -\tau s & s^2+\kappa^2 \end{pmatrix}\right](t)\begin{pmatrix} \vec{t}\,(0) \\ \vec{n}\,(0) \\ \vec{b}\,(0) \end{pmatrix}.$$

As we have

$$\begin{aligned}
\mathcal{L}^{-1}\left[\frac{s^2+\tau^2}{s(s^2+\kappa^2+\tau^2)}\right](t) &= \frac{\tau^2}{\kappa^2+\tau^2}\,e^t + \frac{\kappa^2}{\kappa^2+\tau^2}\cos\sqrt{\kappa^2+\tau^2}\,t, \\
\mathcal{L}^{-1}\left[\frac{1}{s^2+\kappa^2+\tau^2}\right](t) &= \frac{1}{\sqrt{\kappa^2+\tau^2}}\sin\sqrt{\kappa^2+\tau^2}\,t, \\
\mathcal{L}^{-1}\left[\frac{1}{s(s^2+\kappa^2+\tau^2)}\right](t) &= \frac{1}{\kappa^2+\tau^2}\,e^t + \frac{1}{(\kappa^2+\tau^2)^{3/2}}\sin\sqrt{\kappa^2+\tau^2}\,t, \\
\mathcal{L}^{-1}\left[\frac{s}{s^2+\kappa^2+\tau^2}\right](t) &= \cos\sqrt{\kappa^2+\tau^2}\,t,
\end{aligned}$$

we get the conclusion. □

3. Curvatures and torsions of Laplace transformed curves

We here study curvatures and torsions of Laplace transformed curves. Given a smooth regular curve $c : \vec{r}_c = \vec{r}_c(t)$ parameterized by its arclength, we get its Laplace transformed curve $\mathcal{L}[c] : \vec{R}(s)$. We suppose this transformed curve is also regular. That is, $\dot{\vec{R}}\,(s) = \mathcal{L}[t\vec{r}_c(t)](s) \neq \vec{0}$ everywhere. Here, $\mathcal{L}[t\vec{r}_c(t)]$ shows the vector-valued Laplace transformation, and does not show the Laplace transformation of the curve given by $t \mapsto t\vec{r}_c(t)$ because this new curve is not parameterized by its arclength. We use such mixing up of notations from now on.

Since s is not the arc-length parameter for $\mathcal{C} = \mathcal{L}[c]$, its Frenet frame $\{\vec{T}\,(s), \vec{N}\,(s), \vec{B}\,(s)\}$ is given as

$$\vec{T}\,(s) = \frac{\dot{\vec{R}}\,(s)}{\|\dot{\vec{R}}\,(s)\|}, \quad \vec{B}\,(s) = \frac{\dot{\vec{R}}\,(s)\times\ddot{\vec{R}}\,(s)}{\|\dot{\vec{R}}\,(s)\times\ddot{\vec{R}}\,(s)\|}, \quad \vec{N}\,(s) = \vec{B}\,(s)\times\vec{T}\,(s).$$

By use of the rule of differentiation on Laplace transformed functions

$\frac{d^k}{ds^k}\mathcal{L}[f\,](s) = (-1)^k\mathcal{L}\big[t^k f(t)\big](s)$, we obtain

$$\vec{T}\,(s) = -\frac{\mathcal{L}[t\vec{r}\,(t)](s)}{\|\mathcal{L}[t\vec{r}(t)](s)\|},$$

$$\begin{aligned}\vec{B}\,(s) &= \frac{-\mathcal{L}[t\vec{r}(t)](s)\times\mathcal{L}[t^2\vec{r}(t)](s)}{\|\mathcal{L}[t\vec{r}(t)](s)\times\mathcal{L}[t^2\vec{r}(t)](s)\|}\\ &= \frac{-\mathcal{L}[t\vec{r}(t)](s)\times\mathcal{L}[t^2\vec{r}(t)](s)}{\Big\{\big\|\mathcal{L}[t\vec{r}(t)](s)\big\|^2\,\big\|\mathcal{L}[t^2\vec{r}(t)](s)\big\|^2 - \big\langle\mathcal{L}[t\vec{r}(t)](s),\mathcal{L}[t^2\vec{r}(t)](s)\big\rangle^2\Big\}^{1/2}},\end{aligned}$$

Differentiating $\vec{T}$ with respect to the arc-length parameter, we have

$$\begin{aligned}&\frac{1}{\|\dot{\vec{R}}\,(s)\|}\,\frac{d}{ds}\vec{T}\,(s)\\ &\quad= \frac{1}{\big\|\mathcal{L}[t\vec{r}(t)](s)\big\|^4}\Big\{\big\|\mathcal{L}[t\vec{r}(t)](s)\big\|^2\mathcal{L}[t^2\vec{r}(t)](s)\\ &\qquad\qquad - \big\langle\mathcal{L}[t\vec{r}(t)](s),\mathcal{L}[t^2\vec{r}(t)](s)\big\rangle\mathcal{L}[t\vec{r}(t)](s)\Big\}.\end{aligned}$$

Hence we obtain the curvature $\kappa_C(s)$ of the Laplace transformed curve is given by

$$\kappa_C(s) = \frac{\Big\{\big\|\mathcal{L}[t\vec{r}(t)](s)\big\|^2\big\|\mathcal{L}[t^2\vec{r}(t)](s)\big\|^2 - \big\langle\mathcal{L}[t\vec{r}(t)](s),\mathcal{L}[t^2\vec{r}(t)](s)\big\rangle^2\Big\}^{1/2}}{\big\|\mathcal{L}[t\vec{r}(t)](s)\big\|^3},$$

and $\vec{N}\,(s)$ is expressed as

$$\vec{N}\,(s) = \frac{\big\|\mathcal{L}[t\vec{r}(t)](s)\big\|^2\mathcal{L}[t^2\vec{r}(t)](s) - \big\langle\mathcal{L}[t\vec{r}(t)](s),\mathcal{L}[t^2\vec{r}(t)](s)\big\rangle\,\mathcal{L}[t\vec{r}(t)](s)}{\big\|\mathcal{L}[t\vec{r}(t)](s)\big\|^4\kappa_C(s)}.$$

Also, as $\mathcal{L}[t\vec{r}(t)]$ is orthogonal to $\mathcal{L}[t\vec{r}(t)]\times\mathcal{L}[t^2\vec{r}(t)]$, we obtain

$$\begin{aligned}\tau_C(s) &= \frac{1}{\|\dot{\vec{R}}\,(s)\|}\,\big\langle\vec{N}\,(s),\frac{d}{ds}\vec{B}\,(s)\big\rangle\\ &= \frac{\big\langle\mathcal{L}[t^2\vec{r}(t)](s),\mathcal{L}[t\vec{r}(t)](s)\times\mathcal{L}[t^3\vec{r}(t)](s)\big\rangle}{\big\|\mathcal{L}[t\vec{r}(t)](s)\big\|^2\,\big\|\mathcal{L}[t^2\vec{r}(t)](s)\big\|^2 - \big\langle\mathcal{L}[t\vec{r}(t)](s),\mathcal{L}[t^2\vec{r}(t)](s)\big\rangle^2}.\end{aligned}$$

Summarizing up we have

Theorem 3.1. *Let $c:\vec{r}\,(t)$ be a smooth curve parameterized by its arc-length. Suppose its Laplace transformed curve $\mathcal{L}[c]$ is also regular. Then* *the curvature and the torsion of* $\mathcal{L}[c]$ *are given by*

$$\kappa_C = \frac{\sqrt{\left\|\mathcal{L}[t\vec{r}(t)]\right\|^2 \left\|\mathcal{L}[t^2\vec{r}(t)]\right\|^2 - \left\langle \mathcal{L}[t\vec{r}(t)], \mathcal{L}[t^2\vec{r}(t)]\right\rangle^2}}{\left\|\mathcal{L}[t\vec{r}(t)]\right\|^3},$$

$$\tau_C = \frac{\left\langle \mathcal{L}[t^2\vec{r}(t)], \mathcal{L}[t\vec{r}(t)] \times \mathcal{L}[t^3\vec{r}(t)]\right\rangle}{\left\|\mathcal{L}[t\vec{r}(t)]\right\|^2 \left\|\mathcal{L}[t^2\vec{r}(t)]\right\|^2 - \left\langle \mathcal{L}[t\vec{r}(t)], \mathcal{L}[t^2\vec{r}(t)]\right\rangle^2}.$$

Remark 3.1. When a smooth curve is contained in a plane passing through the origin, it is clear that its Laplace transformed curve $\mathcal{L}[c]$ is contained in this plane, hence $\tau_{\mathcal{L}[c]} \equiv 0$.

We now calculate curvatures and torsions for Laplace transformed images of curves given in §2.

Example 3.1. For a ray $c : \vec{r}(t) = \vec{q} + t\vec{p}$ parameterized by its arclength, we have

$$\kappa_{\mathcal{L}[c]}(s) = 2s^3 \sqrt{\frac{\|\vec{q}\|^2 - \langle \vec{p}, \vec{q}\rangle^2}{\left(s^2\|\vec{q}\|^2 + 4s\langle \vec{p}, \vec{q}\rangle + 4\right)^3}}, \quad \tau_{\mathcal{L}[c]}(s) = 0.$$

Example 3.2. We take a circle $c : \vec{r}(t) = \vec{q} + \frac{1}{\kappa}\cos\kappa t\,\vec{v} + \frac{1}{\kappa}\sin\kappa t\,\vec{w}$ of positive curvature κ with an orthonormal pair $\vec{v}, \vec{w}$. When $\vec{q} = \vec{0}$, its transformed curve is $\vec{R}(s) = \{(s/\kappa)\vec{v} + \vec{w}\}/(s^2 + \kappa^2)$, and we have

$$\kappa_{\mathcal{L}[c]}(s) = 2\kappa^2, \quad \tau_{\mathcal{L}[c]}(s) = 0.$$

This curve $\mathcal{L}[c]$ is a half circle-arc centered at $\big(0, 1/(2\kappa^2)\big)$ of radius $1/(2\kappa^2)$. When $\vec{r}(0) = \vec{0}$, which is the case $\vec{q} = -\vec{v}/\kappa$, its transformed curve is $\vec{R}(s) = \{(-\kappa/s)\vec{v} + \vec{w}\}/(s^2 + \kappa^2)$, and we have

$$\kappa_{\mathcal{L}[c]}(s) = 6\kappa s^4(4s^2 + \kappa^2)^{-3/2}, \quad \tau_{\mathcal{L}[c]}(s) = 0.$$

Example 3.3. We take a helix of curvature κ and torsion τ

$$c : \vec{r}(t) = \vec{q} + \frac{t}{\sqrt{2}}\vec{u} + \frac{1}{\sqrt{\kappa^2 + \tau^2}}\left\{\cos\sqrt{\frac{\kappa^2 + \tau^2}{2}}\,t\,\vec{v} + \sin\sqrt{\frac{\kappa^2 + \tau^2}{2}}\,t\,\vec{w}\right\}$$

with an orthonormal triplet $\vec{u}, \vec{v}, \vec{w}$. When $\vec{q} = \vec{0}$, its transformed curve is $\vec{R}(s) = \big(1/(\sqrt{2}\,s^2)\big)\,\vec{u} + \{(2s/\omega)\vec{v} + \sqrt{2}\,\vec{w}\}/(2s^2 + \omega^2)$ with $\omega = \sqrt{\kappa^2 + \tau^2}$. Thus we have

$$\kappa_{\mathcal{L}[c]}(s) = \frac{\omega^2 s^5 \left(32s^8 + 96\omega^2 s^6 + 120\omega^4 s^4 + 56\omega^6 s^2 + 9\omega^8\right)^{1/2}}{\left(2s^6 + 4\omega^2 s^4 + 4\omega^4 s^2 + \omega^6\right)^{3/2}},$$

$$\tau_{\mathcal{L}[c]}(s) = \frac{24\omega^3 s^3 (2s^2 + \omega^2)^2}{32s^8 + 96\omega^2 s^6 + 120\omega^4 s^4 + 56\omega^6 s^2 + 9\omega^8}.$$

When $\vec{r}\,(0) = \vec{0}$, which is the case $\vec{q} = -\vec{v}/\kappa$, its transformed curve is $\vec{R}(s) = \big(1/(\sqrt{2}\,s^2)\big)\,\vec{u} + \big\{(-\omega/s)\vec{v} + \sqrt{2}\,\vec{w}\big\}/(2s^2+\omega^2)$. We then have

$$\kappa_{\mathcal{L}[c]}(s) = \frac{\sqrt{2}\omega s^3\big(288s^8 + 224\omega^2 s^6 + 120\omega^4 s^4 + 24\omega^6 s^2 + \omega^8\big)^{1/2}}{\big(16s^4 + 9\omega^2 s^2 + 2\omega^4\big)^{3/2}},$$

$$\tau_{\mathcal{L}[c]}(s) = \frac{48\omega s^3(2s^2+\omega^2)^2}{288s^8 + 224\omega^2 s^6 + 120\omega^4 s^4 + 24\omega^6 s^2 + \omega^8}.$$

We here reformulate our study on the Laplace transformation acting on smooth curves. We take a smooth regular curve c which is parameterized by its arc-length and that satisfies $\vec{r}_c(0) = 0$ and $\|\vec{r}_c\| \in \mathcal{O}$. We consider a smooth curve $u \mapsto \mathcal{L}[c]\big(\lambda_c + \frac{1}{u}\big)$ defined on the interval $(0, \infty)$. When this curve is also regular, we reparameterize it by its arc-length and obtain a new smooth regular curve $\widehat{\mathcal{C}} : \vec{\widehat{R}}(t)$. Since we can extend this curve continuously by setting $\vec{\widehat{R}}(0) = \vec{0}$, we obtain a correspondence $\widehat{\mathcal{L}} : c \mapsto \widehat{\mathcal{C}}$ of smooth curves emanating from the origin. Example 2.1 shows that rays emanating from the origin are invariant under this correspondence.

References

1. W. Arendt, C.J.K. Batty, M. Hieber & F. Neubrander, *Vector-valued Laplace transformations and Cauchy problems*, Springer (2011).
2. Balth. van der Pol, H. Bremer, *Operational calculus based on the two-sided Laplace integral*, Cambridge Univ. Press, 1950.
3. M. Hristov, Vector-valued Laplace transformation applied to rational Bézier curves, this volume.

Received January 28, 2015
Revised June 4, 2015

Current Developments in Differential Geometry
and its Related Fields *pp. 235 – 241*

VECTOR-VALUED LAPLACE TRANSFORMATION APPLIED TO RATIONAL BÉZIER CURVES

Milen J. HRISTOV *

Department of Algebra and Geometry, Faculty Mathematics and Informatics,
"St. Cyril and St. Methodius" University of Veliko Tarnovo,
5000 Veliko Tarnovo, Bulgaria
E-mail: m.hristov@uni-vt.bg

In this paper we study images and inverse images of some curves which are not parameterized by their arclengths through vector-valued Laplace transformations.

Keywords: Vector-valued Laplace transformation, Laplace inverse transformation, regular curves, rational Bézier curve.

1. Introduction

In his paper [5], the author study Laplace transformed curves from the Euclidean geometric point of view. Since the Laplace transformation is originally an analytical transformation which maps vector-valued functions to vector-valued functions, in order to treat this transformation geometrically, we need to take care that it preserves some action of a subgroup of the congruence group. For example, if we change the parameter of a curve c as $\vec{r}(t)$ to $\vec{r}\big(t(\tau)\big)$, then their images through the Laplace transformation acting on the family of vector-valued functions do not coincide with each other. We therefore restrict ourselves in [5] to curves of unit speed. But as is known, for concrete examples it is difficult to write down their arclength parameters explicitly.

In order to proceed our study on Laplace transformed curves, in this paper we study curves analytically by considering arbitrary parameters. This means that we study vector-valued Laplace transformations of some curves in a Euclidean space. Once we drop the arclength condition, we can treat curves much easier. We give an example of a regular curve whose Laplace transformed image is not regular. When we study transformations, it is interesting to investigating the images and the inverse images of some typical

*Supported by Scientific researches fund of "St.Cyril and St. Methodius" University of Veliko Tarnovo under contract RD-09-422-13/09.04.2014.

objects through these transformations. In this paper as typical objects we take rational Bézier curves which represent curves of second order, and that are important in the field of Computer Aided Geometric Designs. Since we can not express them by their arclength parameters explicitly, we study images of Bézier curves and the inverse images of rational Bézier curves through vector-valued Laplace transformations.

2. Vecror-valued Laplace transformations

A continuous function $f : [0,\infty) \to \mathbb{R}$ is said to be of exponential order λ if $|f(t)|e^{-\lambda t}$ is bounded for $t > 0$. We denote by $\mathcal{O}$ the set of all continuous functions defined on the interval $[0,\infty)$ which are of exponential order for some constant λ. For such a function $f \in \mathcal{O}$ of exponential order λ, we can define a function $\mathcal{L}[f] : (\lambda,\infty) \to \mathbb{R}$ by

$$\mathcal{L}[f](s) = \int_0^\infty e^{-st} f(t)\, dt.$$

We set $\mathcal{O}^n = \{\vec{f} = (f_1,\dots,f_m) \mid f_i \in \mathcal{O} \text{ for } i = 1,\dots,n\}$. For $\vec{f} \in \mathcal{O}^n$ its vector-valued Laplace transformation is nothing but the vector-valued function $\mathcal{L}[\vec{f}\,] = \big(\mathcal{L}[f_1],\dots,\mathcal{L}[f_n]\big)$ defined on the interval $(\lambda_{\vec{f}},\infty)$, where $\lambda_{\vec{f}} = \inf\{\lambda \mid f_i \text{ is of exponential order } \lambda \text{ for all } i\}$. Such transformations are studied many part in analysis (cf. [1, 2]). In this paper we regard a vector valued-functions as curves, and study some curve theoretic properties.

A smooth curve $\vec{r}$ of parameter t is said to be regular if its differential does not vanishes (i.e. $\frac{d\vec{r}}{dt}(t) \neq \vec{0}$) at each point t. In treating curves geometrically, this regularity condition is quite important. Unfortunately, the vector-valued Laplace transformation does not preserve the regularity condition. We here give an example.

Example 2.1. We take a regular curve $\vec{r}(t) = \big(\cos t, \cos(2t+b)\big)$ on a plane $\mathbb{R}^2$, where $\tan b = -3/4$. Its laplace transformed curve is

$$\vec{R}\,(s) = \mathcal{L}[\vec{r}\,](s) = \left(\frac{s}{s^2+1}, \frac{s\cos b - 2\sin b}{s^2+4}\right)$$

which is defined on $(0,\infty)$. As we have

$$\vec{R}\,'(s) = \frac{d}{ds}\mathcal{L}[\vec{r}\,](s) = \left(\frac{1-s^2}{(s^2+1)^2}, \frac{4\cos b + 4s\sin b - s^2\cos b}{(s^2+4)^2}\right),$$

we see

$$\vec{R}\,'(1) = \left(0, \frac{3\cos b + 4\sin b}{25}\right) = (0,0),$$

and find that it is not regular.

We do not have examples of smooth curves parameterized by their arclengths whose images through the Laplace transformation are not regular now. But this suggests us the existence of such curves. We should note that for a smooth curve $c : \vec{r}(t)$ the differential of its Laplace transformed curve $\mathcal{C} : \vec{R}(s)$ is expressed as $\mathcal{L}\big[t\vec{r}(t)\big](s)$ and is smooth. Therefore we can reparameterize it by its arclength.

3. Rational Bézier curves

We devote the rest of this paper to study images and inverse images of rational Bézier curves of order 2 through the Laplace transformation. Bézier curves and rational Bézier curves are quite important tools used in computer aided geometric design (see [3], for example). Moreover such curves generate remarkable objects described in barycentrics on a projectively extended plane of a fixed triangle (see [4]). We hence give some consideration on them. Our porblems in this section and next are the following:

1) Study the vector-valued Laplace transformed images of rational Bézier curves;
2) Study curves whose vector-valued Laplace transformed images are rational Bézier curves.

We shall start by recalling rational Bézier curves. Let $\triangle = (\mathrm{A}_0, \mathrm{A}_1, \mathrm{A}_2)$ be a triangle in a Euclidean 3-space $\mathbb{R}^3$. We put $\vec{a}_\ell = \overrightarrow{\mathrm{OA}_\ell}$. For a parameter u $(u > -1)$ of control, a rational Bézier curve of order 2 with normal parameterization is given as

$$c_{\triangle}^{(u)} : \vec{r}_u(t) = b_u^0(t)\,\vec{a}_0 + b_u^1(t)\,\vec{a}_1 + b_u^2(t)\,\vec{a}_2$$

with the coefficient functions

$$b_u^0(t) = \frac{(1-t)^2}{B_2(u,t)}, \quad b_u^1(t) = \frac{2ut(1-t)}{B_2(u,t)}, \quad b_u^2(t) = \frac{t^2}{B_2(u,t)},$$

$$B_2(u,t) = (1-t)^2 + 2ut(1-t) + t^2.$$

This curve is invariant under the action of the group of affine transformations in projectively extended plane of the triangle $\triangle$. If we restrict the domain of this curve to the interval $[0,1]$, this curve is contained in the triangle $\triangle$ and lies in its interior except the points $\vec{r}_u(0)$ and $\vec{r}_u(1)$. This curve satisfies also $\vec{r}(0) = \overrightarrow{\mathrm{OA}_0}$, $\vec{r}(1) = \overrightarrow{\mathrm{OA}_2}$ and its differential $\vec{r}_u{}'(0)$, $\vec{r}_u{}'(1)$

are parallel to $\overrightarrow{\mathrm{A}_0\mathrm{A}_1}$, $\overrightarrow{\mathrm{A}_1\mathrm{A}_2}$, respectively. Hence it is usually considered on the interval $[0,1]$. It is well known that rational Bézier curves are classified into three classes according to the parameter of control:

i) When $u \in (-1,1)$, it is an ellipse;
ii) When $u = 1$, it is a parabola;
iii) When $u \in (1,\infty)$, it is a hyperbola.

For the sake of expressing these classes clearly, we denote a rational Bézier curve by $\epsilon_{\triangle}^{(u)}$, $\rho_{\triangle}$, $\chi_{\triangle}^{(u)}$ according to this classifications. In the case $u = 1$, it is a Bézier curve of order 2 and its coefficients are so called the Bernstein polynomials.

First we study the image of a Bézier curve $\rho_{\triangle}$ through the Laplace transformation. Since Laplace transformation is a linear transformation, we have

$$\begin{aligned}\mathcal{L}[\vec{r}_1](s) &= \mathcal{L}[(1-t)^2](s)\,\vec{a}_0 + 2\mathcal{L}[t(1-t)](s)\,\vec{a}_1 + \mathcal{L}[t^2](s)\,\vec{a}_2\\ &= \frac{1}{s^3}\big\{(s^2-2s+2)\vec{a}_0 + 2(s-2)\vec{a}_1 + 2\vec{a}_2\big\}.\end{aligned}$$

We should note that this image is obtained by considering $\rho_{\triangle}$ as a curve defined on $\mathbb{R}$. As we usually consider that Bézier curves are defined on the interval $[0,1]$ we here take two curves $\widetilde{\vec{r}}$, $\widehat{\vec{r}} : [0,\infty) \to \mathbb{R}^3$ which is defined as

$$\widetilde{\vec{r}}(t) = \begin{cases}\vec{r}_1(t), & 0 \le t \le 1,\\ \vec{0}, & t > 1,\end{cases} \qquad \widehat{\vec{r}}(t) = \begin{cases}\vec{r}_1(t), & 0 \le t \le 1,\\ \vec{a}_2, & t > 1.\end{cases}$$

We then have

$$\begin{aligned}\mathcal{L}\big[\widetilde{\vec{r}}\big](s) &= \frac{1}{s^3}\big\{(s^2-2s+2-2e^{-s})\,\vec{a}_0 + 2\{s-2+e^{-s}(s+2)\}\,\vec{a}_1\\ &\qquad + \{2-e^{-s}(s^2+2s+2)\}\,\vec{a}_2\big\},\end{aligned}$$

$$\begin{aligned}\mathcal{L}\big[\widehat{\vec{r}}\big](s) &= \frac{1}{s^3}\big\{(s^2-2s+2-2e^{-s})\,\vec{a}_0 + 2\{s-2+e^{-s}(s+2)\}\,\vec{a}_1\\ &\qquad + 2\{1-e^{-s}(s+1)\}\,\vec{a}_2\big\}.\end{aligned}$$

For rational Bézier curves we can not express their transformed images explicitly. We have

$$\begin{aligned}\mathcal{C}_{\triangle}^{(u)} : \vec{R}\,(s) &= \int_0^\infty \frac{(1-t)^2 e^{-st}}{1+2(u-1)t(1-t)}\,dt\,\vec{a}_0\\ &\quad + \int_0^\infty \frac{2ut(1-t)e^{-st}}{1+2(u-1)t(1-t)}\,dt\,\vec{a}_1 + \int_0^\infty \frac{t^2 e^{-st}}{1+2(u-1)t(1-t)}\,dt\,\vec{a}_3.\end{aligned}$$

Though the original rational Bézier curve $c_{\triangle}^{(u)} : \vec{r}_u(t)$ is invariant under the action of the group of affine transformations in projectively extended plane of the triangle $\triangle$ as we mentioned before, its transformed image $\mathcal{C}_{\triangle}^{(u)} : \vec{R}_u(s)$ is not necessarily invariant under this action in general.

4. Laplace inverse images of rational Bézier curves

In this section we study the inverse images of rational Bézier curves through the Laplace transformation. We take a rational Bézier curve $c_{\triangle}^{(u)} : \vec{R}_u(s) = b_u^0(s)\,\vec{a}_0 + b_u^1(s)\,\vec{a}_1 + b_u^2(s)\,\vec{a}_2$ $(u \neq 1)$ defined on $[0, \infty)$. Since the b-normalizer $B_2(u, s)$ is expressed as

$$B_2(u,s) = 2(1-u)\left\{\left(s-\frac{1}{2}\right)^2 + \frac{1+u}{4(1-u)}\right\}.$$

By putting $\lambda = (1+u)/\{4(1-u\}$ we have

$$\begin{cases} b_u^0(s) = \dfrac{1}{2(1-u)}\left\{1 - \dfrac{s-\frac{1}{2}}{(s-\frac{1}{2})^2+\lambda} - \dfrac{4\lambda-1}{4} \times \dfrac{1}{(s-\frac{1}{2})^2+\lambda}\right\}, \\ b_u^1(s) = \dfrac{1}{1-u}\left\{-u + \dfrac{4\lambda-1}{4} \times \dfrac{1}{(s-\frac{1}{2})^2+\lambda}\right\}, \\ b_u^2(s) = \dfrac{1}{2(1-u)}\left\{1 + \dfrac{s-\frac{1}{2}}{(s-\frac{1}{2})^2+\lambda} - \dfrac{4\lambda-1}{4} \times \dfrac{1}{(s-\frac{1}{2})^2+\lambda}\right\}. \end{cases} \tag{4.1}$$

We note that $\lambda > 0$ if and only if $-1 < u < 1$.

First we study the inverse images of an ellipse $\epsilon_{\triangle}^{(u)}$. As $\lambda > 0$ in this case, by putting $\sigma = \sqrt{\lambda}$ we have

$$\mathcal{L}^{-1}[b_u^0](t) = \frac{\delta(t)}{2(1-u)} - \frac{e^{t/2}}{2(1-u)}\left\{\cos\sigma t + \frac{4\sigma^2-1}{4\sigma}\sin\sigma t\right\},$$

$$\mathcal{L}^{-1}[b_u^1](t) = \frac{-u\delta(t)}{1-u} + \frac{4\sigma^2-1}{4\sigma(1-u)}\,e^{t/2}\sin\sigma t,$$

$$\mathcal{L}^{-1}[b_u^2](t) = \frac{\delta(t)}{2(1-u)} + \frac{e^{t/2}}{2(1-u)}\left\{\cos\sigma t - \frac{4\sigma^2-1}{4\sigma}\sin\sigma t\right\}.$$

Here $\delta(t)$ denotes the Dirac impulse function. Thus if we take $\theta_0 \in (-\pi/2, \pi/2)$ which satisfies

$$\sin\theta_0 = \frac{4\sigma}{4\sigma^2+1}, \qquad \cos\theta_0 = \frac{4\sigma^2-1}{4\sigma^2+1},$$

we obtain that the inverse image of an ellipse $\epsilon_{\triangle}^{(u)}$ through the Laplace transformation is given as

$$\begin{aligned}\vec{r}_u(t) &:= \mathcal{L}^{-1}\big[\vec{R}_u\big](t)\\ &= \frac{\delta(t)}{2(1-u)}\big(\vec{a}_0 - 2u\vec{a}_1 + \vec{a}_2\big)\\ &\quad + \frac{e^{t/2}}{2(1-u)}\left\{-\frac{\sin(\sigma t+\theta_0)}{\sin\theta_0}\,\vec{a}_0 + \frac{2\sin\sigma t}{\tan\theta_0}\,\vec{a}_1 + \frac{\sin(\sigma t-\theta_0)}{\sin\theta_0}\,\vec{a}_2\right\}.\end{aligned}$$

We should note that the point P_u given by $\overrightarrow{\mathrm{OP}}_u = (\vec{a}_0 - 2u\vec{a}_1 + \vec{a}_2)/\{2(1-u)\}$ lies on the line containing the median $\mathrm{A}_1\mathrm{M}$ of $\triangle$, where M denotes the midpoint of the edge $\mathrm{A}_0\mathrm{A}_2$, and is determined by the affine ratio $u = \overline{\mathrm{MP}}_u/\overline{\mathrm{A}_1\mathrm{P}}_u$, which is usually denote as $(\mathrm{MA}_1\mathrm{P}_u)$.

Next we study the inverse image of a hyperbola $\chi_{\triangle}^{(u)}$. Since $\lambda < 0$ in this case we set $\sigma = \sqrt{-\lambda}$. By (4.1) we have

$$\mathcal{L}^{-1}[b_u^0](t) = \frac{\delta(t)}{2(1-u)} - \frac{e^{t/2}}{2(1-u)}\left\{\cosh\sigma t - \frac{4\sigma^2+1}{4\sigma}\sinh\sigma t\right\},$$

$$\mathcal{L}^{-1}[b_u^1](t) = \frac{-u\delta(t)}{1-u} - \frac{4\sigma^2+1}{4\sigma(1-u)}\,e^{t/2}\sinh\sigma t,$$

$$\mathcal{L}^{-1}[b_u^2](t) = \frac{\delta(t)}{2(1-u)} + \frac{e^{t/2}}{2(1-u)}\left\{\cosh\sigma t + \frac{4\sigma^2+1}{4\sigma}\sinh\sigma t\right\}.$$

Thus if we take $\theta_0 \in (0,\infty)$ which satisfies

$$\sinh\theta_0 = \frac{4\sigma}{4\sigma^2-1}, \qquad \cosh\theta_0 = \frac{4\sigma^2+1}{4\sigma^2-1},$$

we obtain that the inverse image of an ellipse $\chi_{\triangle}^{(u)}$ through the Laplace transformation is given as

$$\begin{aligned}\vec{r}_u(t) &:= \mathcal{L}^{-1}\big[\vec{R}_u\big](t)\\ &= \frac{\delta(t)}{2(1-u)}\big(\vec{a}_0 - 2u\vec{a}_1 + \vec{a}_2\big)\\ &\quad - \frac{e^{t/2}}{2(1-u)}\left\{\frac{\sinh(\sigma t-\theta_0)}{\sinh\theta_0}\,\vec{a}_0 + \frac{2\sinh\sigma t}{\tanh\theta_0}\,\vec{a}_1 - \frac{\sinh(\sigma t+\theta_0)}{\sinh\theta_0}\,\vec{a}_2\right\}.\end{aligned}$$

Summarizing up, we find the following

Theorem 4.1. *The inverse image* $\widehat{\mathcal{C}}^{(u)} : \vec{r}_u(t)$ *of each rational Bézier curve* $c_{\triangle}^{(u)} : \vec{R}_u(s)$ *is not a smooth curve. When* $u \neq 1$*, the curve* $\widetilde{\mathcal{C}}^{(u)}$:

$\vec{r}_u(t) - \delta(t)\overrightarrow{\mathrm{OP}}_u$ is collinear to the line $t \mapsto t(\vec{a}_0 + \vec{a}_1 + \vec{a}_2)$ at infinity in the projectively extended plane of the referenced triangle.

We note that even the rational Bézier curve $c_{\triangle}^{(u)} : \vec{R}_u$ is invariant under the action of the group of affine transformations in projectively extended plane of the triangle $\triangle$, its inverse image $\widehat{\mathcal{C}}^{(u)} : \vec{r}_u$ is not invariant under this action in general.

References

1. W. Arendt, C.J.K. Batty, M. Hieber & F. Neubrander, *Vector-valued Laplace transformations and Cauchy problems*, Springer (2011).
2. Balth. van der Pol, H. Bremer, *Operational calculus based on the two-sided Laplace integral*, Cambridge Univ. Press (1950).
3. G.E. Farin, *Curves and surfaces for computer aided geometric design: A practical guide*, Academic Press Inc. (1988).
4. M. Hristov, Geometric group structures and trajectories of rational Bézier curves, in *Prospects of Differential Geometry and its Related Fields*, T. Adachi, H. Hashimoto & M. Hristov eds, World Sci. Publ., Singapore, 2013, pp. 129-143.
5. M. Hristov, A geometric study on Laplace transformed curves, this volume.

Received January 28, 2015
Revised June 8, 2015